Thomas Lincoln Casey

Studies in the Ptinidae, Cioidae and Sphindidae of America

Thomas Lincoln Casey

Studies in the Ptinidae, Cioidae and Sphindidae of America

ISBN/EAN: 9783337389284

Printed in Europe, USA, Canada, Australia, Japan

Cover: Foto ©berggeist007 / pixelio.de

More available books at **www.hansebooks.com**

JOURNAL

OF THE

Ⓝⓔⓦ Ⓨⓞⓡⓚ Ⓔⓝⓣⓞⓜⓞⓛⓞⓖⓘⓒⓐⓛ Ⓢⓞⓒⓘⓔⓣⓨ.

Vol. VI.   JUNE, 1898.   No. 2.

STUDIES IN THE PTINIDÆ, CIOIDÆ AND SPHIN-
DIDÆ OF AMERICA.

By Thos. L. Casey.

The term "America," in the above title, is employed to designate that portion of the American continent embraced within the boundaries of the United States. There should be no more ambiguity in designating the United States of America as America simply, than in calling the United States of Brazil, or the United States of Colombia by the last word of their respective titles. It may perhaps be considered egotistical for us to appropriate to ourselves the name characterizing the continents of the western hemisphere, but as we have no other title to distinguish us among the numerous aggregates of united states which compose these continents, there is no reasonable motive for avoiding the apparent conceit.

PTINIDÆ.

PTILININI.

The insects of this tribe form an appropriate introduction to the Bostrichinæ, for they are evidently a connective bond with the Anobiini. Our species have not been studied for many years. We have two genera as follows:—

Antennæ strongly flabellate in the male, serrate in the female...........**Ptilinus**
Antennæ slender and feebly serrate in the male, shorter and somewhat stouter but
    rather more strongly serrate in the female.................**Euceratocerus**

The eyes are rather larger and more convex in *Euceratocerus* than in *Ptilinus*, but are much smaller in the female than in the male. Selecting the apparent males by this character I have been unable to observe any pectination of the male antenna in *pleuralis*, though it may exist in *horni*, of which I have not seen the male.

Ptilinus *Geoff.*

The long slender pubescent appendages of the male antenna are not an extreme development of the usual serriform structure, but project from the base of the joints, the joints themselves being slender and sometimes cylindrical.   In the female, however, the joints are prolonged outwardly and in an obliquely anterior direction, forming a truly and quite strongly serriform antenna.   The males differ from the females not only in the structure of the antennæ but in the much denser sculpture of the entire body, and frequently to a very great degree in the form of the prothorax.   In the following table of the species included within my cabinet, all the discriminating characters refer to the female, except in the case of *flavipennis*, of which the only known example is a male :—

Color uniform throughout or very nearly, the elytra not paler.................. 2
Color blackish, the elytra flavate... 8
2—Prothorax sinuate at the sides subapically, the apex more or less prominent in a
    rounded or feebly sinuate lobe...................... 3
Prothorax arcuately oblique subapically, the apex angulate and slightly prominent 6
Prothorax broadly and evenly arcuate at apex, feebly sinuate in the middle...... 7
3—Elytra distinctly punctured only toward the base, the punctures nearly obselete
    toward tip; thoracic lobe feebly and evenly crenulate. California. . **basalis** *Lec.*
Elytra distinctly punctured throughout..................................... 4
4—Thoracic lobe narrowly rounded and with a few closely approximate serrules at
    tip. Pennsylvania to Indiana .... **ruficornis** *Say*
Thoracic lobe broadly rounded, with a feeble cuspiform emargination and minutely
    and evenly serrulate throughout...5
5—Scutellum elongate, finely and densely rugose ; epipleuræ gradually wider at base.
    Female.—Body rather stout, cylindrical, dark piceo-castaneous in color throughout, the legs scarcely, the antennæ much, paler ; lustre rather dull, the pubescence very short and dense on the elytra.   Head convex, minutely and densely granulate, the transverse frontal impression distinct.   Prothorax distinctly wider than long, widest at about the middle, narrowed slightly to the base, rapidly and just visibly sinuate to the apical lobe ; surface minutely and densely granulose toward base, more coarsely, sparsely and irregularly so toward apex, the median line finely impressed.   Elytra three-fourths longer than wide, equal in width to the prothorax, minutely and densely subgranulose in texture, with the punctures rather strong, sparse and distinct, feebler toward tip, where they are more distinctly intermingled with small granules and the ground lustre is more shining.   Length 4.0 mm.; width 1.5 mm. Nebraska...... ....**lobatus**, sp. nov.
Scutellum quadrate, coarsely rugose ; epipleuræ rapidly wider at base.   *Male.*—Cylindrical, opaque, black, the legs scarcely paler, the antennæ pale rufous, the rami infuscate ; pubescence extremely minute and not very conspicuous.   Head convex, dull, minutely subgranulose, the eyes convex, separated on the front by about four times their own width.   Prothorax about a third wider than long, broadly, evenly arcuate at base, gradually narrowed and arcuate at the sides from

the broadly rounded basal angles to the apex, the latter much narrower than the base and evenly and more strongly arcuate ; surface densely granulato-rugose throughout, the median line very fine and subobsolete.   Elytra three-fifths longer than wide, rather wider than the prothorax and much more than twice as long, very densely granulato-rugose and opaque, the punctures distinct throughout, with about two impressed series along the lateral margin.   *Female.*—A little larger and paler in color than the male, with more shining elytra, upon which there are more distinct traces on each of three or four feeble ridges.   Prothorax larger, fully as wide as the elytra, rounded at the sides, widest just behind the middle, the surface more sparsely and decidedly granose toward tip.   Elytra scarcely three-fifths longer than wide, twice as long as the prothorax.   Length 3.0–3.5 mm.; width 1.1–1.2 mm.   Indiana....................**pruinosus**, sp. nov.

6—Elytral punctures only distinct near the base, where they are fine.   *Female.*— Evenly cylindrical, piceous, the legs and antennæ paler ; pubescence very short, even, extremely dense, yellowish in color and conspicuous on the elytra ; lustre rather shining.   Head evenly convex, minutely granulato-rugose, the epistomal impression small and rather feeble.   Prothorax not quite as long as wide, the sides parallel and feebly arcuate ; apex broadly and evenly ogival ; surface minutely granulose, becoming nearly smooth at the sides toward base, the granules coarse and pronounced toward apex except laterally.   Scutellum quadrate, feebly convex, dull.   Elytra fully three-fourths longer than wide, about twice as long as the prothorax, smooth and alutaceous, without trace of impressed lines at any part.   Abdomen rather convex, the second segment somewhat longer than the first.   Length 2.8–4.0 mm.; width 0.9–1.4 mm.   California (Sta. Cruz Mts.).
acuminatus, sp. nov.

7—Elytra with fine, even and somewhat impressed striæ in both sexes.   *Male.*— Cylindrical, blackish, the elytra generally a little paler ; legs and antennæ pale, the flabellum infuscate ; surface dull, the humeral callus more shining.   Head short, inserted to the eyes which are well developed and strongly convex ; surface but feebly convex, densely scabrous and opaque ; antennal joints very short, the rami very long and slender.   Prothorax a little shorter than wide, parallel and straight at the sides, broadly and evenly rounded in apical third or fourth, with a minute sinus at the middle ; surface coarsely, densely and roughly granulato-scabrous throughout the width, becoming much more finely so and smoother toward base.   Scutellum moderate, subquadrate.   Elytra three-fourths longer than wide, a little more than twice as long as the prothorax and rather wider, densely dull and finely granulato-rugose, the second and fifth intervals uniting and rather convex near the declivity, the ninth also becoming broader and slightly convex behind.   *Female.*—Rather shining and dark rufo-testaceous throughout, the prothorax similar in form but rather shorter and fully as wide as the elytra, with the rugulosities more distinct and isolated, nearly smooth toward base, the head more elongate, narrower and with the eyes small and distant from the prothorax ; elytra rather flattened on the posterior declivity, with the intervals slightly uneven.   Length 3 0–4.2 mm.; width 0.9–1.4 mm.   California (Sta. Cruz Mts)......................................**ramicornis**, sp. nov.

8—Elytra with rather strong punctures unevenly arranged throughout.   *Male.*— Cylindrical, blackish and opaque, the elytra flavate and less dull ; legs paler, the antennæ pale flavate; pubescence short, fine and moderately distinct. Head rather short, inserted nearly to the eyes, which are well developed and very convex; surface moderately convex, dull and subscabrous; antennæ moderate in length, the rami unusually short and gradually thickened from their bases, the ramus of the fourth joint three times as long as the joint. Prothorax distinctly shorter than wide, the outline broadly parabolic from the base continuously around the apex, the sides becoming almost parallel near the base, which is broadly arcuate, finely margined toward the middle; surface densely granulato-scabrous, larger individual granules but slightly evident toward tip. Scutellum longer than wide, dull, obtuse at tip as usual. Elytra three-fourths longer than wide, twice as long at the prothorax and scarcely wider, the punctures equally visible throughout, rather large but sparse and with but the vaguest suggestion of lineal arrangement. Length 2.4 mm.; width 0.8 mm. California (Los Angeles Co.) .... **flavipennis**, sp. nov.

The female in this genus generally has a short acute transverse ridge near the apex of the last ventral segment, but in *ramicornis* the fifth segment is simple in that sex, having merely a very small and shallow impression at the apex. The male usually has the fifth ventral simple or slightly more convex at the apex, where it is broadly and evenly rounded.

Acuminatus is represented before me by seven females varying greatly in size, and the male is apparently rare; on the other hand *ramicornis* is represented by nine specimens, only two of which are females. *Thoracicus* Rand., is not known to me at present and is therefore omitted from the table.

Euceratocerus Lec.

The fifth ventral segment is generally impressed in the female of *Euceratocerus* and is rather shorter than in the male, where it is simple. The species are all elongate and subcylindrical, though rather less convex than in *Ptilinus*, the head minutely and densely granulose, the prothorax less minutely and very clearly and evenly granulate throughout the disk, but rather more densely at the summit of the more convex median parts near the base. The elytra have very fine, scarcely impressed striæ, which extend nearly to the apex in *horni*, and that species is well distinguished from any of the California representatives by the two basal impressions of the pronotum. The species are very much more closely allied among themselves than those of *Ptilinus*, and the male appears to be very rare in comparison with the female. The four species in my cabinet may be identified as follows from the female :—

Basal joint of the hind tarsi very much shorter than the entire remainder, the second joint relatively more elongate; pronotum impressed at each side near the basal margin; elytra more elongate; eyes separated by rather less than three times their own width. Texas .... **horni** *Lec.*

Basal joint but slightly shorter than the entire remainder, the second joint relatively
 much shorter; pronotum not impressed sublaterally at base................2
2—Pleural sulcus below the humeri deep and strongly marked; elytra twice as long
 as wide; eyes separated by evidently more than three times their own width.
 Body rather stout, the elytra subdilated near the tip, blackish-piceous above,
 the legs and antennæ dark rufous or rufo-piceous; surface feebly shining, the
 pubescence extremely short, pale, dense and conspicuous on the elytra. Head
 short, inserted to the eyes, dull, the epistomal impression well marked. Prothorax three-fifths wider than long, rounded at apex, the sides thence strongly diverging and feebly sinuate, becoming parallel and broadly rounded in basal half;
 basal angles rounded. Elytra twice as long as wide, more than three times
 as long as the prothorax and fully as wide, a little wider at apical third; humeral
 angles rounded. Length 3.7-4.5 mm.; width 1.3-1 7 mm. California (Sta.
 Cruz Mts.) .. .. **pleuralis**, sp. nov.
Pleural sulcus narrow and feeble.. 3
3—Elytra fully twice as long as wide; eyes small, separated by fully four times their
 own width. Body very slender, nearly as in *pleuralis* but narrower, the elytra
 not distinctly wider behind and fully three and one-half times as long as the prothorax, the latter nearly similar in outline but still more transverse, the sides becoming parallel in less than basal half, with the median line similarly finely impressed anteriorly. Length 2.9 mm.; width 1.0 mm. California (locality not
 indicated) ....'. .... **macer**, sp. nov.
Elytra much shorter, three-fourths longer than wide; eyes more convex and better
 developed, separated by three times their own width. Body suboblong, moderately convex, dull, blackish, almost similar throughout to *pleuralis* but shorter,
 the prothorax relatively rather smaller and the elytra much shorter, not distinctly
 dilated subapically, and with the minute subgranuliform rugulosity still finer
 and the pubescence a little denser, the fine striæ distinct to the summit of the
 convex declivity. The hind tarsi are longer than in *pleuralis*. Length 3.4
 mm.; width 1.25 mm. California (locality not indicated)..**saginatus**, sp. nov.

The descriptions are derived throughout from the female, the only
male accessible to me being one of the four examples of *pleuralis*. This
male is very much smaller and narrower than the female, with the eyes
better developed and separated by slightly more than twice their own
width; the prothorax is more transverse and almost semicircular in
outline from the base around the apex, near which it is perhaps more
correctly broadly parabolic; the last ventral segment is simple, rounded
ta apex and not quite as long as the two preceding combined; the antennæ do not differ essentially in structure from those of the female, but
are rather more slender.

BOSTRICHINI.

The genera of Bostrichini have not been considered in their mutual
relationships for twenty years, when a review of them was published by
Dr. Horn. I find it necessary to increase the genera recognized by

that author by five, the species hitherto placed in *Sinoxylon* being quite heterogeneous and in no single instance truly a member of that genus. *Sinoxylon ·dinoderoides*, *Amphicerus fortis* and *Dinoderus brevis* are also types of distinct genera. The genus proposed for the last named species is called *Patea* in the table. The genera known thus far may be thus distinguished :—

Tarsi long, with the last joint relatively shorter, the second joint usually elongate ;. claws and tibial spurs stouter, grooved beneath, the edges of the groove minutely crenulate. 2

Tarsi short, the four basal joints subequal among themselves and together nearly equal to the fifth ; claws smaller and more slender, not at all crenulate within. .9·

2—Funicular joints of the antennæ very short and closely united, together never longer than the first joint of the club, the latter long, loose and strongly compressed 3

Funicular joints more elongate and less closely united, together generally much longer than the first joint of the club .... ̇.....................6

3—Antennal club 3-jointed...4

Antennal club 4 jointed......................... ....**Tetrapriocera**

4—Antennæ 10-jointed, the elytral declivity with well-marked spines or tubercles. .5

Antennæ 9-jointed, the elytral declivity without spines or tubercles at the sides, or with very rudimentary tuberculiform irregularities, excavated near ·the suture, the latter with a spiniform elevation ; antennal club moderate in development, its joints decreasing in width and only very feebly serriform .....**Xylopertha**

5—Antennal club moderately developed, its first and second joints nearly similar in form ; elytral declivity sexspinose.......................... **Xylobiops**

Antennal club greatly elongated, its first and second joints dissimilar in form, the former more or less outwardly produced at apex ; elytral declivity quadrituberculose. ....**Dendrobiella**

6—Joints of the antennal club strongly compressed and deeply bistriate from the apical margin ; front simply tumid ; hind angles of the prothorax rounded.

Amphicerus

Joints of the antennal club strongly compressed but not striate, the two sensitive patches near the apices of the joints rounded and feebly marked ; front lamellarly prominent behind the clypeus ; hind angles of the prothorax not rounded. . **Apatides**

Joints of the antennal club but feebly compressed and generally quite convex, the first two more or less transverse, with the two sensitive patches rounded and subapical ; front not transversely prominent.................................7

7—Tibiæ dentate externally ; claws abruptly bent at base, not distinctly crenulate ;. size very large ....**Dinapate**

Tibiæ not dentate ; claws evenly arcuate ; size moderate or small...............8

8—Front margined at the sides ; eyes well developed .... **Bostrichus**

Front not margined, convex ; eyes small ....**Micrapate**

9—Antennæ with the two basal·joints relatively smaller, the funicle well developed ;. club rather short, 3-jointed.

Antennæ 10-jointed ; body elongate, the head exserted.............**Dinoderus**

Antennæ 11-jointed ; body short and stout ; head deeply inserted.........**Patea**

In *Tetrapriocera* and *Patea* the antennæ are 11-jointed. In all the others they are 10-jointed, except in *Xylopertha* and in one species of *Bostrichus*, where they have but nine joints. *Tetrapriocera longicornis* (= *schwarzi* Horn) is the only known species of that genus. *Xylopertha* is confined, as might be expected, to the subsiberian fauna of the Pacific coast, where it is represented by *bidentata, declivis* and *suturalis*, hitherto placed in *Sinoxylon*, which genus has the two basal joints of the antennal club short and transverse. *Xylobiops* is proposed for the *Sinoxylon basillare, texanum, sextuberculatum* and *floridanum* of the present lists. *Dinapate wrighti*, the type and only known species of the genus, is the largest bostrichid known ; it will probably soon become extinct by reason of the destruction of its food-plant for commercial purposes.

Dendrobiella, gen. nov.

This genus inhabits the warmer parts of the North American continent and also the West Indies; the species known to me may be identified by the following characters :—

Elytral punctures distinct throughout the disk, except at the sides, where they are obsolete, finer toward base, coarser posteriorly to the brink of the declivity, the latter smooth and impunctured as usual throughout the genus.

Larger species, 5.5–6 mm. in length, blackish in color...........**sericans** *Lec.*

Smaller species, 4 mm. in length, rufo-piceous in color.....**quadrispinosa** *Lec.*

Elytral punctures rather fine and sparse but distinct toward base, becoming wholly obsolete toward the declivity. *Male.*—Head well developed, the surface flattened, polished, nude and finely, sparsely punctulate, bituberculose at the base of the vertex ; eyes moderate, very prominent ; antennæ pale, longer than the width of the head, the first seven joints together scarcely longer than the first joint of the club, the tenth joint long and narrow. Prothorax wider than long, slightly narrowed anteriorly, broadly truncate at apex, the sides becoming parallel behind the middle ; apical asperities moderately coarse and obtuse at the sides ; disk polished, finely, subimbricately punctulate toward the middle in more than basal half. Elytra shining, the pubescence rather long, fine, decumbent, fulvous and conspicuous ; apical truncature flat and shining, the tubercles moderate, the lower more obtuse. *Female.*—Smaller than the male but nearly similar, except that the head is smaller, more convex, less shining, feebly convex, punctured, pubescent and devoid of tubercles. Length 4.3–5 0 mm.; width 1.75–2 1 mm. Texas (Brownsville)..............................**pubescens**, sp. nov.

Elytral punctures wholly obsolete, being feebly traceable only very near the base. *Male.*—Head moderate, flattened, becoming concave behind the frontal margin. minutely, sparingly puberulent, slightly shining, finely and rather closely punctulate throughout ; two small tubercles of the vertex on a transverse line through the posterior limits of the eyes; antennæ but little longer than the width of the head, nearly as in *pubescens.* Prothorax much wider than long, feebly narrowed in apical half, very broadly truncate at apex, the apico-lateral serrules acute, about three in number ; sculpture nearly as in *pubescens*, except that the disk is finely, sparsely punctulate toward base, without trace of imbricate sculpture.   Elytra smooth, conspicuously pubescent; apical tubercles small and rather feeble, Length 5.0 mm.; width 2.0 mm.   Island of Jamaica.......**sublævis**, sp. nov.

It is probable that *pubescens* is the species identified as *sericans* by Gorham in the ''Biologia.''

Amphicerus *Lec.*

This is a rather large and important genus among our bostrichids, not at all closely allied to *Apate* as is said to be the case by Mr. Gorham in the '' Biologia,'' the two differing radically in the form of the antennal club among other characters.   The species known to me are as follows :—

Elytra deeply margined at apex..... 2
Elytra not strongly margined at apex, smaller species, brown or testaceous in color, with the sculpture toward the base of the pronotum less broadly granulose and more nearly strigose......... 7
2—Elytra 4-tuberculate at the summit of the apical declivity, less distinctly so in the female ; body generally black throughout.................................3
Elytra bituberculate at the summit of the apical declivity, rudimentarily so in the female ; color dark brown, the elytra less coarsely punctate.................6
3—Pubescence of the elytra wanting or not distinguishable under low power......4
Pubescence of the elytra conspicuous but decumbent..........................5
4—Elytra of the female more elongate, distinctly more than twice as long as wide, with coarse and close-set punctures.   Southern Texas to Honduras.
                                       punctipennis *Lec.*
Elytra of the female distinctly shorter, about twice as long as wide, with smaller and sparser punctures.   *Female.*—Body cylindrical, black, polished and glabrous. Head two-thirds as wide as the prothorax, with the eyes rather large, very convex and prominent; vertex transversely tumid and pubescent; sculpture coarsely granulato-rugose; antennæ as long as the width of the head, dark rufo-piceous.   Prothorax as long as wide, parallel and broadly arcuate at the sides, narrowed and serrate at the sides anteriorly, the apex sinuato-truncate, with the apical teeth small; surface coarsely asperato-granose anteriorly, smooth with flattened contiguous tubercles posteriorly.   Elytra more than twice as long as the prothorax and a little wider, the punctures coarser and closer toward the sides and strongly and coarsely confluent on the apical declivity; tubercles rudimentary.   Abdomen with whitish pubescence, minutely and densely punctulate, with coarse punctures interspersed.   Length 12.0 mm ; width 4.0 mm.   Texas (Galveston) .... **maritimus**, sp. nov.
5—*Female.*—Body very slender, cylindrical, shining, black with a feeble piceous tinge.   Head three-fourths as wide as the prothorax, the eyes very convex and prominent; vertex moderately tumid, the surface granulato rugose with a smooth

median spot posteriorly ; antennæ rather stout, not quite as long as the width of
the head. Prothorax obviously shorter than wide, narrowed somewhat from
very near the base, more rapidly and arcuately and with moderate serrules anteriorly, the apical sinuation narrow with the teeth small ; surface coarsely as-
perato-tuberculate anteriorly, smoother in basal two thirds, the sculpture becoming coarsely subimbricate in the middle toward base, with the surface shining
and the median line finely impressed. Elytra much more than twice as long as
wide, nearly three times as long as the prothorax and just visibly wider, the
punctures coarse and close-set, subserial in arrangement, coarse, contiguous and
subconfluent on the declivity, the tubercles feeble, especially the inner. Abdomen finely punctulate, pubescent, the scattered larger punctures rather small.
Legs quite slender. Length 9.0 mm.; width 2.5 mm. Kansas, Iowa and
North Carolina. ....**gracilis**, sp. nov.
6—Pubescence distinct, decumbent. Sutural series rather impressed, the suture elevated on the declivity. Head rather small, the eyes moderate in size. Elytral
punctures not serial in arrangement, but with traces of three fine raised lines.
Indiana and Kansas ....**bicaudatus** *Say.*
7—Prothorax emarginate at apex, fully as wide as and with the usual terminal
teeth of the lateral series. *Male.*—Rather stout, cylindrical, shining, dark testa-
ceous-brown in color; antennæ pale; surface virtually glabrous. Head moderate,
nearly two-thirds as wide as the prothorax, broadly, almost evenly convex, with
a large median impunctate area; transverse impression behind the clypeus deep
and distinct ; eyes small and but moderately prominent; antennæ fully as long as
the width of the head. Prothorax fully as long as wide, the sides broadly arcuate,
becoming parallel only very near the base, converging anteriorly where the serrules are prominent and close-set in less than apical half ; apex narrowly sinuate.
surface tuberculose anteriorly, becoming smooth and polished in basal half and
almost sculptureless toward the sides but sparsely imbricato-strigose toward the
middle. Elytra short, one-half longer than wide, equal in width to the prothorax, strongly but not very closely, confusedly punctate, more closely but
scarcely coalescently behind, the declivity very steep, more convex at each side
above but not tuberculate, the suture elevated. Abdomen finely, strongly and
densely punctulate,the scattered coarser punctures not visible,the pubescence even,
decumbent and rather dense ; last segment shorter than any of the preceding.
Hind tarsi very much longer than the tibiæ. Length 6.7 mm ; width 2.2 mm.
Texas (El Paso)..**grandicollis**, sp. nov.
Prothorax truncate at tip, with the angles obtuse and rounded, without trace of processes..8
8—Larger species, the prothorax much wider than long and trapezoidal in form;
Female.—Rather slender, cylindrical, shining, subglabrous, dark rufo-testaceous
in color. Head well developed, nearly three-fourths as wide as the prothorax,
the surface granose throughout, tumid posteriorly, the epistomal suture just beyond the middle of the length and impressed toward the middle, the epistoma
large ; eyes very large, convex and prominent ; antennæ obviously shorter than
the width of the head, with the club relatively very long, the five joints of the
funicle together barely equal in length to its first joint. Prothorax much wider
than long, the sides parallel and feebly arcuate nearly to the middle, then

strongly convergent to the truncate apex, the latter not visible from above but narrow and feebly sinuate ; declivity coarsely asperate above, smoother near the apex, subserrate laterally, the teeth not extending to the apex; basal half rather dull in lustre and with short strigiform lines not densely placed.   Elytra about twice as long as wide, between two and three times as long as the prothorax and rather wider, rather coarsely, deeply and irregularly but uniformly and quite densely punctate, very densely and perforately so behind, the declivity moderately steep, more convex at each side but not tuberculate, the suture elevated. Abdomen closely punctulate, the pubescence moderately abundant.   Tarsi very long.   Length 6.5–7.0 mm. ; width 2.0–2.2 mm.   Texas (El Paso).

brevicollis, sp. nov.

Small species, 4.5–5.5 mm. in length, the prothorax as broad as long.   Body elongate, cylindrical, sparsely clothed with moderately long semi-erect hair; elytra coarsely and seriately punctate; under surface sparsely punctate.   California (Fort Yuma) ....**teres** *Horn*

Grandicollis is described from what appears to be the male, but the eyes are very small when compared with those of *brevicollis*, of which the four homogeneous examples before me seem to be females ; both of these species and probably *teres* also, which I have not seen, have the funicle of the antennæ much shorter than in the others; in *grandicollis* the five joints together are however quite distinctly longer than the first joint of the club ; in *brevicollis* they are barely as long as the first joint but do not have the closely crowded structure observed in *Sinoxylon* and *Tetrapriocera*.   In *brevicollis* there are a few erect hairs observable near the sides of the elytra especially behind, but otherwise the surface is glabrous and the punctures are only feebly subseriate in arrangement.

Apatides, gen. nov. [*Bostrichopsis Lun*]

This genus is amply distinct from *Amphicerus* in the characters of the table.   We have the following three species:—

Inner margin of the epipleuræ continuous and obliquely ascending at base to the humeral angles in the female ; basal angles of the prothorax acute and prominent; head and abdomen finely punctate, the former slightly tumid or subcarinate along the middle toward the frontal margin.   Lower California and California (Yuma)..**fortis** *Lec.*

Inner margin of the epipleuræ discontinuous at base in the female, basal angles of the prothorax not at all rounded but at the same time not distinctly prominent, the surface less impressed before the angles.................................2

2—Vertex gradually ascending to the prominent frontal margin, finely and sparsely punctate, the abdomen minutely punctulate throughout ; thoracic processes separated by rather more than a third of the total width.   *Male.*—Head three-fifths as wide as the prothorax, the latter nearly as long as wide, with the apical processes long and obliquely convergent; surface with the usual isolated tubercles toward base. Elytra twice as long as wide, just visibly wider than the prothorax, the apical declivity flattened, becoming alutaceous in lustre and almost impunctate toward the suture, which is elevated   *Female.*—Similar to the male but larger, with the thoracic processes short and parallel, the elytra rather more than twice as long as wide and more distinctly wider than the prothorax, the apical declivity convex and coarsely perforato-punctate throughout, the suture moderately elevated.   Length 13.0–15.5 mm.; width 4.3–5.1 mm.   Texas (El Paso)..**robustus**, sp. nov.

Vertex more tumid and convex, less finely and quite strongly punctured throughout; abdomen strongly though sparsely punctured toward base; thoracic processes more approximate, separated by but little more than a fourth of the total width.   *Female.*—Head moderate in size, the eyes very convex and prominent as usual. Prothorax not quite as long as wide, nearly as in *robustus* but less devoid of sculpture toward the basal angles.   Elytra not at all mcre than twice as long as wide, the apical declivity rather more convex at each side than in *robustus*, steeper and a little less coarsely punctured.   Abdomen polished as usual, the punctures becoming finer and denser toward apex.   Length 12.5 mm.; width 4.3 mm.   Arizona (Locality not specified—Levette Cabinet.)

puncticeps, sp. nov.

The male of *fortis* has the apical processes more convergent and longer than the female, but there seems to be no modification of the elytral declivity near the suture.   Individuals vary much in size as usual in the Bostrichinæ.

Bostrichus *Geoff.*

The genus *Bostrichus*, as represented in America, differs remarkably from *Amphicerus* in the structure of the antennal club, the joints being short, subglobose, and with the sensitive spaces small and circular; it also differs in having the basal angles of the prothorax acute and prominent, but in that respect resembles *Apatides*, from which it differs in turn in the structure of the antennal club and frontal parts of the head.   The following table comprises all the species known to me at present : —

Prothorax narrowly and deeply sinuate at apex, with the limiting processes prominent and generally unciform ; elytra each with two ridges more or less distinct or interrupted...2
Prothorax sinuato-truncate at apex, with the limiting angles acute and somewhat prominent ; vestiture hair-like, decumbent and unevenly distributed ; elytra without trace of ridges ; species smaller and more slender 5
2—Hind tarsi fully as long as the tibiæ ; unciform processes of the prothorax more prominent.   Atlantic regions...3
Hind tarsi shorter than the tibiæ ; unciform processes shorter, not differing in form from the lateral serrules.... 4

3 —Vestiture of the elytra squamiform ; inner ridge strong and continuous to the
   apical declivity ....**bicornis** *Web.*
Vestiture more hair-like and still more unevenly disposed in clusters ; inner ridge
   feeble and much interrupted, the outer almost obsolete ....**armiger** *Lec.*
4—Elytral vestiture long and hair-like, very sparse and almost evenly disposed ;
   ridges fine, feeble and subobsolete ....**californicus** *Horn*
5—Antennæ 10-jointed as usual........ ....**truncaticollis** *Lec.*
Antennæ 9-jointed.   Evenly cylindrical, black, the antennæ and tarsi paler ; vestiture coarsely hair-like, fulvous in color, dense and conspicuous, somewhat uneven on the elytra but much less nucleated than in *truncaticollis*.   Head
   moderate, opaque, pubescent, the eyes well developed ; antennæ as long as the
   width of the head, the funicle 4-jointed.   Prothorax nearly as long as wide,
   roughly tuberculose, pubescent, the basal angles acutely prominent ; median
   line somewhat depressed.   Elytra slightly wider than the prothorax, two and
   one-half times as long as wide, coarsely, densely, unevenly punctured and finely
   tuberculose.   Legs rather short and slender, the hind tarsi longer than the tibiæ.
   Length 6.4 mm.; width 1.8 mm.   New Jersey (Woodbury).

                                                        angustus, sp. nov.

In the males the elytral apices are minutely spinulose throughout,
but there is very little sexual difference otherwise, except that the male
is generally smaller and with the elytra less elongate.   It will probably
prove necessary to generically separate the American species of *Bostrichus* from the European forms, when the family is monographed as a
whole.

Micrapate, gen. nov. [*Bostrychulus Leun*]

This genus is founded upon the *Sinoxylon dinoderoides* of Horn,
and its allied species, and I have ventured to include also the *S. simplex*
of that author, although the size is so much greater that renewed observation would possibly disclose some divergencies of a generic nature.   I
should have been disposed to refer the specimens described above under
the name *Amphicerus brevicollis* to *S. simplex*, were it not for the fact
that the basal parts of the pronotum are said to be "densely punctate,"
which language it would be impossible to apply to *brevicollis*, where the
sculpture of that part consists of short, isolated and longitudinal raised
lines, as in the *Amphicerus teres* of Horn.   It is a peculiarity of *Micrapate* that the basal parts of the pronotum are truly and simply punctate, and not in any way asperate, granose or tuberculose.   Our species
are as follows :—

Pronotum less densely or rather sparsely punctured toward base ; size much smaller,
   never materially exceeding 4 mm. in length.............................2
Pronotum densely punctate toward base.......................................3
2—Surface "feebly shining;" suture moderately and evenly elevated on the apical
   declivity.   Arizona and Texas (Brownsville) ....**dinoderoides** *Horn*

Surface strongly shining; sutural elevation on the declivity strong, its summit for a short distance at the middle of the declivity, still more elevated, dilated and canaliculate. *Female*.— Similar to *dinoderoides* but smaller, the epistomal suture more deeply impressed and more remote from the apical margin. Prothorax nearly as long as wide, similar to *dinoderoides* but still more sparsely punctate toward base. Elytra rather coarsely, strongly punctured and very densely so, the punctures rather sparser toward the suture except on the declivity, but not as sparse as in *dinoderoides*, the surface unevenly rugose by anteriorly oblique light. Under surface finely and densely punctulate, confluently so on the sterna. Length 3.4 mm.; width 1.15 mm. District of Columbia .....**cristicauda**, sp. nov.

♂—Size larger, 6.5 mm. in length. Body piceous, the elytra brownish; head opaque, tuberculate, the maxillary palpi with the last two joints equal; prothorax wider than long; elytra not wider than the prothorax, very coarsely and closely punctate, the punctures of the declivity coarser and denser, the sutural region slightly elevated, especially in the apical declivity. Body beneath moderately densely punctate, sparsely pubescent. Texas (southwestern) ....**simplex** *Horn*

I have here regarded the specimens recently taken by Mr. Wickham in the extreme southern part of Texas, near Brownsville, as representing the true *dinoderoides*, but actual comparison will be necessary to decide, as these examples are certainly strongly shining.

Dinoderus *Steph.*

The rather numerous species of this genus may be outlined in the table which follows. *Punctatus* and *truncatus* are the only discordant elements after eliminating *brevis*, and they may have to be separated at some future time.

Apex of the elytra convex, the suture only very rarely somewhat prominent, the apical margin not concave or prominently margined; pubescence erect...... 2
Apex of the elytra more abruptly truncate, concave and prominently margined at tip; · pubescence decumbent. 13
2—Pronotum with granuliform and separated tubercles toward base............ 3
Pronotum with flattened and generally subcontiguous tubercles toward base; side margins almost devoid of serrulation except at apex; body more cylindro-convex 8
3—Elytra polished or strongly shining.................................... 4
Elytra opaque; color dark brown or blackish-piceous........................ 5
4—Elytra with very close-set perforate punctures, larger than the width of the intervals, the latter tuberculose; color dark brown throughout. Michigan, Canada and Europe......................... ....**substriatus** *Payk.*
Elytra with less coarse and impressed punctures, not larger than the width of the intervals, the latter less elevated and more feebly but distinctly tuberculose; color black or blackish. Head moderate, exserted, with a polished constriction at base as usual; surface subopaque, granulose, the epistomal suture distinct; apex sinuate; eyes small, convex; antennæ stout, dark rufous, not as long as the

width of the head. Prothorax not quite as long as wide, the apex broadly arcuate, the sides becoming parallel and feebly arcuate near the base, serrate throughout, rather strongly at the rounded basal angles and still more coarsely around the apex; base broadly lobed; surface with small, strong and isolated granules throughout, intermingled anteriorly with some larger sparse asperities. Scutellum small. Elytra not quite twice as long as wide, more than twice as long as the prothorax and slightly wider; surface with series of moderately coarse punctures, confused near the suture and smaller and less seriate on the flanks; intervals asperate; apex evenly convex, with the punctures confused and asperate. Abdomen shining, sparsely punctulato-rugose and finely, sparsely pubescent. Length 4.0–4.8 mm ; width 1.3–1.6 mm. California (Calaveras Co.), Colorado and Idaho (Cœur d'Alène)..........................**pacificus**, sp. nov.
Elytra with less coarse and more impressed punctures, nearly as in *pacificus* and not larger than the width of the intervals, the latter perfectly even, polished and devoid of tubercles or asperities throughout. Body deep black, the erect hairs of the elytra rather short. Head dull, sparsely pubescent, the epistoma broadly sinuate. Prothorax not quite as long as wide, arcuately swollen toward base, broadly rounded and asperato-tuberculose at apex; disk granose toward base. Elytra slightly wider than the prothorax, rather short, four-fifths longer than wide, the punctures seriate in arrangement, densely confused near the suture, more broadly toward base, small and irregular in arrangement toward the side margins, the apical declivity evenly convex and not at all granulose though more closely and unevenly punctate. Abdomen shining, sparsely punctulate. Length 2.7–3.7 mm.; width 0.8–1.2 mm. Wyoming (Laramie) and Arizona.
sobrinus, sp. nov.
5—Elytral granules strong and well defined, arranged in even single series along the intervals........................ 6
Elytral granules subobsolete except on the declivity, the punctured series contiguous, with the intervening ridges narrow and alternately slightly stronger 7
6—Elytra roughly and densely punctate on the declivity, the tuberculose intervals equal throughout, finely and confusedly on the flanks. Head short and transverse, granose, the basal constriction exposed as usual; eyes small; antennæ short, the club paler; epistomal suture subobsolete. Prothorax slightly shorter than wide, nearly as in *pacificus*. Elytra not quite twice as long as wide, rather wider than the prothorax, the lustre dull, the sculpture coarse and rough, the punctures of the series large, deep and approximate but circular and well defined, except at the sides. Abdomen minutely, sparsely punctulate, feebly pubescent. Length 4.0 mm.; width 1.4 mm. New Mexico (Fort Wingate)...**asperulus**, sp. nov.
Elytra finely, evenly and strongly granose on the declivity; intervals separating the punctured series equal in elevation ; punctures of the series coalescent and not well defined. Head short and transverse, finely granose, the labrum declivous, the eyes and antennæ moderate. Prothorax not quite as long as wide, broadly rounded and strongly asperate anteriorly, the sides feebly diverging to the rounded and asperate basal angles; disk with the granules equal, strong and isolated toward base. Elytra but little wider than the prothorax, scarcely twice as long as wide, densely sculptured in even series, except near the suture and

more broadly on the flanks, the elevations polished. Length 5.2 mm.; width
1.6 mm. Arizona (Seligman).. .. ..**amplus**, sp. nov.
Elytra rather sparsely and strongly granose on the declivity; intervals separating the
punctured series alternating in prominence; punctures of the series subtransverse, subcoalescent and not well defined. North Carolina...**porcatus** *Lec.*

7—Punctures of the elytral series confluent, opaque and not well defined. Head
transverse, opaque and granulose; eyes small; antennæ short, dark rufous, the
club not paler. Prothorax nearly as in *pacificus*, the basal angles less rounded.
Elytra about twice as long as wide, slightly wider than the prothorax and much
more than twice as long; sculpture very dense, the surface densely opaque;
erect hairs moderate in length, stiff and fulvous. Abdomen rather dull, finely,
sparsely punctulate. Length 3.0–4.0 mm.; width 0.9–1.2 mm. Virginia (Norfolk)......................... ....**opacus**, sp. nov.

8—Elytral punctures confused in arrangement, at least toward the sides and suture..9
Elytral punctures forming perfectly even series throughout the width, the intervals
even...12

9—Apical declivity of the elytra granulose, the punctures more close-set throughout. 10
Apical declivity simply punctate................................. 11

10—Elytral punctures distinctly asperate throughout. Body and legs blackish, the
antennæ rufo-piceous; surface moderately shining. Head short, not very
densely granose. Prothorax not quite as long as wide, the sides feebly convergent from near the broadly rounded basal angles, merging gradually into the
broadly rounded and moderately serrulate apex; surface sparsely, rather strongly
asperate anteriorly, more closely granulate toward base, the granules flattened,
less dense laterally. Elytra about two-thirds longer than wide, twice as long as
the prothorax and scarcely wider; punctures not very coarse, serial in arrangement, the intervals flat and even; apex evenly convex, strongly grano-tubercu-
lose. Abdomen shining, sparsely punctulate. Length 3.7 mm.; width 1.2
mm. New Jersey......................... ....**hispidulus**, sp. nov.
Elytral punctures circular, not asperate on the disk and toward the suture, feebly
granuliferous on the convex declivity; elytra polished, the intervals flat; serial
arrangement of the punctures only observable along the middle of each elytron.
South Carolina............................. ....**densus** *Lec.*
Elytral punctures abnormal, not rounded but somewhat dilated at their posterior
limits, serial in arrangement and well separated, more confused near the suture
and broadly toward the sides, not granulose except posteriorly and on the declivity. Body evenly cylindrical, shining, dark piceous, the elytral vestiture
sparse, stiff and erect. Head nearly smooth, constricted at base as usual. Prothorax nearly as long as wide, oval, asperulate anteriorly, the basal angles
rounded; disk with the flattened and nearly contiguous tubercles toward base
small. Elytra perfectly cylindrical, barely twice as long as the prothorax and
perceptibly wider, not quite twice as long as wide, polished. Length 2.4 mm.;
width 0.7 mm. Pennsylvania.;.............. ....**parvulus**, sp. nov.

11—Dark rufo-piceous, the elytra blackish, highly polished with rather small and
simple punctures, which are only feebly subserial in arrangement, becoming
very small and feebler on the flanks and simple on the declivity. Indiana and
South Carolina.......................... ....**cribratus** *Lec.*

12—Body small, narrow, subglabrous, highly polished and pale rufo-testaceous throughout; apical margin of the prothorax rather crenulate than serrulate. Iowa (Keokuk).   Cosmopolitan and introduced............**pusillus** *Fabr.*

13—Antennæ with the second joint nearly as slender as the third, the funicle bristling with long coarse hairs anteriorly; ridge of the apical declivity short; head strongly, transversely tumid behind the epistoma.   New York, Indiana and South Carolina.. **punctatus** *Say*

Antennæ with the second joint stout, the funicle not more setose in front; declivity more abrupt and flat, with the marginal ridge long.   California.

truncatus *Horn*

I have not been able to compare *substriatus** of the table with European examples, and the identification is taken from the books; it is referred to the genus *Stephanopachys* by Heyden, Reitter and Weise, who separate also *pusillus* under the generic name *Rhizopertha* (*Rhyzopertha*).   The differences seem to be scarcely generic in value. *Truncatus* of Horn, I have not seen.

CIOIDÆ.

The Cioidæ are intimately related to the Bostrichinæ, as shown by general organization, and particularly by the two small rounded sensitive areas near the apices of the joints of the antennal club, greatly developed in the genus *Plesiocis*; but, at the same time, they are closely allied also to some groups at present assigned to the Clavicornia, such as the Cryptophagidæ and Mycetophagidæ.   In fact, the assemblages which are at present collectively known as the Clavicornia, are so heterogeneous among themselves as to indicate that they do not form a natural division of the Coleoptera at all, but are in many cases the extreme developments of various types of Serricornia or Adephaga, and the Heteromera belong near them in immediate succession.   *Berginus* has a purely serricorn habitus, and yet has been placed with the Mycetophagidæ.   I believe that the Cryptophagidæ and Mycetophagidæ should not be widely separated from Cioidæ and Sphindidæ, and I am in favor of removing them from the so-called Clavicornia and placing them in the Serricornia near Cioidæ.   This would be far more natural than to remove the Cioidæ to the Clavicornia.   The Cucujidæ, consisting of the subfamilies Passandrinæ, Colydiinæ, Monotominæ, Rhysodinæ, Lyctinæ, Silvaninæ, Brontinæ, Cucujinæ and Hemipeplinæ should also be removed from the Clavicornia and follow Cioidæ, Cryptophagidæ, etc., in the Serricornia.   The Hemipeplinæ form a natural transition to the Heteromera.

---

* *Dinoderus substriatus* is said by Mannerheim (Bull. Mosc., 1853, p. 233), to inhabit also the Kerai Peninsula, in Alaska.

The Cioidæ consist of two subfamilies, Cioinæ and Rhipidandrinæ, distinguished by clavate and compactly serrate antennæ respectively. The American genera of Cioinæ are as follows :—

Antennæ 10-jointed 2

Antennæ 9-jointed............................ 7

Antennæ 8-jointed ; body glabrous........ 8

2—Prosternum well developed before the coxæ; lateral edges of the prothorax acute to the apex 3

Prosternum very short and transversely excavated before the coxæ; lateral edges of the prothorax becoming subobsolete at the apex..........................6

3—The prosternum simple or nearly so..4

The prosternum tumid or carinate along the middle 5

4—Body setose or pubescent the vestiture erect and bristling, the anterior tibiæ finely produced and dentiform externally at apex, sometimes simple ....**Cis**

Body glabrous, the anterior tibiæ wholly unarmed at apex; elytral suture margined toward tip; body elongate, the head rather less deflexed than usual, the head and prothorax simple in the male, the latter with a deep rounded setigerous fovea at the centre of the first ventral segment....................**Orthocis**

5—Body glabrous or with very short decumbent pubescence or inclined setæ.

Xestocis

6—Body very short, oblong-oval in form, with stiff erect pubescence as in *Cis.*

Brachycis

7—Body stout, convex, coarsely cribrate and setose; anterior tibiæ strongly, obliquely produced and acute externally at apex....................**Plesiocis**

Body narrow, cylindrical, feebly sculptured and glabrous, the anterior tibiæ thickened and externally rounded and spinulose at apex.................**Ennearthron**

8—Anterior tibiæ swollen, rounded and spinulose externally at apex as in *Ennearthron*; head and prothorax strongly modified in the male............ **Ceracis**

Anterior tibiæ narrowly triangular, the external edge straight throughout and minutely spinulose ; head and prothorax not modified in the male. . **Octotemnus**

The term glabrous, as used above, signifies the absence of distinct pubescence ; with high power each puncture can be seen to bear a very small hair. Many of Mellié's species are still unknown to me, and the localities of some of them may be open to doubt ; a few may possibly be synonyms, as, for example, *atripennis*, which may have been founded upon a damaged specimen of *fuscipes*. It is possible that the *Cis pumicatus* of Mellié may prove to be an *Octotemnus*. *Ceracis* is very closely allied to *Ennearthron*, and was indeed considered to be more properly a subgenus by Mellié. The figure of *C. sallei*, on plate 4 of the monograph, seems to have been taken from a specimen of *Ennearthron mellyi*.*

---

* I am indebted for several very interesting species of Cioidæ to my friend, P. Jerome Schmitt, of Westmoreland county, Pa., and Mr. Wickham has also contributed a number of interesting species in Bostrichinæ, Cioidæ and Sphindidæ.

Cis *Latr.*

Only those species represented before me are included in the following table:—

Elytra with shallow, variolate and nude punctures, intermingled with others smaller and deeper which bear the setæ...........2
Elytra deeply punctured throughout, all the punctures bearing hairs or setæ......11
2—Body stout and convex, the elytra confusedly rugulose; maxillary palpi very stout; anterior tibiæ acute and feebly everted externally at tip; apical angles of the prothorax right and somewhat prominent, the apex prolonged and broadly rounded over the head; base not distinctly margined......................3
Body narrowly elongate-oval and more depressed, the surface less rugose but with the elytral series more distinct; maxillary palpi variable but generally less stout; antennal club smaller, with the two basal joints wider than long; apical angles of the prothorax obtuse, the apex broadly, evenly rounded over the basal parts of the head, the base finely margined; scutellum transversely oval............8
3—Elytral bristles moderate in length, more or less distinctly serial in arrangement, the antennal club long and loose, with the two basal joints as long as wide; head and pronotum finely, evenly punctured, the elytral punctures fine, not very distinct and rather sparse; male sexual characters very feeble................4
Elytral bristles extremely short, distributed uniformly but without order; antennal club shorter, with the two basal joints wider than long; male characters pronounced......................... 7
4—Third and fourth joints of the antennæ elongate and equal, each as long as the fifth and sixth together. Body stout, shining, castaneous in color, the head moderate, with the clypeal margin feebly reflexed and broadly subtruncate; eyes rather well developed; prothorax distinctly wider than long, the sides reflexed, feebly convergent and feebly, evenly arcuate from the obtuse basal angles to the apex; elytra one-half longer than wide, twice as long as the prothorax and just visibly wider. Length 2.4–2.8 mm.; width 1.05–1.25 mm. North Carolina (Asheville) .... **carolinæ**, sp. nov.
Third joint much longer than the fourth, the latter distinctly shorter than the fifth and sixth combined...5
5—Concave side margin of the pronotum not at all inwardly prolonged at base; body large, generally pale in color, shining; prothorax more than one-half wider than long, the basal angles very obtuse and rounded; sides slightly convergent and very feebly, evenly arcuate throughout; elytra barely one-half longer than wide, nearly two and one-half times as long as the prothorax and slightly wider, the humeral callus small but pronounced. Length 2 8–3.0 mm.; width 1.1–1.3 mm. Montana (Missoula) ....**pallens**, sp. nov.
Concave margin more or less distinctly prolonged inwardly at base; color black or piceous-black, the size smaller...6
6—Male with the prothorax simple throughout. Atlantic regions, from Massachusetts to Iowa and Texas (Houston) ....**fuscipes** *Mell.*
Male with the prothorax broadly impressed at apex. Body moderately stout, strongly convex, blackish in color and shining, the elytral punctures generally stronger

than those of the pronotum, with the impressed lines distinct; head with the
clypeal margin moderately reflexed and broadly sinuato-truncate in both sexes;
prothorax two-fifths wider than long, the sides feebly convergent, rather more
rounded near the base; elytra one-half longer than wide, fully twice as long as
the prothorax and just visibly wider posteriorly. Length 1.8–2.75 mm.; width
0.8–1.2 mm.   Pacific coast—Vancouver Island, Washington State (Tacoma) and
California (Humboldt Co. and Alameda)............... **impressa**, sp. nov.

7—Body rather stout, strongly convex, oblong-suboval, shining, blackish· in color
throughout, the legs and antennæ dark rufous; vestiture very short and almost
scale-like, erect as usual; head moderate, the eyes well developed, convex and
prominent; prothorax one half to three-fifths wider than long, the sides rather
widely reflexed, slightly convergent and broadly, evenly arcuate throughout, the
basal angles very obtuse ; surface finely, closely punctured but polished; elytra
more than one-half longer than wide, nearly two and one-half times as long as
the prothorax and very slightly wider, the humeral callus small; surface confusedly rugulose, finely punctate and with slightly evident longitudinal lines and
short transverse rugæ.   *Male.*—Head concave, the clypeal margin reflexed and
broadly bidentate; prothorax impressed transversely at the apical margin, the
latter moderately reflexed, with a small rounded sinuation at the middle.   *Female.*—Head flat, the clypeal margin very slightly reflexed, broadly, feebly
sinuato-truncate, the prothorax rounded and unmodified at apex.   Length 2.2–
2.5 mm.; width 0.9–1.1 mm.   Rhode Island (Boston Neck)..**pistoria**, sp. nov.

8—Anterior tibiæ finely everted and acute externally at apex.................9
Anterior tibiæ simple at apex 10

9—Pronotum not impressed at the apical angles, the flanks deeper, the side margin
feebly reflexed, more strongly about the basal angles.   *Female.*—Body elongate-
oval, moderately convex, piceous, the elytra black; legs and antennæ paler,
rufous, shining, the bristles short and pale, moderately abundant, not arranged
in definite series on the elytra though with feeble suggestion of such arrangement at certain parts; head moderate, the eyes small, the clypeal margin broadly
arcuate; prothorax nearly as long as wide, circularly arcuate in apical third, the
sides thence nearly straight and parallel to the basal angles, which are very obtuse; base arcuate; punctures rather fine, strong and close-se ; elytra three-
fourths longer than wide, two and one-third times as long as the prothorax and
scarcely wider; punctures fine, strong, close-set, the impressed lines distinct and
with rather coarser irregular punctuation.   Length 2.3 mm.; width 0.8 mm.
Colorado (Salida)...... ....**striolata**, sp. nov.
Pronotum impressed at the apical angles, the side margins strongly, narrowly and
equally reflexed throughout.   *Female.*—Nearly similar to *striolata* but shorter,
the prothorax fully one third wider than long, with the sides subparallel, evenly
and feebly arcuate throughout, the apex broadly, evenly arcuate; punctures fine,
strong and rather close-set; elytra two-thirds longer than wide, two and one-half
times as long as the prothorax, the surface polished, with distinctly impressed
lines of much coarser punctures, which are shallow, nude and variolate as usual,
the bristles arranged more definitely in series.   *Male.*—Smaller than the female
and more slender, the clypeal margin rather strongly rounded near the eyes and
remotely and feebly bituberculate at the middle; prothorax only slightly shorter

80   JOURNAL NEW YORK ENTOMOLOGICAL SOCIETY.   [Vol. VI.

than wide, the sides feebly convergent and evenly and feebly arcuate from the
base, the apex circularly rounded, the surface dull; elytra polished, nearly as in
the female; first ventral segment foveate at the centre.   Length 2.0–2.2 mm.;
width 0.65–0.75 mm.   Utah (southwestern)............**fraterna,** sp. nov.
10—Body more slender, picious black throughout.   *Female.*—Narrowly elongate-
oval, moderately convex, shining; legs and antennæ rufous; bristles short, pale
as usual, arranged in almost regular series on the elytra; front feebly convex;
eyes moderate in size; clypeus broadly arcuate, very short before the eyes; prothorax nearly one-third wider than long, the sides feebly convergent, evenly and
feebly arcuate from base to the rather pronounced apical angles, which are not
rounded, the apex circularly arcuate, the punctures fine but deep, moderately
close; elytra two-thirds longer than wide, nearly two and one-half times as long
as the prothorax and somewhat wider, the humeral callus minute; series well
impressed, almost regular but not much more coarsely punctate, the intervals
sparsely punctulate.   Length 1.9 mm; width 0.7 mm.   California (Lake Tahoe).
                                                        macilenta, sp. nov.
Body stouter and more cylindric, bicolored, the head and prothorax rufous, the elytra
black.   *Female.*—Oblong-subcylindric, moderately convex, slightly dull in
lustre; bristles short, feebly subserial on the elytra; head feebly convex, the
clypeus broadly arcuato-truncate, oblique at the sides to the eyes, which are
small; prothorax fully one-third wider than long, nearly as in *macilenta,* the
basal angles more broadly rounded; elytra scarcely more than one-half longer
than wide, but little more than twice as long as the prothorax and not wider,
the impressed lines feeble and somewhat irregular, more coarsely punctured.
Length 1.4–1.8 mm.; width 0.55–0.75 mm.   California (Calaveras, Humboldt,
Lake and Los Angeles Cos.)......................**versicolor,** sp. nov.
11—Vestiture of the elytra stiff and bristle-like ,.............. 12
Vestiture of the elytra long, slender and hair-like but erect and conspicuous; elytral
punctures arranged without order, not at all seriate at any point; last joint of the
maxillary palpi acutely pointed........ 25
12—Vestiture more or less distinctly serial in arrangement 13
Vestiture not at all serial at any point, the punctures evenly distributed..........15
13—Body strongly cylindro-convex, the elytral punctures differing among themselves
in size, the larger forming more or less indefinite series; bristles unusually
long......... 14
Body narrow, parallel, distinctly depressed, the punctuation dense, the elytral punctures more uniform in size, the bristles moderate in length, forming close and
nearly even series.   Pennsylvania to Texas ....**creberrima** *Mell.*
14—Sides of the prothorax becoming straight and parallel behind the middle.   *Male.*
—Body subcylindric, somewhat shining, castaneous in color, the bristles coarse,
erect, longer than the width of the scutellum, subserial on the elytra; head moderate, the front flat, the eyes small; clypeal margin feebly reflexed, remotely and
feebly bituberculate, a small sinus just without each tubercle and thence strongly
oblique for some distance to the eyes; prothorax nearly as long as wide, circularly rounded at apex, narrowly subsinuate at the middle; angles obtuse; base
finely margined; surface very obsoletely, transversely impressed at apex; punctures uneven in size, small, deep, not very close-set; scutellum pointed behind ;

elytra two-thirds longer than wide, equal in width to the prothorax and barely twice as long, obtuse at apex; series of coarse punctures scarcely impressed. *Female.*—Nearly similar to the male, the clypeal margin evenly arcuato-truncate, the prothorax not modified. Length 2.4–2.9 mm.; width 0.9–1.1 mm. Utah (southwestern)........................ ....**mormonica**, sp. nov.
Sides of the prothorax subparallel and evenly arcuate throughout. *Male.*— Similar to *mormonica* in the modifications of the clypeus and prothorax, pale piceous, polished, the bristles long, stiff and erect, subserial on the elytra; eyes small; prothorax fully one-third wider than long, the angles obtuse; punctures moderately fine, deep, somewhat uneven in size, rather close-set; elytra less than twice as long as wide, as wide as the prothorax and barely twice as long; punctures rather coarse and close-set, the larger only partially forming indefinite and scarcely at all impressed series. Length 2.0 mm; width 0.85 mm. Pennsylvania (Westmoreland Co.)............. ....**horridula**, sp. nov.
15—Body obese and strongly convex, suboval; male sexual characters pronounced, the female also having the apex of the prothorax at least feebly bilobed; apical angles of the anterior tibiæ everted and acute externally...................16
Body subcylindric, convex; male sexual characters feeble, the clypeus finely bituberculate; maxillary palpi slender; prothorax margined at base, the angles obtuse........................ 21
16—Clypeus angulate at each side near the eyes in both sexes.................17
Clypeus emarginate in the middle and bidentate, not angulate near the eyes......20
17—Elytra very nearly one-half longer than wide..........................18
Elytra very short, scarcely one-third longer than wide 19
18—Elytral punctures rather close-set. *Male* with the clypeal margin reflexed and quadridentate, the apex of the prothorax with two broad porrect triangular processes, separated by a rounded sinuation. California..........**vitula** *Mann.*
Elytral punctures rather sparse, the integuments more shining. *Female.*—Body elongate-oval, very convex, polished, castaneous, the legs, antennæ and sometimes the anterior parts paler; bristles of the prothorax very small and rather fine, not conspicuous, of the elytra coarse, moderately long and rather sparse; head concave apically, the clypeus broadly rounded and obscurely quadrangulate; eyes rather small; prothorax one-fourth wider than long, the sides feebly convergent and very feebly, evenly arcuate from base to the rather obtuse but somewhat prominent apical angles; base transverse, very feebly lobed at the middle, very finely margined; apex advanced, rounded and feebly bilobed; punctures fine and moderately close; scutellum obtuse, wider than long; elytra as wide as the prothorax and slightly less than twice as long, perfectly even, the punctures deep, very much larger than those of the pronotum. Length 2.3–2.5 mm.; width 1.15 mm. California (Humboldt Co.)............**illustris**, sp. nov.
19—*Female.*—Body stout, oval, strongly convex, pale in color, polished, the elytral bristles very short, those of the prothorax rather inconspicuous; head nearly as in *illustris*, less concave anteriorly, the eyes very small; prothorax nearly as in *illustris* but shorter, nearly one-half wider than long, the punctures very small and rather sparse; surface occasionally with a very obsolete median canaliculation near the apex; elytra very short, scarcely two-thirds longer than the prothorax, strongly convex, obtusely rounded behind, the punctures rather

coarse but feebly impressed and quite sparse. Length 2.1 mm.; width 1.0 mm.
Louisiana .... **congesta**, sp. nov.
20—*Male.*—Cylindric-oval, not very stout. strongly convex, pale in color probably
from immaturity, rufo-testaceous, shining; bristles very stout but short, distinct
and rather close on the prothorax, somewhat sparse on the elytra ; head and
eyes rather well developed, the front flat ; clypeus strongly reflexed, triangularly
bidentate ; prothorax two-fifths wider than long, the sides rather strongly convergent and arcuate from base to apex, the latter reflexed and triangularly bidentate ; base truncate ; punctures quite coarse, deep and close-set ; elytra less
than one-half longer than wide, four-fifths longer than the prothorax, the punctures about equal in size to those of the pronotum but sparser. Length 1.4
mm. ; width 0.6 mm. California...................... **duplex**, sp. nov.
21—Prosternum normally convex ; anterior tibiæ externally everted and acute at
apex ; scutellum small, not wider than long ; prothorax rounded at the apex ;.
male with the first ventral simple. *Male.*—Body narrowly cylindric-oval,
moderately convex, piceous-black, with the legs and antennæ pale ; surface
shining ; bristles coarse, pale, erect, moderately sparse, even in length on the
elytra ; head and eyes small ; clypeal margin feebly reflexed, bituberculate ; prothorax nearly as long as wide, parabolically rounded anteriorly, with a small and
very feeble median sinuation, the sides becoming straight and parallel toward
base ; punctures fine but perforate, rather close-set ; elytra rather more than one-
half longer than wide, as wide as the prothorax and twice as long, the punctures rather coarse, well separated and subeven in size. Length 2.0 mm. ;
width 0.7 mm. California (Lake Tahoe) ....'.....**hystricula**, sp. nov.
Prosternum broadly and feebly biconcave ; anterior tibiæ thickened and rounded externally at apex ; scutellum larger ; prothorax feebly sinuate from above at the
converging sides of the apex ; maxillary palpi with the last joint more acutely
pointed ; male with the first ventral foveate at the middle..................22
22—Eyes small, the body more elongate and cylindric......................23
Eyes large and well developed ; body stouter and more cylindric-oval...........24
23—Antennal funicle longer than the club. *Male.*—Moderately convex and shining,
rather pale castaneous, the bristles stiff, moderately long and rather abundant ;
head rather well developed, the clypeal tubercles small and separated by a fourth
of the entire width ; prothorax nearly a fourth wider than long, rounded and somewhat lobed at apex, the sides becoming nearly straight and parallel toward
base ; punctures rather strong and close-set though not very coarse ; elytra one-
half longer than wide, as wide as the prothorax and rather more than twice as
long, the punctures quite coarse, impressed and somewhat close-set. Length
1.75–1.8 mm. ; width 0.75 mm. Montana (Missoula).....**montana**, sp. nov.
Antennal funicle equal in length to the club. *Male.*—Dark rufo-piceous, the elytra
black, the legs and antennæ pale, shining, the bristles stiff, erect and pale but
rather sparse throughout ; head well developed, the minute tubercles of
the clypeus separated by a little more than a fourth of the width ; prothorax
nearly as in *montana* but nearly a third wider than long, with the punctures
much less close-set ; elytra one-half longer than wide, as wide as the prothorax
and rather more than twice as long, the punctures moderately coarse, deep and

not very close-set. Length 1.5–1.7 mm.; width 0.65–.75 mm. Vancouver
Island, Washington State and Northern California............**soror**, sp. nov.
24—*Male.*—Black and shining, the anterior parts picescent; legs and antennæ pale;
bristles rather sparse, short and somewhat inconspicuous anteriorly, longer on
the elytra; head well developed, the minute clypeal tubercles separated by a
fifth of the width; prothorax nearly as in *soror* but fully two-fifths wider than
long, the punctures strong and well separated; elytra suboval, not more than
two-fifths longer than wide, rather wider than the prothorax and distinctly more
than twice as long; punctures only moderately coarse but deep and quite
sparse. Length 1.6–1.75 mm.; width 0.75 mm. New York.

curtula, sp. nov.

25—Anterior tibiæ everted and acute externally at apex; hairs very long, a fifth or
sixth as long as the entire width of the elytra. *Male.*—Body stout, cylindric,
polished, piceous in color, the vestiture very long and bristling, abundant; head
and eyes moderately developed; front feebly concave; clypeus with two long
slender erect and widely separated processes; prothorax slightly wider than
long, the sides just visibly convergent and nearly straight from base nearly to
the apex, then rounding and strongly convergent to a trapezoidal and obliquely
upturned lamina, which is subtriangularly emarginate at tip; surface with
rather strong and close-set punctures, feebly impressed at apex behind the
lamina; elytra short, cylindric, obtuse at apex, two-fifths longer than wide, as
wide as the prothorax and three-fifths longer, the punctures rather coarse, even,
moderately close-set, the surface not at all rugose. *Female.*—Smaller than the
male, the clypeus feebly reflexed at each side; prothorax shorter and more transverse, simple. Length 1.75–2.1 mm.; width 0.75–0.9 mm. Florida (Lake
Worth)..**hirsuta**, sp. nov.
Anterior tibiæ simple at apex, not dilated or produced; hairs shorter, about an eighth
as long as the entire width of the elytra. *Female.*—Oblong oval, moderately
convex, shining, pale rufo-testaceous, the vestiture only moderately abundant
and not dense; head moderate, the eyes small; clypeal margin feebly reflexed
at each side; prothorax nearly one-half wider than long, parabolically rounded
at apex, the sides becoming parallel and nearly straight behind the middle;
punctures fine and rather sparse, elytra suboval, rather ogival at tip, two-fifths
longer than wide, rather wider than the prothorax and more than twice as long,
the punctures somewhat coarse but feeble, well separated; humeral callus rather
small and feeble. Length 1 4 mm.; width 0 65 mm. Alabama.

ursulina, sp. nov.

Fuscipes is our most abundant species, and the west coast *impressa*
resembles it very much in external appearance. Mellié states that the
anterior margin of the head in the male of *fuscipes* is surmounted by
two very small tubercles; this is not the case in the representatives before me, but as Mellié included with his American specimens some
from Madeira, it is probable that he had one of these under observation, and that it is a species different from *fuscipes*. *Vitula* of Mannerheim, is assigned to *Ennearthron* in the Henshaw list, but without

reason, as it is in no way related to that genus, and the *Cis dichrous*, of that list, is a manuscript name, appearing only in the LeConte list of Coleoptera. *Hirsuta* and *ursulina* are remarkable in having long fine and bristling pubescence. *Hystricula* seems to have the elytral bristles vaguely inclined to serial arrangement, and it is undoubtedly more closely allied to *mormonica* than to the three species immediately following it in the table. The Alaskan *Cis ephippiatus*, of Mannerheim, (Bull. Mosc., 1853, p. 234), is omitted from our lists. It is unknown to me, but seems to be peculiar in having the elytra profoundly and remotely, subseriately punctate, red, with a large common transverse black spot at the middle, which attains the margin at each side.

Orthocis, gen. nov.

This genus is very closely allied to *Cis*, but differs in the more parallel form of the body, in its glabrous surface, margined elytral suture and absolutely simple apex of the anterior tibiæ. The maxillary palpi are rather stout, the ligula large and corneous, the antennæ long, with the two basal joints of the funicle elongate and the club rather small and loose. The head and clypeus are absolutely simple in the male, and the only visible male sexual character is a small oval opaque and densely pubescent area at the centre of the first ventral segment, at the point occupied by a deep circular fovea in some species of *Cis*. Our two species greatly resemble each other but may be distinguished as follows :—

Ligula broader and flat ; third antennal joint nearly or quite as long as the next two combined ; body rather less elongate, the sides of the prothorax somewhat less rectilinear, otherwise similar to the following. New York....**punctata** *Mell.*
Ligula narrow and convex ; third antennal joint distinctly shorter than the next two combined. *Male.*—Oblong, parallel, moderately convex, polished, black, the legs and antennæ rufous, glabrous, each puncture of the elytra with a very minute simple silvery hair ; head well developed, convex, the eyes small and prominent ; clypeal margin perfectly simple, evenly arcuate from side to side ; prothorax two-fifths wider than long, the sides parallel and straight, rather widely reflexed ; apex broadly arcuate and slightly advanced ; angles obtuse ; base finely margined ; punctures rather fine but deep, well separated ; elytra two thirds longer than wide, rather wider than the prothorax and nearly two and one-half times as long, obtusely ogival at apex, the sides very feebly arcuate ; punctures confusedly arranged, rather small but deep and somewhat sparse, the surface smooth ; elytral suture margined toward tip. Length 2.3-2.5 mm.; width 0.85 mm. California (Alameda Co.) ....**aterrima**, sp. nov.

Xestocis, gen. nov.

A few species of peculiar facies are separated under this name, because of the prosternal carination. The antennæ are of the normal
structure, with the club well developed and the second funicular joint
only slightly longer than wide. The anterior tibiæ are strongly oblique
and acute externally at apex, except in *opalescens*, where the external
angle is slightly thickened and rounded. The maxillary palpi are usually
rather slender. The first ventral segment is subfoveate and densely pubescent at the centre of the disk in the males. Our five species are
strongly differentiated among themselves, and may be described as follows :—

Body glabrous...2
Body clothed with short pubescence or bristles...............................3
2—Clypeus bidentate in the male, the prothorax simple and rounded at apex in both
   sexes. *Male.*—Suboval,very dark rufo piceous, sometimes paler, polished; head
   and eyes moderately developed ; prothorax two-fifths wider than long, the sides
   just visibly convergent from base to the obtuse apical angles, rather distinctly and
   evenly arcuate ; lateral margin very fine, the base finely margined, more distinctly in the middle ; punctures minute and rather sparse ; elytra less than one-
   half longer than wide, twice as long as the prothorax and barely wider, rather
   narrowly rounded at apex, very feebly subrugulose, minutely, sparsely punctate.
   Length 1.5–1.9 mm.; width 0.7–0.85 mm. Canada (Toronto), New Jersey,
   Pennsylvania, Indiana and Iowa.......................**levettei**, sp. nov.
Clypeus monocerate in the male, the prothorax with two long slender porrect processes. *Male.*—Oblong-oval, convex, testaceous, polished, glabrous, each puncture with an excessively minute hair ; head and eyes moderately developed, the
   front impunctate, broadly concave, the clypeus reflexed, with a long erect parallel process at the middle, the latter very feebly expanding toward apex, the
   emarginate ; prothorax distinctly wider than long, the sides moderately convergent and evenly arcuate throughout, the apex prolonged over the head and with
   two long remote straight and porrect processes, the surface behind their separating
   sinus broadly impressed ; punctures fine, not very sparse ; elytra short, one third
   longer than wide, twice as long as the prothorax at the median line and somewhat wider, rather rapidly and narrowly rounded at apex, the sides arcuate ;
   surface nearly smooth, minutely, rather sparsely punctate. *Female.*—Nearly
   similar to the male, the clypeus broadly sinuate at the middle, the prothorax
   broadly rounded at apex. Length 1.2–1.35 mm.; width 0.55–0.6 mm. Pennsylvania (Westmoreland Co) and Rhode Island (Boston Neck)..**miles**, sp. nov.
3—Upper surface normal ; vestiture distinct, even but arranged without order ; punctures of the elytra intermingled with larger sparse punctures, which are sometimes disposed in vague series ; clypeus bidentate in the male...............4
Upper surface covered with a waterproof crust, through which the extremely minute
   simple hairs protrude...5
4—Vestiture composed of small simple and subdecumbent hairs *Male.*—Oblong-
   oval, moderately convex, rufo testaceous, feebly shining ; prothorax wider than

long, minutely but strongly, closely punctate ; elytra two-fifths longer than wide,
ogivally rounded at apex, finely, rather closely punctured and vaguely subrugose.
Alaska (Sitka) and Queen Charlotte Islands (Massett).....**biarmata** *Mann.*
Vestiture composed of coarse stiff and suberect squamules. *Male.*—Slightly smaller
than *biarmata* but similar, oblong-oval, moderately convex, testaceous, feebly
shining, the bristles short and abundant ; head rather small, feebly concave, the
eyes moderate ; clypeus triangularly reflexed at each side ; prothorax nearly one-
half wider than long, the sides just visibly convergent, feebly and evenly arcuate ; angles obtuse ; apex subangularly produced and rounded, with the tip very
narrowly sinuato-truncate ; punctures moderately fine, deep and quite dense ;
elytra not quite one-half longer than wide ; less than twice as long as the prothorax and somewhat wider, feebly rugulose, minutely and rather closely punctate, the punctures smaller than those of the prothorax ; apex ogivally rounded.
Length 1.6 mm.; width 0.7 mm.   Pennsylvania (locality not recorded).
insolens, sp. nov.
5—*Male.*—oblong oval, rather stout, only moderately convex, polished, dark piceo-
testaceous ; head moderate, the eyes small but prominent ; front broadly concave,
the clypeus acutely, bitriangularly reflexed ; prothorax one-half wider than
long, the sides very feebly convergent, rather strongly, evenly arcuate ; apex
subtriangularly prolonged, with the tip minutely emarginate and feebly reflexed ;
punctures rather fine but distinct, slightly separated ; lateral margins narrowly
reflexed ; elytra short, scarcely a third longer than wide, as wide as the prothorax
and four-fifths longer, the apex rather acutely ogival ; surface not very finely,
evenly and rather closely punctate.   Length 1.4-1.6 mm ; width 0 6-0 75 mm.
Pennsylvania (Westmoreland Co.)...................**opalescens**, sp. nov.

Biarmata is misprinted " *bicarinatus* " in the Henshaw list.

Brachycis, gen. nov.

The chief peculiarities of the single type of this genus are the short
and suboval form, very short, transversely excavated prosternum and obsolescent side margin of the prothorax at the rounded and obtuse apical
angles.   The antennal club is strongly developed, as long as the preceding six joints combined and has the sensitive apical pores small but
deep and bristling with white setæ, thus leading up to the remarkable
Plesiocis which follows.   The maxillary palpi are rather slender and
acutely pointed, and the anterior tibiæ are finely, acutely and almost
perpendicularly produced externally in a well-marked process.   Sexual
characters are wanting in the single specimen before me, which is probably a female :—

Broad, moderately convex, oblong-oval, piceous, the elytra, legs and antennal shaft
paler rufo-testaceous ; body clothed above with stiff pale and erect setæ, moderate in length and density, uniformly distributed ; head rather small, the eyes moderate ; clypeus simple, subtruncate ; prothorax two-thirds wider than long ;

angles rounded, the sides strongly convergent and broadly arcuate from base to apex, the latter broadly subparabolic, not extending much over the head ; punctures not very coarse but deep, rather close-set ; elytra oblong, very broadly, obtusely rounded at apex, one-fourth longer than wide, two and one-half times as long as the prothorax, and, at apical third or fourth, visibly wider ; sides nearly straight ; punctures rather coarse, deep and somewhat close set, the surface nearly smooth. Length 1.4 mm.; width 0.75 mm. New York (Ithaca)

brevicollis, sp. nov.

Plesiocis, gen. nov.

This genus, which is also represented at present by a single species, is remarkably distinct in the structure of the antennæ, which are 9-jointed, with the club large and well developed, more corneous than usual, dark in color and with the two sensitive subapical pores on each side large, rounded and filled with white spongy pubescence ; the club is nearly as long as the entire basal portion, with its first two joints transverse and obtriangular.   The maxillary palpi are well developed, but rather slender, the prosternum normal, the process however rather thin and sublamellar.   The anterior tibiæ are strongly, obliquely produced and acute externally at apex.   The male has very simple characters, the clypeus having two minute and rather approximate tubercles, but the first ventral segment has, as in so many other cases, a small pubescent fovea at the centre of the disk.   The type resembles a large subcylindrical *Cis* :—

Body cylindric-oval, strongly convex, piceous-brown, shining, the legs and antennal shaft rufous ; vestiture rather sparse, the hairs coarse, pale, stiff and erect, moderate in length ; head rather small, the eyes moderate, convex ; front flat, finely punctate ; prothorax two-fifths wider than long, the sides feebly convergent and slightly arcuate, the basal angles rather broadly rounded, the apical obtusely subprominent ; lateral margin narrowly reflexed ; base finely margined ; apex broadly rounded, slightly produced over the head ; punctures coarse, perforate and close-set ; elytra one-third longer than wide, twice as long as the prothorax and scarcely at all wider, the sides nearly straight, the apex very broadly and obtusely rounded ; punctures coarse, perforate, close-set and arranged evenly but without order, the interspaces smooth.   Length 2.1–2.3 mm ; width 1.0–1.1 mm.   California (Mokelumne Hill, Calaveras Co.)........**cribrum**, sp. nov.

Ennearthron *Melliè*.

The small cylindrical species which compose this genus may be readily identified by the characters given in the table.   The antennæ are slender, with the club rather feeble, the first joint of the funicle generally equal to the next two, which, with the last are equal and moniliform.   Prosternum moderately developed before the coxæ, with

the process very narrow and sublamellar. The male characters are
always pronounced and generally affect both the clypeus and thoracic
apex, but these characters greatly diminish in degree in the smaller and
less developed males, these depauperate individuals not differing much
from the females in either the present genus or *Ceracis*. In both of
these genera the male also has a small deep pubescent fovea, not at the
centre, but near the posterior margin, of the first ventral segment. The
characters of the following table are taken throughout from what appear
to be fully developed males only:—

Male with the clypeal margin broadly and strongly reflexed in a trapezoidal process, the
    thoracic process bidentate. **2**
Male with a long slender erect clypeal process, the thoracic apex simple and rounded;
    species very small. **11**
2—Elytra without trace of impressed lines. **3**
Elytra with very feebly and unevenly impressed lines, the punctures feebly sub-
    serial in arrangement. **10**
3—Male with the thoracic processes longer, narrower and more approximate ; punc-
    tuation feeble, the elytral punctures always confused in arrangement. Atlantic
    and Gulf regions. **4**
Male with the thoracic processes shorter, more widely separated and more lamellarly
    triangular ; punctuation stronger, the elytral punctures generally confused but
    occasionally very feebly subserial. Pacific Coast regions. **6**
4—Apex of the pronotum rather feebly impressed behind the processes. **5**
Apex of the pronotum strongly, transversely impressed behind the processes. Mod-
    erately slender, polished, piceo-rufous in color ; head well developed, concave,
    the clypeal process large and well developed, with the apex feebly sinuate at
    the middle : prothorax slightly wider than long, the sides parallel and nearly
    straight, the angles all rounded ; processes long, slender and distinctly diverging
    as usual ; base and sides finely margined ; punctures fine and sparse ; elytra less
    than one-half longer than wide, as wide as the prothorax and two-thirds longer,
    the surface very feebly subrugulose, sparsely and very minutely punctate, the
    punctures much more minute than those of the pronotum ; apex evenly rounded.
    Length 1.2–1.5 mm.; width 0.45–0.6 mm. Texas (Columbus) and Louisiana.
                                                                    piceum, sp. nov.
5—Elytra fully one-half longer than the prothorax, slender, cylindric oval, black,
    rather strongly shining ; legs, mouth parts and antennæ pale ; punctures fine and
    rather sparse ; thoracic processes rather short. Canada to Pennsylvania (*mell, i*
    Mell., *unicolor* Csy.). . . . . **thoracicorne** *Zieg.*
Elytra very short and quite strongly cuneiform, very much less than one-half longer
    than the prothorax. Rather stout, the head polished and concave, the eyes
    small ; clypeal process well developed but with the sinuate sides rather rapidly
    converging, the apex a little less than half as wide as the head, feebly sinuato-
    truncate ; prothorax large, not quite as long as wide, the sides parallel and
    broadly arcuate, the corneous processes moderately long and rather stout ;
    punctures fine but distinct, only moderately sparse ; elytra at base as wide as the

prothorax, the sides nearly straight and distinctly convergent, the apex narrowly rounded ; punctures rather sparse and very minute, the surface feebly rugulose. Length 1.3 mm.; width 0.53 mm.   Louisiana (Morgan City).

laminifrons, sp. nov.

6—Thoracic process of the male abruptly formed............................7

Thoracic process gradually formed, its sides merging gradually and obliquely into the sides of the prothorax ; thoracic punctures quite dense....................9

7—Elytral punctures rather coarse and closer ; thoracic process one-half as wide as the elytra.   Body rather stout, blackish, the elytra piceous ; legs and oral organs pale ; lustre moderately shining ; head and eyes moderately developed ; clypeal process very broad, only moderate in length, almost transversely truncate ; prothorax only slightly wider than long, the sides parallel and broadly arcuate ; apical process very broad, deeply sinuate ; punctures strong and close-set ; elytra quite distinctly narrower than the prothorax and scarcely more than one-half longer ; one-half longer than wide, the sides parallel ; apex broadly rounded ; surface feebly rugulose, strongly punctured, less closely than the prothorax.   Length 1.4–1.7 mm ; width 0.55–0.7 mm.   California (southern)...........grossulum, sp. nov.

Elytral punctures fine but distinct, rather sparse, the thoracic process much less than one-half as wide as the elytra...8

Thoracic punctures close-set ; angles of the clypeal process scarcely at all rounded. Blackish, the elytral punctures generally very feebly subserial in arrangement. California (especially northern coast regions) ....californicum Csy.

Thoracic punctures sparse ; angles of the clypeal process rounded.   Body evenly cylindric, moderately shining, more or less rufo-testaceous in color, the elytra sometimes blackish toward base ; head and eyes moderate, the clypeal process moderately strong, with the sides rather strongly convergent and the apex broadly sinuate ; prothorax slightly transverse, the sides parallel and almost straight, rounding anteriorly, the process rather feebly developed, sinuate at apex ; elytra scarcely one-half longer than wide, as wide as the prothorax and three-fifths longer, the sides parallel and straight, the apex broadly rounded ; punctures fine and sparse, the surface almost smooth.   Length 1 2 mm.; width 0.5 mm.   California (Sonoma Co.)................................discolor, sp. nov.

9—Narrowly cylindric, blackish, the elytra rufescent at tip ; legs, trophi and antennæ pale ; surface moderately shining ; head well developed, the clypeus moderate in length, with the sides strongly convergent, the apex broadly sinuato-truncate and the angles blunt ; prothorax but little wider than long, the sides subparallel and very slightly arcuate ; process rather short, sinuate at tip ; elytra fully one-half longer than wide, as wide as the prothorax and fully three-fourths longer, somewhat parabolically rounded at tip, the punctures fine but strong, but little smaller than those of the prothorax and much less close-set.   Length 1.1–1.3 mm.; width 0.4–0 5 mm.   California (Los Angeles)...convergens, sp. nov.

10—Moderately stout, piceous to blackish in color, with the elytra paler ; legs, trophi and antennæ pale, the club dusky ; surface polished ; head well developed, broadly concave, the clypeus only moderately reflexed, the apex broadly truncate; prothorax but little wider than long, the sides subparallel and broadly, distinctly arcuate ; process rather short, lamelliform, with a triangular incisure at the middle, the process abruptly formed ; punctures very fine and sparse ; elytra short, less than one half longer than wide, scarcely as wide as the prothorax and one-half longer, evenly rounded at apex, the punctures minute and sparse, those of the series larger and closer. Length 1 0–1.2 mm ; width 0.4–0 5 mm. Florida.

pullulum, sp. nov.

11 - -Narrowly cylindric-oval, moderately shining, pale flavo-testaceous throughout ; head and eyes well developed, the front concave; clypeal process narrow, long, the apex strongly rounded ; prothorax distinctly wider than long, the apex evenly and circularly rounded, the sides becoming parallel toward base; punctures very minute, sparse and feeble ; base distinctly margined as usual ; elytra fully as wide as the prothorax and two-thirds longer, not quite one-half longer than wide, parabolically rounded at tip, margined at base, the punctures sparse, very feeble and extremely minute, even smaller than those of the prothorax ; surface smooth. Length 1.0–1.1 mm.; width 0.4–0.45 mm.   Florida......**unicorne,** sp. nov.

Unicorne is evidently closely related to the Brazilian *corniferum* of Mellié, but in that species the cephalic process is said to be broad, recurved and narrowed to the acute apex.

Ceracis *Mellié.*

This genus is scarcely distinct from *Ennearthron*, agreeing in facies and in every structural feature except the antennæ, which are 8-jointed, one of the small joints of the funicle being eliminated.   The two species before me are as follows :—

Rufo-testaceous, the elytra blackish toward base ; punctures very minute and sparse, the remaining characters nearly as in *punctulata.*   Louisiana [Mellié], North Carolina and Pennsylvania (Westmoreland Co.)...... ....**sallei** *Mell.*

Black throughout, the head and prothorax sometimes picescent ; legs, trophi and antennæ pale ; surface polished, the elytra nearly smooth ; head and eyes well developed ; clypeal process rather well developed, with its sides but slightly converging, the apex broadly sinuato-truncate ; prothorax slightly shorter than wide, the sides feebly convergent and broadly arcuate from the base to the rather pronounced but obtuse apical angles ; process very abruptly formed, moderate in length, the exterior angles somewhat everted and the apex deeply sinuate ; punctuation quite deep and close-set but rather fine ; base finely margined ; elytra at base as wide as the prothorax, less than one-half longer than wide ; the sides nearly straight and feebly convergent ; apex broadly rounded, base not in the least margined ; punctures finer and sparser than those of the prothorax, confusedly arranged.   Length 1.25–1.3 mm.; width 0 55–0.6 mm. Florida.

punctulata, sp. nov.

The species recently described from Lower California by Dr. Horn, under the name *similis*, appears to resemble *sallei*, but differs from both the above in having the elytral punctures coarser than those of the prothorax, a very exceptional character in *Ennearthron* and *Ceracis*.

Octotemnus *Mellié.*

This is a very pronounced and distinct genus, differing from *Ennearthron* in the oval outline of the body, absence of male sexual characters of the head and prothorax, and in tibial structure. The maxillary palpi are slender and pointed, the antennal club well developed and very loose, the joints being attached by very slender pedicels and with the sensitive pores approximate at each side of the apex. The prosternum is short and somewhat concave before the coxæ, with the process thin and laminate. There is no fovea on the first ventral segment of the male, but the surface is feebly and approximately bi-impressed near the base, the intervening area elevated and prolonged backward in an isolated triangular point, a structure not suggested elsewhere in the family. The surface is glabrous, but the elytra have a few widely dispersed erect setæ. Our two species are very closely allied; they may be described as follows from the male : —

Form more narrowly oval ; size larger, the basal abdominal process of the male very acute at apex, pale testaceous, polished throughout ; head and eyes well developed, the latter convex ; front broadly, evenly convex, very minutely, sparsely punctate ; clypeal margin slightly thickened for a short distance from the eyes ; prothorax but little wider than long, circularly rounded at apex, the sides diverging slightly to the base ; angles all very broadly rounded ; base very minutely margined ; punctures very minute, feeble and sparse ; elytra fully one-half longer than wide, a little wider than the prothorax and nearly twice as long ; sides feebly arcuate, the apex rather narrowly parabolic ; base not margined ; humeral callus very small, feeble ; surface feebly rugulose, the punctures extremely minute and sparse. Length 1.4–1.75 mm.; width 0.55–0.75 mm. Pacific coast (from Vancouver to San Francisco).... ....**denudatus**, sp. nov.

Form rather shorter and more broadly oval, polished, the pronotum more or less alutaceous, blackish to pale flavo-testaceous throughout ; head and prothorax nearly as in *denudatus*, the elytra barely one-half longer than wide, distinctly wider than the prothorax and barely twice as long, the surface nearly smooth, very minutely, sparsely punctate. Length 1.35–1.6 mm.; width 0.6–0.65 mm. Rhode Island, New York and Pennsylvania.................**lævis**, sp. nov.

Both of these species are very common, and it is remarkable that they have not been heretofore described. Perhaps the *Cis pumicatus* of Mellié may be the same as *lævis*, but that species, taken apparently near New Orleans, is said to have the prothorax longer than wide and the elytra only one-half longer than the prothorax, which language agrees rigorously also with the figure and in no way suits either of the above species.

SPHINDID.E.

This family forms a very good transition to the Cryptophagidæ. The antennæ are of a more perfectly clavicorn type than in Cioidæ, and have a large compact club, with the ninth joint variable in size. The mentum is very large, in striking contrast to the Cioidæ, where it is unusually minute. The maxillary palpi are small, slender and pointed and the anterior coxæ rather widely separated. The clypeus is convex, continuous with the front, narrowed and continued over the larger part of the mandibles, the labrum being small, almost atrophied in *Odontosphindus*, the epistomal suture fine and posteriorly arcuate. The eyes are large, convex and coarsely faceted. The two genera before me may be distinguished as follows:—

Tarsi heteromerous; antennæ 11-jointed, the ninth joint wider than the eighth, but very short, the club virtually 2 jointed, large and nearly cylindr,c; prothorax denticulate at the sides, the elytra with impressed series of coarse punctures; body glabrous ....**Odontosphindus**

Tarsi pentamerous; antennæ 10-jointed, the club variable, 2 or 3-jointed; prothorax not denticulate, the elytra with unimpressed series of fine punctures; body sparsely pubescent.................................·.............**Sphindus**

Another genus, *Eurysphindus*, has been described by LeConte, but I have seen no example; the inferior flanks of the prothorax are said to be deeply concave and the body clothed with erect hair.

Odontosphindus *Lec.*

These species are much larger than those of *Sphindus* and are distinguishable at once by the characters of the table; the two species are as follows:—

Sides of the prothorax scarcely at all reflexed, finely margined.   Atlantic regions.

denticollis *Lec.*

Sides of the prothorax distinctly though not very broadly explanato-reflexed, more strongly and quite irregularly denticulate.   Body subparallel, moderately convex, pale rufo-testaceous in color, shining though somewhat alutaceous in lustre; head moderate in size, the epistoma polished and impunctate; eyes moderately large, convex; antennæ as long as the width of the head, the first joint large, the second more slender, contorted at base as usual; prothorax quite transverse, parallel, the sides feebly arcuate; punctures rather coarse and close set; elytra only just visibly wider than the prothorax, three-fourths longer than wide, the serial punctures obsolete toward tip.   Length 3.0–3.5 mm.; width 1.25 mm.   California (Sonoma Co.)................................**clavicornis**, sp. nov.

Clavicornis is materially larger than *denticollis*, and has a larger, more transverse and more coarsely punctured prothorax.

Sphindus *Chev.*

The species of this genus are small and oblong, with duller surface lustre and moderately long, rather sparse pubescence, serial in arrangement on the elytra. The three species before me may be thus separated :—

Antennal club 2 jointed...2

Antennal club purely 3-jointed...3

2—Pronotum minutely and rail er closely punctured, more or less rufo-piceous in color. Atlantic regions to Iowa ....**americanus** *Lec.*

Pronotum more coarsely and quite sp rsely punctured. Body black, stouter, dull in lustre ; head and eyes moderate in size, the epistoma polished ; antennæ a little longer than the width of the head, the club only moderately stout, the tenth joint twice as long as the ninth ; prothorax nearly twice as wide as long, the sides just visibly convergent from base to apex, feebly arcuate ; apex broadly arcuate ; elytra scarcely a fourth longer than wide, barely wider than the prothorax and two and one-half times as long ; serial punctures rather fine, the intervals dull and minutely shagreened. Length 1.9 mm.; width 0.9 mm. Colorado (Buena Vista—8000 feet)....................................**crassulus**, sp. nov.

3—Narrowly oblong, more shining, piceous black, the elytra, legs and antennæ pale testaceous ; head moderate, the eyes large and convex, separated by about three times their own width ; antennæ moderate in length, the last three joints forming a compact subcylindric club; prothorax much smaller than in the two pre-ceding species, transverse, the sides subparallel ; surface evenly convex, very minutely and not very closely punctulate ; elytra two-fifths longer than wide, about a fifth wider than the prothorax and three times as long ; serial punctures feeble but distinct, the intervals smooth and aluaceous ; apex obtusely rounded as usual. Length 1.7 mm.; width 0.7 mm. Canada (Toronto).

trinifer, sp. nov.

Americanus varies greatly in size as usual in this and neighboring families ; it is quite abundant and occurs in fungi of various species.

THE LIFE-HISTORIES OF THE NEW YORK SLUG CATERPILLARS.—XV.

PLATE VI, FIGS. I–II.

BY HARRISON G. DYAR, A.M., PH.D.

Heterogenea flexuosa *Grote.**

1880—*Limacodes flexuosa* GROTE, North Am. Ent. I, 60.
1880—*Limacodes cæsonia* GROTE, North Am. Ent. I, 60.
1894—*Heterogenea cæsonia* and *flexuosa* NEUMOEGEN & DYAR, Journ. N. Y. Ent. Soc. III, 74.

LARVA.

1878—GLOVER, Ill. North Am. Ent. pl. 95, fig. 19.
1893—PACKARD, Proc. Am. Phil. Soc. XXXI, 105 (as "full grown larva of *Heterogenea* sp.").
1895—DYAR & MORTON, Journ. N. Y. Ent. Soc. III, 146 (in synopsis).
1896—DYAR, Journ. N. Y. Ent. Soc. IV, pl. VI, figs. 3 and 4 (as *Tortricidia pallida*).

SPECIAL STRUCTURAL CHARACTERS.

Dorsal space narrow, of even width, scarcely narrowing at the ends, gently arched; joint 13 rounded prominent. Lateral space broad, oblique, narrowing to the extremities; subventral space small, retracted. Subdorsal ridge slight; indicating the change in direction between back and sides; lateral ridge rather prominent, overhanging the subventral space. Outline elliptical, joint 13 only slightly notched on the sides, not forming a quadrate tail. Depressed spaces (1)–(8) present, the subventral ones (7) and (8) only indicated, the others sharp edged and deep, large, dividing the surface into latticed ridges as in *Tortricidia pallida*, (4) the largest, transversely elongated, the lower segmental (5) moderate, the intersegmental (6) very small, alternating exactly in line with the lower edge of (5). Skin surface covered with coarse clear granules, the depressed spaces finely granular in the base. In the first stage the setæ are arranged as in *T. pallida*, but disappear at the first molt when all the structural characters are assumed nearly in their mature form. Coloration of the pattern and colors of *T. pallida*, modified in detail.

---

* This is not a *Heterogenea;* but I reserve generic corrections till the end of these articles.

A REVISION OF THE AMERICAN COCCINELLIDÆ.

BY THOS. L. CASEY.

The object of the following pages is to give a short outline or sketch of every species occurring within the limits of the United States accessible to me at the present time, and also to invite attention to certain features in the taxonomy of the family which do not seem to have been hitherto brought to notice.   In an appendix a list of African species is given, containing quite a number of novelties, and the descriptions of certain new species from other parts of the world are also appended.

COCCINELLIDÆ.

The separation of this family into two parts based upon mandibular structure has never seemed entirely satisfactory to me ; first, because of the difficulty of observing the character, causing the classification of Chapuis to be unpractical, and, secondly, because *Epilachna* and related genera are merely pubescent halyziids, slightly modified by reason of perverted food habits and attendant environments.   Many of the Harpalini of the Carabidæ are known to be either wholly or partially phytophagous, but no one has proposed to divide the Carabidæ on these lines, and would scarcely do so even if a minute structural divergence in the mandibles existed, and it has never been demonstrated that the mandibular teeth serving as the basis of the Chapuisian classification are not found elsewhere in the family.   The Epilachnini, in fact, resemble the Psylloborini in all external structures, including the long antennæ, a character of more importance than has apparently been conceded.   In view of these facts I have not employed the classification of Chapuis in the following pages.

The latter author appeared also to be constantly striving to reduce the generic groups hitherto proposed, but this cannot be done with propriety, and many more will be needed, both of genera and tribes, before the taxonomy of the family can be made entirely clear.   This is well shown by some small species which we had held to belong to the genus *Pentilia*, until Weise recently proved that they were in no way related, and separated them under the name *Smilia ;* as a matter of fact they do not resemble *Pentilia* at all, and are much more closely allied to *Scymnus.* Again, our representatives of *Cryptognatha* are likewise widely separated from the *Cryptognatha* of Mulsant, and form in reality one of the most isolated types of the family, the special character relating to the prosternum, which caused LeConte to associate them, being of subordinate value and liable to appear in any tribe ; it exists, for instance, in *Stethorus* of the Scymnini, and in *Nipus* of the Cranophorini, though not the distinguishing feature of that remarkable type. In *Zagloba* of the Scymnillini it also tends to reappear. Again the genus *Rhyzobius* is tribally distinct from *Scymnus* in the structure of the eyes, antennæ and epipleuræ.

The character relating to the anterior coxal cavities, announced by LeConte, is apparently of no significance even if wholly true, as it would bring together genera with no special affiliation otherwise, and the character made use of by Mulsant to separate Coccinellini from Cariini is of no value, there being no tribal difference between *Coccinella* and *Synonycha*, in spite of their general dissimilarity of habitus.

The abdomen is composed throughout of five segments, but the genital armature sometimes becomes distinct and assumes the form of a sixth segment. This character is very useful in the classification of the tribes related to Chilocorini, and of the compact Coccinellidæ having narrow epipleuræ, as will appear ; it generally affects both sexes and is particularly developed in the Hyperaspini. The Hyperaspini of Chapuis include several distinct tribes, and those with but five ventral segments should be removed, the retractility of the legs and epipleural depressions not being tribal characters necessarily, but appearing in several tribes with the legs generally free.

The tarsi in this family are in reality 4-jointed, the third small and generally forming a rigidly anchylosed basal lobe of the last, but it is sometimes free or partially so. The second is lobed beneath, the lobe truncate at tip and hollowed on its upper surface, not bilobed as stated by Crotch (Rev. Cocc., p. 53).

In the following pages I have made use of all generic types, foreign and native, which have been accessible to me, and regret that my exotic material might not have been more extensive. Where names not belonging to the fauna of the United States are introduced they are preceded by an asterisk.

Crotch employs the name *affinis* Rand., for the species *venusta* and *notulata*, but in error, as *affinis*, of Randall, is simply a synonym of *Hyperaspis binotata* Say.

The family may be divided into numerous tribes, as follows : —

Middle coxæ narrowly separated; body glabrous, elongate-oval, the epipleuræ moderately wide, horizontal ; legs long, free, the femora extending beyond the sides of the body ; abdomen with the genital or sixth segment visible in both sexes ; head not deeply inserted, the prothorax strongly sinuate but not covering the eyes ; epistoma, eyes and antennæ as in Coccinellini . . . . HIPPODAMIINI

Middle coxæ widely separated ; legs shorter, the femora generally not extending beyond the sides of the body ; head deeply inserted, the pronotum covering a considerable part of the eyes except in certain rare cases such as *Selvadius*. 2

2—Eyes finely faceted. 3

Eyes coarsely faceted ; antennæ long, with the club loose ; body pubescent ; abdomen with the sixth segment visible in both sexes. 18

3—Epipleuræ wide, concave, strongly descending externally ; body loosely articulated, generally rounded in form. 4

Epipleuræ narrow, generally horizontal, flat or feebly concave; body compact, generally oval in form 14

4—Fourth joint of the maxillary palpi securiform. 5

Fourth joint narrow, elongate with circular section, finely acuminate at tip. 13

5—Epistoma narrowed from the base, sometimes expanded slightly at apex, the antennal fossæ more or less exposed 6

Epistoma broadly dilated, concealing the antennæ and subdividing the eyes. 11

6—Legs free ; antennæ more or less elongate ; sixth ventral segment small but visible in both sexes. 7

Legs retractile and lodged in moderately deep to shallow depressions ; antennæ short ; abdomen with five segments, the fifth longer, the sixth always invisible. 10

7—Upper surface of the body glabrous. 8

Upper surface pubescent. 9

8—Epistoma more or less sinuate at apex and obliquely dentiform at the sides, the sinus generally more or less closed by a semi-corneous additional piece united to the front without visible suture ; antennæ more or less approximate to the eyes, which are narrowly and rather deeply emarginate, the fossæ large, with distinctly overreaching superior ridge ; prothorax deeply emarginate ; body moderate to large in size. . . . . COCCINELLINI

Epistoma narrower, truncate, without semi-corneous additional piece and not obliquely denticulate at the sides, the antennæ more frontal in insertion and more distant from the eyes, which are broadly and more feebly sinuate, the fossæ small, more exposed frontally and with very slight superior ridge ; body smaller, with thinner integuments, the head small, the prothorax smaller, very feebly sinuate at apex, with broadly rounded apical angles ; antennæ slender, with the last joint elongate . . . . PSYLLOBORINI

9—Antennæ long, with loosely articulated club, inserted within very small and completely exposed subfrontal foveæ remote from the eyes, nearly as in Psylloborini, the eyes not or only very feebly sinuato-truncate ; epistoma truncate, not denticulate at the sides ; prothorax deeply emarginate at apex ; mandibles bifid at tip and denticulate within ; body rounded or elongate-oval, the legs free.

EPILACHNINI

10—Epistoma feebly sinuate, with rounded lateral angles and coriaceous margin within the sinus, the sides sinuate above the moderate exposed antennal foveæ, the eyes deeply but very narrowly emarginated by the post-antennal canthus; mandibles simple and finely acumiuate at tip; body rounded, very convex, the prothorax very deeply emarginate and formed as in Chilocorini ....*PENTILIIŃ[1]

11—Upper surface glabrous; body very convex or subcompressed, rounded, the abdomen with five segments, a small genital segment visible in the males; antennæ very short, more or less bent, the club with fóur connate joints; legs free or feebly retractile...CHILOCORINI

Upper surface pubescent; legs retractile within shallow depressions; antennæ very short, bent.............. 12

12—Abdomen composed of six segments nearly as in Hyperaspíni, the fifth short.

*PLATYNASPINI

Abdomen composed of five segment, the fifth large and rounded, the sixth wholly invisible in both sexes; body very small, rounded...............*TELSIMIINI

13—Body rounded or oval, very convex, pubescent or partially so; epistoma large but not dilated, broadly rounding from the base into the apex, which is feebly sinuate medially; eyes entire, the antennæ short and slender, straight, inserted in small exposed foveæ very close to the eyes; prosternum widely separating the coxæ, bicarinate, flat; abdomen composed of but five segments, the fifth large, rounded; basal node of the last tarsal joint free; legs retractile, the impressions feebly concave; prothorax deeply emarginate.....................*PHARINI

14—Abdomen composed of only five segments, the genital segment wholly obsolete in both sexes, the fifth segment large, ogival or rounded.................. ...15

Abdomen with the sixth segment well developed and distinct in both sexes, the fifth shorter...16

15—Legs strongly retractile within deep concavities of the under surface; prosternum widely separating the coxæ, strongy deflexed at tip, forming a protection to the mouth in repose; eyes entire; antennæ with exposed insertion; body oval, moderately convex, glabrous or only partially pubescent..............ŒNEINI

Legs free; prosternum flat, remotely separating the coxæ, the apex not deflexed or with feeble tendency thereto; antennal foveæ shallow, the eyes narrowly and deeply emarginate; body rounded or oval, moderately convex, pubescent or partially so..................,........................SCYMNILLINI

16—Body glabrous; epipleuræ generally slightly descending externally but relatively narrow; legs moderately retractile or free; eyes emarginate or entire.

HYPERASPINI

Body pubescent; epipleuræ generally flat and horizontal; legs always free........ 17

17—Pronotum covering the head, rounded or feebly truncate in front; body oval or elongate-oval, moderately convex, subglabrous in *Nipus*........CRANOPHORINI

Pronotum deeply sinuate at apex and never produced; body oval or oblong-oval.

SCYMNINI

18—Prothorax narrowed anteriorly from the base; epipleuræ moderately wide and more or less concave, descending externally.....................RHYZOBIINI

Prothorax narrowed at base; body elongate; epipleuræ very narrow, flat and horizontal.......................... .... COCCIDULINI

The Rhyzobiini are not marked with an asterisk as they have been to some extent acclimated in California; they are not however, as far as known, endemic.

HIPPODAMIINI.

The characters heretofore used to distinguish this tribe from the Coccinellini are of little or no value, as the sternal and ventral post-coxal plates or arcs are frequently both as distinct in the former as in the latter, but the ventral plates are always short, as in those Coccinellini allied to *Adalia*.

The Hippodamiini are not relatively very numerous and are almost essentially American. They may be distinguished at once from the Coccinellini by the elongate-oval form of the body, narrowly separated intermediate coxæ and the other characters given in the table. The frequently obsolete or ill-defined post-coxal lines are the obvious result of long disuse, as the legs are unusually developed for the present family and perfectly non-retractile. The genera before me may be distinguished as follows :—

Tarsal claws simple, being evenly arcuate, slender and very acutely pointed, with a more or less slight bulbiform enlargement at base..........................2

Tarsal claws acutely pointed, with a large quadrate basal tooth within, separated from the slender apical part by a deep acute fissure—a very usual structure in Coccinellidæ...6

Tarsal claws slender, bifid within behind the apex, the two lobes unequal in length and both acutely pointed.. 7

2—Sternal and ventral coxal plates both dis inct ; basal angles of the prothorax obtuse but distinct and not rounded.................................... 3

Sternal plates distinct, the abdominal obsolete..................................4

Sternal and ventral plates both completely obsolete......................... ...5

3—Body oval, the elytra maculate and strongly punctate ; side margins all strongly and quite broadly reflexed................................. **Anisosticta**

Body elongate and subparallel, the elytra vittate and finely punctate ; side margins very narrowly reflexed .... **Macronæmia**

4—Basal angles of the prothorax broadly rounded........ ....**Næmia**

5—Basal angles broadly rounded as in *Næmia*........ .. ..**Paranæmia**

6—Body nearly as in *Næmia*, the elytra and pronotum almost similarly ornamented ; sternal and ventral plates both completely obsolete..... ....**Megilla**

7—Base of the prothorax rounded in the middle ; sternal and ventral plates variously developed or wanting...............................:..........**Hippodamia**

Another genus of our fauna,—*Ceratomegilla* of Crotch,—is unknown to me but is said to differ from *Megilla* in having the third joint of the antennæ dilated and triangular. *Eriopis*, which is said to occur here, differs from *Hippodamia* only in having the base of the prothorax sinuate at the middle. *Anisosticta* is represented within our confines by *hitriangularis* Say (=*multiguttata* Rand.), related to the European *19-punctata*, and still more closely to *strigata*, but distinct from either. *Macronæmia* (gen. nov.) has for its unique representative the *Coccinella episcopalis* of Kirby, assigned to *Næmia* by Mulsant. *Næmia* has for its type, and only species within the United States, the *Coccinella seriata* of Melsheimer (=*litigiosa* Muls.).

Paranæmia, gen. nov.

The type of this genus is the *Hippodamia vittigera*, of Mannerheim, assigned to *Næmia* by Mulsant. The specimens in my cabinet may be grouped in the two following closely allied species or perhaps subspecies:—

Form short and broadly suboval, the prothorax twice as wide as long and broadly rounded at base ; elytra rather shining and distinctly punctate. Length 4.8–5.2 mm.; width 2.9–3.0 mm. California. ....**vittigera** *Mann.*

Form more elongate but broad and subparallel, larger, though similarly ornamented with black, the prothorax much less than twice as wide as long and more strongly rounded at base ; elytra strongly alutaceous and more finely and very much more sparsely punctate. Length 5.2–6.4 mm.; width 2.9–3.4 mm. Colorado and Arizona. . . . . **similis**, sp. nov.

Megilla *Muls.*

The type assumed by Mulsant is the *M. maculata*, of De Geer (Spec., p. 24), but this name was applied by its author to one of the large South American forms, which are in all probability specifically distinct from our familar and very constant modification, and it is therefore proper to apply the name *fuscilabris* to the latter. The material before me indicates three species or subspecies as follows :—

Head finely and feebly punctured ; surface lustre alutaceous ; pronotum narrowly reflexed at the sides. 2

Head strongly and closely punctured ; lustre much more shining, the pronotum more broadly reflexed at the sides. *3*

2—Prothorax less than twice as wide as long. Length 4.7–6.2 mm.; width 2.7–3.4 mm. Delaware, North Carolina, Iowa, Arizona and California (Yuma).
fuscilabris *Muls.*

Prothorax twice as wide as long ; body larger and much more broadly oval. Length 5.2–7.2 mm.; width 3.0–4.0 mm. Texas (Brownsville)...**strenua**, sp. nov.

3—Body in form and size nearly as in *fuscilabris*, the ground color of the type yellowish, the discal transverse spot of the elytra posteriorly angulate ; punctures of the elytra fine and rather close. Length 5.7 mm.; width 2.9 mm. Honduras. . . . . **medialis**, sp. nov.

These forms are all virtually similar in ornamentation to the common *fuscilabris*.

Hippodamia *Chev.*

The species of this genus are rather numerous, and constitute by far the larger part of the tribe; they are frequently closely allied among themselves and are common to the arctic and subarctic faunas of both hemispheres, although poorly represented in the palæarctic provinces. The sternal and ventral plates lose all value in a generic sense, and the *Adonia* of Mulsant, must consequently be suppressed, as suggested by Crotch. Sometimes, as in *parenthesis* and *apicalis*, both the sternal and ventral plates are distinct and as perfect as in *Anisosticta*. In *obliqua* and *convergens*, also, they are similar, though more feebly outlined. In *lecontei, quinquesignata*, with related species, and in the *sinuata* group, the sternal plates become obsolete or very indistinct, but the ventral are still complete or very nearly. In *glacialis* the sternal plates are completely obliterated and the ventral are only represented by an oblique and isolated external line, and finally in *tredecempunctata*, the type of the genus, both plates become obsolete.

Hippodamia (Adonia) variegata of Goeze, (*constellata* Laich.), is a European species which is said to occur within the United States ; this is probably an error, however, and it is omitted from the following table of the American species known to me by actual examples. The sternal and ventral plates are exactly as in *parenthesis* and *apicalis*, but in habitus and ornamentation it agrees with the majority of species much better than they :—

Pronotum with a broad pale lateral border enclosing an isolated black dot or dot-like spur from the central black area, the latter without trace of the usual white discal diverging lines; elytra each very constantly with six rounded black dots, and also a small common scutellar spot; femora black, the tibiæ and tarsi pale throughout; claws rather thicker and more feebly arcuate than usual. Length 4.3–5.3 mm.; width 2.4–3.3 mm. Europe, Siberia and the United States. [*tibialis* Say]... **13-punctata** *Linn.*

Pronotum with a narrower white lateral margin which is intruded upon by a more or less pronounced angulation of the central black area, occasionally completely dividing the white area, in which case the white near the basal angles also frequently disappears; legs black throughout, the anterior sometimes in part pale, especially in those species with distinctly formed sternal and ventral plates.....2

2—Pronotum without trace of a median white spot at the basal margin; sternal and abdominal plates very variable in development...................... · 3

Pronotum with a white or whitish median spot at the basal margin; sternal and abdominal plates both distinct, the latter complete but short, extending to about the middle of the segment..21

3—Elytra completely black, with two small and obsolescent transverse whitish spots
at the basal margin and one on each elytron, larger and triangular, at the lateral
margin and apical fourth. Length 6.0 mm. ; width 4.0 mm. California to Van-
couver Island. . . . . **mœsta** *Lec.*
Elytra red, with a transverse basal fascia of black, either complete and constant, or
formed occasionally and in certain individuals by the coalescence of the small
scutellar and two post-scutellar spots with the two humeral. 4
Elytra never with a transverse basal fascia, the two post-scutellar points when present never coalescent with the scutellar spot, the latter always very small or obsolete ; elytra frequently immaculate, generally very finely and inconspicuously
punctured 11
Elytra never with a tranverse basal band or post-scutellar spots, the scutellar spot
larger and more or less elongate-oval or rhomboidal, sometimes involving almost
the entire suture ; discal and humeral spots tending to unite to form a black
vitta ; marginal white area of the pronotum narrow and subequal in width
throughout, the diverging discal lines distinct, the outer post-median spot when
disconnected always small, the inner large. 17
4—Subapical black spot of the elytra constantly large and distinct ; body generally
more broadly oval 5
Subapical black spot constantly wanting or extremely rudimentary ; body generally
more narrowly oval ; lateral angulation of the pronotal black area pronounced,
the white margin very broad anteriorly, frequently interrupted in the middle, the
basal part sometimes obsolete as in typical *extensa*. 8
5—Lateral angulation of the black pronotal area strong, frequently dividing the white
marginal area, the apical and basal parts of the latter wider, the basal becoming
obsolete in typical examples of *5-signata* ; body larger and more broadly oval,
the pronotal punctures very fine and not close-set. 6
Lateral angulation of the central black area very obtuse, the marginal white area nar-
row throughout but entire. 7
6—Basal band of the elytra broad, very constant and almost equally wide throughout,
obtusely truncate at its lateral limits on the callus and angularly involving the
scutellum ; post-median black spot large, somewhat obliquely transverse,
straight, even, extending nearer to the side margin than the suture. Length
6.2 mm. ; width 4.0 mm. Colorado, Lake Superior and Hudson Bay [*mulsanti*
Lec.] . . . . **5-signata** *Kirby*
Basal band of the elytra rarely entire and then very irregular, the scutellar and post-
scutellar points generally coalescent, forming a trilobed star, which is generally
isolated from the humeral spots ; post-median black spot transversely arcuate or
sinuate, evidently formed by the amalgamation of two transverse spots, the subsutural slightly the more basal. Length 4.9–6.0 mm. ; width 3.2–4.0 mm.
New Mexico, Colorado, Utah and Oregon. . . . . **lecontei** *Muls.*
7—Pronotum more strongly and quite densely punctate ; basal band of the elytra
strongly developed and entire, the humeral dilatation well marked ; post-
median spot composite, consisting of a large, outwardly and anteriorly oblique
spot, united behind its anterior limit, with a smaller external, inwardly and anteriorly oblique spot ; subapical spot transversely oval, with an internal posterior
angulation ; body smaller. Length 4.7 mm.; width 2.9 mm. Canadian
Rocky Mts. . . . . **puncticollis**, sp. nov.

8—Pronotum closely punctulate; basal band of the elytra equally broad throughout, with a scutellar angulation as in *5-signata*; post-median spot broad, slightly oblique and oval, the subapical wholly obsolete; surface of the elytra strongly alutaceous and rugulose; body small and more depressed. Length 4.5 mm.; width 2.7 mm. Colorado...............................**dispar,** sp. nov.

Pronotum minutely and sparsely punctulate, more convex and polished; basal band of the elytra crescentiform, acuminate at the callus, with an anterior scutellar angulation; elytra polished 9

9—Elytra undulato-rugulose externally and toward apex, without trace of black spots behind the basal band. Length 5.1 mm.; width 3.1 mm. California (Alameda)...**extensa** *Muls.*

Elytra smooth throughout............................... 10

10—Elytra closely punctate; post-median feebly oblique line narrow and composed of two slightly confluent transverse spots; subapical spot of *5-signata* and allies visible as a minute and feeble point. Length 4.75 mm.; width 3.1 mm. California?............................... ....**subsimilis,** sp. nov.

Elytra sparsely punctate, the post-median spot almost transverse, narrow and subentire, the subsutural part not more basal—as it evidently is in *subsimilis*—the subapical spot completely obsolete; surface very highly polished throughout. Length 5.8 mm.; width 3.6 mm. Wyoming—Mr. Wickham...**vernix,** sp. nov.

11—Pale lateral margin of the pronotum wider anteriorly and posteriorly, the angular extension of the black area strongly marked.........................12

Pale margin narrower and much less unequal in width from apex to base, the angular extension of the black area more obtuse; diverging discal pale spots distinct; elytra each with six black spots nearly as in *13-punctata*, the three posterior generally more developed and constant, the lustre faintly alutaceous............16

12—Subapical spot of the elytra large, constant and conspicuous, the two post-median spots large and obliquely coalescent; anterior spots always wanting, the scutellum alone black; body large and rather broadly oval. Length 5.9–7.0 mm.; width 3.75–4.7 mm. New Jersey and Indiana.............**glacialis** *Fabr.*

Subapical spot of the elytra invariably wanting.............................13

13—Elytra very feebly alutaceous, being distinctly microreticulate under sufficient amplifying power...14

Elytra very highly polished and rather more distinctly, though not more closely, punctate, the punctures rather more impressed, the interspaces devoid of distinct microreticulation...15

14—Form broadly oval, the elytra wholly devoid of black spots, excepting a small scutellar sutural dash; pronotum frequently devoid of diverging discal pale spots. Length 5.2–6.6 mm.; width 3.6–4.5 mm. Coast regions of California from San Diego to Sonoma [*punctulata* Lec.]................**ambigua** *Lec.*

Form narrowly oval, the elytra generally with a small subsutural transverse spot behind the middle which is sometimes joined to another external and more posteior, frequently wholly immaculate or with only a small scutellar dash and, rarely, exhibiting very minute post-scutellar points; scutellum always black; discal diverging lines of the pronotum always very fully developed, sometimes coalescing anteriorly with the lateral pale area. Length 4.2–5.0 mm.; width 2.5–3.5 mm. California (Sonoma Co.)................................**obliqua,** sp. nov.

15—Form rather short and broadly oval, the prothorax relatively small, with largely developed pale diverging discal spots ; elytra wholly immaculate, the scutellum alone dark. Length 4.7 mm.; width 2.8 mm. California (Monterey Co.)
.... **politissima**, sp. nov.
16—Three posterior spots of each elytron invariably isolated among themselves. Length 4.6–6.4 mm.; width 2.7–4.4 mm. New Jersey to California (Sonoma Co.), Texas (Brownsville) [*obsoleta* Cr.]...............**convergens** *Guér.*
Three posterior spots much larger and coalescent ; humeral spot distinct, the two at basal fourth equal and extremely minute, the two post-median very large and slightly coalescent, the subapical also large and joined to the inner—not the outer as usual—of the post-median spots by a short straight vitta parallel to the suture. Length 5.2 mm.; width 3.2 mm. California (Sonoma Co.)..**juncta**, sp. nov.
17—Scutellar spot shorter and broad, abruptly terminating at or before basal third. .18
Scutellar spot narrower and elongate....................................... 19
18—Elytra opaque, finely rugose and minutely punctate, each with a black vitta from the callus abruptly ending in a bifurcation at three-fifths from the base, the inner branch not truncate opposite the suture, also with a detached transversely triangular subapical spot ; pronotum polished, minutely punctate. Length 5.1 mm.; width 3.0 mm. California (Lake Co.)................... .**crotchi**, sp. nov.
Elytra more convex, nearly smooth, shining though feebly alutaceous and more distinctly, though not strongly, punctate ; elytra each with a very irregular continuous vitta from the callus to apical sixth or seventh, the vitta strongly constricted just behind the callus, then much dilated inwardly just behind the middle, this part presenting a very broad rectilinearly truncate face opposite and close to the suture, also slightly dilated externally at three-fifths from the base, thence curving in almost regular arc, becoming transverse, and ending at a short distance from the suture at a considerable distance from the apex, this apical part probably being isolated in less fully developed specimens. Length 4.9 mm.; width 3.0 mm. Vancouver Island............... .... **complex**, sp. nov.
19—The scutellar spot extending to about basal third ; elytra more elongate and more acutely rounded behind, the spots four in number, one at the callus, one larger and anteriorly angulate slightly post-median, another, very small and more external, at three-fifths, and the fourth transverse, rather small, submedian, and at apical fifth or sixth, the first and second of these doubtless frequently connected. Length 5.6 mm.; width 3.2 mm. Colorado....................**spuria** *Lec.*
The scutellar spot very elongate, extending to apical fourth or fifth, with a slight rhomboidal enlargement near the base.................................20
20—Elytral spots generally not greatly tending to confluence ; inner post-median sometimes uniting with the spot on the callus to form the usual broad vitta, the subapical always isolated and distant from the apical angles ; lustre of the elytra generally dull, but with the surface almost smooth, the punctures fine, but distinct and rather close-set. Length 5.0 mm.; width 3.1 mm. New Mexico (Fort Wingate)... **americana** *Cr.*
Elytral spots all confluent, forming a broad and nearly even straight vitta from the callus to within a very short distance of the apical angles, slightly angulate externally behind the middle, and thence moderately oblique nearly to the sutural angle, the entire design nearly as in *Paranœmia. vittigera ;* lustre of the elytra

alutaceous, the punctuation sparse and almost obsolete. Length 4.5 mm.;
width 2.7 mm. California (Sonoma Co..).............**trivittata**, sp. nov.
21—Subapical arcuate spot of the elytra not attaining the suture or apical angles.
Length 3.8–5.0 mm.; width 2.3–3.2 mm. New Jersey to Puget Sound [*tridens*
Kirby ; *lunatomaculata* Mots.].....................**parenthesis** *Say*
Subapical arcuate spot flexed posteriorly and inwardly, attaining the suture and apical
angles ; body smaller and more distinctly punctate. Length 3.7–4.75 mm.;
width 2.25–2.6 mm. Nevada and California (valley of the Truckee River).
apicalis, sp. nov.

Of the described species not included above, *15-maculata*, of Mulsant, has a scutellar dash and generally six spots on each elytron, the
anterior juxtasutural dilated and apparently formed of two ; it is said
by Crotch to occur in Missouri and may be inserted after *convergens ;*
leporina Muls., has a subbasal band from one callus to the other and
the elytra each two black spots, the anterior transverse and almost triangular, the posterior smaller, obtriangular and joined to the anterior;
it is described from California and may be placed after *vernix. Sinuata,*
of Mulsant, has the elytral suture black for three-fourths and the elytra
each a vitta from the callus for five-sixths the length, almost semicircularly curved in its posterior half and dilated opposite the suture
near the anterior limit of the arcuate portion ; its dimensions are said
to be 5.9 × 3.3 mm., which is larger than any of the allied species
known to me ; it belongs near *trivittata* in the table ; *interrogans* is
placed as a synonym of *sinuata* by Crotch. Finally, *oregonensis*, of
Crotch, is similar to *spuria*, but lacks the discal white spots of the pronotum and *falcigera* is allied to *trivittata*, but is also devoid of the
discal diverging lines.

The sexual characters are well marked, the anterior and middle
tarsi being distinctly dilated and the abdomen emarginate at apex in
the males. *Extensa, subsimilis* and *vernix*, together with *leporina*
Muls., may all be subspecies of the last, but I have no means of stating this with certitude. *Mæsta* is said to be a variety of *lecontei* by
Crotch, but in my opinion there is no reason for this assumption, as
there is no individual known to me which can be considered a connective bond, my series of both being quite homogeneous ; the elytra in
mæsta are more elongate and more pointed behind than in *lecontei*.
The last two species of the table are almost generically distinct from
the others.

Eriopis connexa Germ., of our lists, is a South American species
which is said by Crotch to occur also in California and Vancouver

Island, but is not recorded from Mexico or any other intervening region. It should be removed from the lists, as there is almost certainly some error of indentification or locality.

COCCINELLINI.

This is by far the most extensive tribe of the family, containing also the largest species and is the most difficult to treat taxonomically, because of the slight amount of structural variety and the evidently great number of groups, which must be accorded generic rank because of habitus or summation of minor characteristics. Type of ornamentation has not been regarded as a generic character hitherto, but is in reality one of the most important, especially that of the pronotum. All of our numerous species of *Coccinella*, for instance, have precisely the same type of pronotal ornamentation and this is true also of *Adalia*, *Cycloneda*, *Anatis*, and all others which comprise enough specific forms to admit of generalization. Where two forms exist, therefore, which seem to belong to different generic types but which do not differ structurally to any decisive extent, I have regarded the general scheme of pronotal ornamentation, and, to a less degree, that of the elytra, as the deciding criterion.

In the following table all the genera accessible to me are included, the exotic ones having an asterisk affixed :—

Metacoxal lines arcuate or feebly angulate, continuous, not quite entire, the plates distinctly shorter than the first ventral segment; body oval (Subtribe ADALIÆ)..2
Metacoxal lines curving outward to the sides of the body along the first suture, the included area frequently divided by an oblique line, which may or may not join the curve posteriorly; body rounded, rarely oval or suboblong..............5
2—Tarsal claws simple, long and well developed; body broadly oval, distinctly punc. tured, pale, maculate with black spots, the scutellum moderate in size; antennæ moderately short, with a rather broadly obtriangular compressed 3-jointed club; metacoxal lines arcuate, the plates slightly shorter than the segment; basal node of the last tarsal joint partially free. Palæarctic .... ***Bulæa**
Tarsal claws with a large subquadrate basal tooth; antennæ slightly longer, with an obtriangular and more closely connate club, the last joint as wide as long. 3
3—Scutellum very small and equilaterally triangular; body distinctly punctate; prosternal process not distinctly bicarinate. Subarctic of both hemispheres. **Adalia**
Scutellum slightly larger, acutely pointed and longer than wide. Austral Africa. . .4
4—Body oval, subimpunctate; prosternal process not evidently bicarinate.
                                                                      *Lioadalia
Body more rounded, finely punctate; prosternal process very narrow, with two strong parallel carinæ extending almost to the apex.......................***Isora**

5— Tarsal claws with a large subquadrate internal tooth at base................6
Tarsal claws cleft within...21
6—Scutellum very minute ; body small, rounded, pale with black spots, the metacoxal plates without an oblique dividing line ; prosternal process very narrow,
strongly bicarinate to apical third or fourth ; antennæ with a narrow, obtriangular club, the last joint rather longer than wide ; claws slender, the basal
tooth but slightly developed transversely.   Africa..............*Micraspis
Scutellum not extremely minute or punctiform ; basal tooth of the claws large and
conspicuous.................................... 7
7—Epistoma truncate or subtruncate at the apex of the coriaceous or semi-corneous
margin (Subtribe COCCINELLÆ).......... 8
Epistoma deeply sinuate. (Subtribe CYDONIÆ)................................18
8—Metacoxal plate divided by an oblique line joining the bounding arc at about its
middle point, forming an angulate inner plate 9
Metacoxal plate not or only partially divided, the oblique line either wholly obsolete
or feeble, or, when more distinct, not joining the boundary curve posteriorly..14
9—Oblique line meeting the bounding curve at a point which is but little beyond the
middle of the segment ; body oval, rather depressed, with coarse and unequal
punctuation, the side margins abruptly but very finely reflexed ; prosternal process
concave along the axial line ; mesosternum with a very small, circularly rounded
median notch ; antennal club large, obtriangular, compact, the last joint nearly
as long as wide and obliquely truncate................... ....Agrabia
Oblique line meeting the bounding curve at or very near the hind margin of the segment..10
10—Mesosternum transversely truncate anteriorly ; body strongly convex, oval, more
or less finely and equally punctate, the side margins very finely reflexed ; pronotum solidly black, with a more or less subquadrate pale spot at the apical
angles ; hind angles rather narrowly rounded...................Coccinella
Mesosternum broadly sinuate at the anterior margin ; side margins more broadly reflexed.............. 11
11—Pronotum solidly black, with broad pale side margins ; body oval, rather strongly
convex, the elytra sometimes having a transverse subapical plica ; punctures
fine and subequal.   Palæarctic.........................* Ptychanatis
Pronotum variegated throughout its extent with black and pale markings, or pale
with small black spots.......................................12
12—Body globularly convex and very broadly rounded, minutely and equally punctate, the pronotum pale with small black points, the elytra with transverse series
of spots on a pale ground, or, by extension, of pale spots on a dark ground ;
prosternum with two fine carinæ converging anteriorly and extending slightly beyond the middle.   Africa.......................................* Stictoleis
Body moderately convex or somewhat depressed, oval in form ; pronotum pale, variegated with black.......................................13
13—Elytral punctures strong and unequal ; prosternum not bicarinate.
Neoharmonia
Elytral punctures finer and equal; prosternum with two fine approximate carinæ, converging slightly in front and extending to about the middle of the length.   Africa.
* Œnopia

14—Elytral punctures very minute and inconspicuous, equal; side margins distinctly
reflexed..15
Elytral punctures strong, conspicuous and more or less unequal..............16
15—Mesosternum truncate anteriorly; body broadly rounded and very convex; pronotum black with pale lateral markings, the elytra immaculate as in *Œnopia*;
metacoxal plates very rarely with a distinct trace of the dividing line.

Cycloneda

Mesosternum broadly and rather feebly sinuate; body as in *Cycloneda* and similarly
punctulate, but having a feeble longitudinal submarginal furrow somewhat as in
Chilocorus, disappearing behind the middle and particularly pronounced in the
black forms; ornamentation dimorphous; oblique line of the metacoxal
plates distinct but not united with the bounding curve posteriorly........Olla
Mesosternum truncate but with a very small, shallow and circularly rounded median
notch; body broadly rounded but rather depressed; pronotum pale, variegated
with black, the elytra pale, usually with black vittæ. Africa.....* Verania
16—Mesosternum truncate, with a very minute shallow rounded notch at the middle
as in *Verania*; body oblong-oval, moderately convex; pronotum pale, variegated with black markings, the elytra pale, with an irregular dark design.

Cleis

Mesosternum broadly and deeply sinuate; body more or less broadly oval, moderately
convex...17
17—Prosternal process narrow, strongly bicarinate; pronotum with two large subquadrate black spots, narrowly and rectilinearly separated; elytra spotted with
black, or dark with pale spots.........................Anisocalvia
Prosternal process broad, strongly convex in a transverse direction and prominent at
the apical margin; pronotum black, with pale lateral or sublateral and basal
areas, the elytra generally pale with black spots or immaculate; body large in
size..Anatis
18—Hypomera with a well-marked but shallow rounded antennal depression; pronotum ornamented almost exactly as in *Coccinella*; body moderate in size, very
broadly rounded..19
Hypomera without an antennal depression; body more broadly oval, the pronotum
nearly as in *Anatis*..20
19—Antennæ inserted very close to the eyes, the latter broadly and feebly sinuated by
the large antennal cavity; epistoma without a semi-corneous margin at the bottom
of the sinus; body moderately convex, the elytra pale with black vittæ. South
Africa. (Type *4-lineata*.).....................................* Cydonia
Antennæ not quite so close to the eyes, which are more deeply and narrowly sinuated
by the post-antennal canthus; epistoma with the usual semi-corneous apical
margin at the bottom of the sinus; body strongly convex, the elytra black, irregularly ornamented with large red areas. Africa. (Type *lunatus*.)

* Cheilomenes

20—Antennæ and eyes as in *Cydonia;* epistoma with a narrow coriaceous apical
margin at the bottom of the sinus; elytra very finely punctulate, black, ornamented with large irregular red blotches; sides gradually less declivous to the
edge, which is not reflexed or thickened; prosternum narrowly excavated along
the median line to beyond the middle. Siberia. (Type *hexaspilota*.)...* Ithone

21—Body very broadly rounded, minutely punctulate, the elytra very broadly explanate at the sides, the edge not thickened, pale, spotted with black, the epipleuræ very broad, continuing to the sutural angles, with a large deep impression
internally at about basal third ; prosternum transversely convex along the median
line, not bicarinate ; metacoxal plates as in *Cyclomeda ;* epistoma feebly emarginate, with coriaceous margin, the sides strongly dentate ; antennæ and eyes as in
Cydonia.   Asia and East Indies.   (Subtribe SYNONYCHÆ.).....* **Synonycha**
Body oval, rather strongly convex, minutely punctulate ; epistoma obliquely denticulate
at the sides, the extreme margin subtruncate ; antennæ and eyes as in *Coccinella ;*
prosternum ornamented nearly as in *Anatis,* the apex less deeply sinuate and the
apical angles less pronounced ; elytra pale, or ornamented with irregular or interrupted dark vittæ, the side margins very narrowly reflexed, with the edge
thickened, the epipleuræ narrower and simple ; metacoxal plates as in *Cycloneda ;*
prosternum feebly convex along the median line.   (Subtribe MYSIÆ.)

Neomysia

Adalia *Muls.*

The type of this genus is the *Coccinella bipunctata* of Linné, which
is now distributed very widely over the world through commerce.
The species before me are as follows :—

Elytra without transverse series of spots ; metacoxal plates rounded or parabolic ;
elytral punctures fine.. 2
Elytra with transverse series of spots or transverse bands ; metacoxal plates frequently
somewhat angular postero-externally ; pronotum pale, with an M-shaped black
design and a submarginal black spot................................ 3
2—Elytra red, each with a rounded or oval black spot at the centre of the disk ; pronotum with a broad M-shaped median black design, the broad pale margins immaculate ; metacoxal plates rounded, extending but slightly beyond the middle
of the segment.   Length 3.8–5.2 mm.; width 2.9–3.9 mm.   United States (except Pacific Coast)............................... ...... **bipunctata** *Linn.*
Elytra red throughout and immaculate, the reflexed lateral margins usually yellowish ;
pronotum with M-shaped design and a black point at the centre of the broad
yellow margin ; metacoxal plates rounded, extending nearly to apical fourth of
the segment.   Length 3.2–4.3 mm.; width 2.3–3.2 mm.   California.

melanopleura *Lec.*

Elytra black with fine yellow side margins, each with a large oblong yellow spot at the
humerus and another, smaller and rounded, at three-fifths and close to the suture ;
pronotum black with narrow apical and side margins pale ; metacoxal plates parabolic, extending nearly to apical third.   Length 3.9–4.6 mm.; width 2.8–3.25
mm.   Utah to California (Siskiyou Co.)....................**humeralis** *Say*
3—Submarginal black spot of the pronotum rounded and isolated, or only connected
to the black design by a narrow isthmus............ 4
Submarginal black spot broadly amalgamated with the central black design, forming a
parallel-sided lateral extension of the latter ; elytra reddish-yellow with black
bands...5

4—Elytra red, coarsely punctured, each with two small black points arranged transversely a little before the middle, the outer one on the median line and not quite
so basal ; metacoxal plates evenly parabolic, extending nearly to apical fourth.
Length 4.8 mm.; width 3.6 mm.   Nebraska............**ophthalmica** *Muls.*

Elytra pale reddish-yellow, rather feebly punctured, each with a small oblique black
dash from the scutellum and two small subbasal spots, the inner the larger and
both oblique and uniting on the humeral callus, also with three widely isolated
black spots in a transverse line just before the middle, the inner more basal and
the outer very close to the margin, and two, very small, on a transverse line at
apical fourth, very near the margin and at inner third ; metacoxal plates extending nearly to apical fourth, obtusely angulate postero externally.   Length 4.5
mm. ; width 3.0 mm.   California (Sonoma Co.)........**ovipennis**, sp. nov.

Elytra reddish-yellow, rather sparsely and moderately strongly punctate, each with a
longitudinal posteriorly pointed dash at each side of the suture from the
base, and two subbasal spots generally disconnected, the outer more basal and on
the callus, also with a transverse series of three rather large spots just before the
middle, the outer two generally connected, and two at apical fourth nearly as
large, the outer slightly more apical, transverse and very close to the margin ;
metacoxal plates rounded though a little more narrowly so postero-externally, extending nearly to apical fourth.   Length 3.9-4.6 mm. ; width 2 8-3.5 mm.
Colorado..**annectans** *Cr.*

5—Elytra coarsely and closely punctured, each with a transverse basal spot acuminate
externally and extending from the suture to inner third, and a large triangular
subbasal spot involving the callus, also with a transverse uneven band near the
middle of the length, not interrupted at the suture, extending to lateral ninth or
tenth, the outer two-thirds straight and transverse, the inner third more basal and
posteriorly oblique toward the suture, and a transverse, somewhat bilobed spot at
apical fourth, equidistant from the suture and margin ; metacoxal plates but
slightly angulated, extending fully to apical fourth.   Length 4.0 mm ; width
3 1 mm.   New Mexico (Las Vegas)...............**transversalis**, sp. nov.

Elytra coarsely but rather less closely punctured, completely devoid of any trace of
basal or subbasal black spots, each with an irregular transverse band just before
the middle, extending from inner sixth to outer third, and a small rounded spot
in the same line at outer fourth or fifth, also with an uneven transverse spot at
apical fourth, extending from inner fifth or sixth very nearly to the margin ; metacoxal plates parabolic, extending to apical third.   Length 3.8 mm. ; width 2.7
mm.   Colorado............. ....**ornatella**, sp. nov.

Humeralis is said to be a variety of *bipunctata* by Crotch, and is
even omitted entirely from the Henshaw list, but my ample series of
each is perfectly homogeneous and without trace of any evidence of relationship, the only variation from the normal being a small red point
in one example just behind the middle and near the side margin ; it is
smaller and more narrowly oval than *bipunctata*, has a differently
formed metacoxal plate, and inhabits a different geographical region.
The last five species of the table are related closely to *frigida*, but they

are distinct among themselves and therefore probably not mere varietal forms of that species. *Annectans* is quite unaccountably placed in *Coccinella* by Crotch. *Ludovicæ* of Mulsant, cannot be identified and has a different type of pronotal ornamentation from any noted in the table. The *Coccinella disjuncta* of Randall, is evidently an *Adalia*, allied to *frigida*, but I have not been able to identify it ; it must resemble *ornatella* very closely.

Agrabia, gen. nov.

The species given below, together perhaps with the Mexican *viridipennis* Muls., is the only known representative of this genus, which resembles *Adalia* in the oval, moderately convex form of the body. The side margins are exceedingly narrowly and finely reflexed :—

Oval, moderately convex, pale rufo-testaceous throughout above and beneath, except the elytra which are bright blue, sometimes with a feeble greenish tinge, the side-margins very narrowly testaceous from the humeral angles to apical four-fifths, where the pale margin is inwardly dilated, forming an elongate, internally arcuate spot, which narrows and disappears completely very near the sutural angles ; punctures strong and rather close-set, somewhat unequal. Length 5.5 mm. ; width 3.9 mm. New Mexico ....**cyanoptera** *Muls.*

The description of Crotch is very inexact, especially in regard to the antennæ, which are not unusually short for the Coccinellini, and the mesosternum, also in stating that the body is "subhemispherical."

Coccinella *Linn.*

This genus is still a receptacle for many discordant elements ; *venusta*, which is assigned to it by Crotch (Trans. Am. Ent. Soc., 1873), is the type of a distinct genus, named *Neoharmonia* in the table, and, in the "Revision," *picta* belongs to *Cleis* and not to *Harmonia*, where is was subsequently placed, and *cyanoptera* to *Agrabia* and not to *Harmonia*. Even as restricted in the present essay, however, the genus is still a large one and our species may be conveniently separated as follows :—

Elytra without trace of a basal fascia, the spot on the callus wanting or moderately developed ; body large, usually broadly oval or elliptic......................2

Elytra with a transverse subbasal fascia, sometimes disintegrating ; body large, strongly convex and broadly oval...................... 11

Elytra with a transverse subbasal fascia, sometimes disintegrating into three spots ; body smaller and generally more narrowly oval, polished ; pronotum with the apical margin and a subquadrate externally broader spot at each apical angle pale in color...................................,.............................12

Elytra with a broad sublasal fascia, not quite attaining the side margins, broadly sinuate medially at its posterior margin and deeply emarginate at each side at base
by two triangular pale areas; body very small, narrowly elliptic.............13
2—Scutellar spot small and oblong or rhomboidal.............................3
Scutellar spot large, transversely suboval or elliptical, the subhumeral always wanting; suture never black...8
3—Pronotum distinctly margined with yellowish-white along the apical margin; each
elytron with four spots, no one of which is ever altogether wanting, that on the
callus and the post-humeral small, the medio-juxtasutural and subapical large;
suture finely black...4
Pronotum without a pale apical margin toward the middle...................5
4—Elytral spots well developed, the juxtasutural rounded or oval and subequal to the
subapical, the subhumeral and post-humeral sometimes connected by a fine line
extending from the outer side of the former to the inner side of the latter, which
rarely shows also a tendency to extend forward externally in a fine line; under
surface and legs black, the meso- and met-epimera white. Length 5.5–6.7 mm.;
width 4.2–5.0 mm. New York, New Jersey, Virginia, Indiana and Iowa.
                                    9-notata *Host.*
Elytral spots very small and feebly developed, the subhumeral and post-humeral reduced to small points, the juxtasutural transversely linear and much smaller than
the transverse subapical, which is the largest; coloration as in *9-notata*, the body
smaller. Length 4.7–6.3 mm.; width 3.8–5.0 mm. New Mexico (Fort Wingate), Arizona (Cañon of the Colorado River) and Colorado.
                                    degener, sp. nov.
5—Elytral suture not at all darker in color; body broadly oval, strongly convex, the
pronotum black with a subquadrate pale spot at each apical angle, the punctures
fine and unusually close-set, giving a feebly alutaceous lustre; elytra immaculate,
except a small black scutellar spot flanked at each side by a paler spot at the
basal margin, the punctures fine and rather close-set, becoming quite strong laterally; abdominal plates strongly defined, broadly ogival in form internally.
Length 5.8 mm.; width 4.5 mm. Nevada (Reno)........**nevadica**, sp. nov.
Elytral suture darker in color but extremely finely so, the scutellar spot, when well
developed, sharply rhomboidal; elytral punctures very fine, sparse, the elytra
frequently immaculate...6
Elytral suture broadly black from the rhomboidal scutellar spot to the apex, toward
which the vitta is noticeably broader...................................7
6—Base of the prothorax very strongly arcuate, the sides scarcely more than two-
thirds as long as the median length, the apical angles very obtuse and broadly
rounded, with the pale spot large, transverse, somewhat prolonged and sharply
angulate at its inner posterior limit; elytra with spots nearly as in *9-notata*, but
smaller, the median discal rather more transverse, and the subhumeral frequently
wanting. Length 6.4 mm.; width 4.7 mm. Utah........ **prolongata** *Cr.*
Base of the prothorax very broadly arcuate, the sides but slightly shorter than the
median length; apical angles more prominent and narrowly rounded, the pale
spot small and subquadrate; elytra generally wholly immaculate, but in rare instances when spots are present they are rounded and disposed nearly as in *9-no-*

tata. Length 5.2–6.2 mm.; width 3.9–4.7 mm. California (Coast regions from
Sonoma to San Diego) [*franciscana* Muls.]............**californica** *Mann.*

7—Body more narrowly oval than usual in this group and very much less convex, the
pronotum finely but strongly and closely punctured, with the pale spot at the
apical angles small and subquadrate; elytra with an even oblique band just before the middle, terminating at equal distances from the suture and margin, and
also with a short transverse spot at apical fourth or fifth; subhumeral spot completely obsolete, the punctures rather strong and close-set. Length 5.7 mm.;
width 4.1 mm. Colorado.........................**suturalis**, sp. nov.

8—Pronotum polished, the minute punctures well separated, the pale spot at the apical
angles moderate in size and subquadrate; elytra each with a long oblique spot
just before the middle and another shorter near the apex....................9

Pronotum strongly alutaceous, the minute punctures deep and close-set, the pale spot
at the apical angles large, extending to basal third. 10

9—Submedian oblique fascia broad, entire and very conspicuous; pronotum evenly convex toward the sides. Length 6.0 mm.; width 4.6 mm. Vancouver Island [*lacustris* Lec]...**monticola** *Muls*

Submedian oblique fascia tending to disintegrate into an outer smaller and inner and
larger spot; body more broadly oval, polished, strongly punctured toward the
sides of the elytra, the impression along the side margin of the pronotum extending arcuately inward just before the middle, disappearing at some distance from
the edge; inner part of the abdominal plates acutely angulate behind. Length
6.4 mm.; width 5.0 mm. California...................**impressa**, sp. nov.

10—Body oval, very strongly convex, the elytra dull, finely and feebly punctate, each
with a transverse spot at the middle as in *5-notata*, and a small rounded spot
near the margin and somewhat more anterior, the two sometimes subunited, the
subapical transverse spot nearer the margin than the suture. Length 6.7 mm.;
width 5.2 mm. New Mexico........ ....**alutacea** sp. nov.

11—Elytral punctures rather strong, moderately close and conspicuous, finer toward the
suture; besides the common subbasal fascia,.each elytron has a transverse spot
from the center of the disk to inner fifth, and a similar or rather wider transverse spot near the apex; submarginal spot before the middle extremely rare;
suture always pale; pronotum with a subquadrate pale spot at each apical
angle. Length 5.8–7.5 mm.; width 4.5–5.8 mm. Colorado, Utah, Wyoming,
Montana and northward, and probably also northern California; [*ransverso-
guttata* Cr. nec Fald., *nugatoria* Muls.].................**5-notata** *Kirby*

12—Elytra with a broad subbasal fascia, equally wide throughout and but little prone to
disintegration, each also with a broad oblique fascia at the middle and another
near the apex; punctures strong and close-set. Length 4.9 mm.; width 3.75 mm.
Rhode Island and Wisconsin [*trifasciata* Cr. nec Linn.]..... **perplexa** *Muls.*

Elytra with a narrower and more irregular subbasal fascia tending to disintegrate
into three spots, and each also with two oblique bands as in *perplexa* but narrower and frequently altogether obsolete, the punctures fine, sparse and feeble.
Length 4.0–5.2 mm.; width 2.9–3.8 mm. California (northern and middle
coast regions); [*harda* Lec.] ....**juliana** *Mus.*

Elytra with a small and evenly equilatero-triangular black scutellar spot widely de-

tached from the subhumeral spots, which are well developed, and each also
with the two oblique bands of the preceding species, which are here rather narrow; body more narrowly oval, the elytra sparsely but more strongly punctured. Length 4.5 mm.; width 3.2 mm.   California (Siskiyou Co).

                                                    eugenii *Muls.*
13—Body strongly convex, the pronotum with a transverse pale spot at each anterior
angle extending narrowly across the median parts of the apical margin; elytra
rather strongly punctured, each with a large irregular transverse discal spot at
apical third in addition to the broad basal fascia.   Length 4.1 mm.; width 2.9
mm.   Lake Superior [*kirbyi* Cr.]. ....**tricuspis** *Kirby*

Perplexa, juliana and eugenii are related to trifasciata, but are all
distinguishable at once by the form of the white apical area of the pronotum, which is expanded into a larger transverse spot in the American forms, but only narrowly and nearer the edge in the European.
Californica is in no wise related to 5-notata or transversoguttata as
stated by Crotch and others, the occurrence of the very rare spotted
examples showing that it is more closely allied to 9-notata. Quinque-
notata is certainly distinct enough from transversoguttata to be entitled
to specific rank, and the variety transversalis seems to be identical
with nugatoria; at any rate the name must disappear as it is preoccu-
pied by Fabricius. Difficilis Crotch, I have failed to identify; it appears to resemble prolongata completely and may be synonymous.
Subversa Lec., is probably allied closely to degener, but the author
states "elytris distincte et subtiliter punctulatis, scutello nigro, et
præcipue macula obliqua ad medium nigra notatis," which will not
agree, as the most conspicuous spot in degener is the subapical; Crotch
states that it is a variety of trifasciata, and that the elytra are spotless.
Mulsant describes eugenii as being subhemispherical; this would be
very inexact for the example before me, which seems to be typical in
every other way; the appearance of the spots indicates that they never
coalesce to form the subbasal fascia of perplexa and juliana.

Neoharmonia, gen. nov.

The genus Harmonia is not considered sufficiently distinct by European authors, and Crotch, while admitting the name to the American lists, assigned to it a number of species belonging to several different genera, no one of which appears to be a true Harmonia. The
Harmonia of Mulsant is also composed of numerous dissimilar elements. In view of this confusion of judgment, it seems best to separate our two species as a distinct genus, allied to Harmonia, but differing apparently in the more widely reflexed side margins. The

form and ornamentation of the body, more broadly reflexed side margins, more depressed surface and emarginate mesosternum are all departures from *Coccinella*, to which these species have been attached, and the two genera are not even closely allied. The genus *Neoharmonia* probably includes also the Mexican *ampla* Muls., which I have not been able to examine. Our two species are the following : —

Broadly rounded, feebly convex, relatively strongly and unequally punctate ; head black, yellow along the eyes ; pronotum pale, with a large oblique fascia of black at each side extending from near the sides to the scutellum, gradually narrowing inwardly and departing slightly from the basal margin externally, also with two approximate median spots before the middle, which are sometimes united with the basal fasciæ at about their medial points ; scutellum black ; elytra pale yellow or reddish, each with two large subquadrate subbasal black spots and one still larger just before the middle, subtriangular and near the margin, also a large subquadrate spot near the margin at apical fourth, extending to inner third, where it is united with a common sutural vitta extending from near the apex to just behind the middle, also with a rounded spot just before the middle and near the suture, prolonged internally obliquely forward meeting—but not quite amalgamating with—its duplicate of the other elytron at the suture some distance behind the scutellum, forming two oblique inverted commas ; under surface and legs blackish. Length 6.0 mm.; width 4.7–5.0 mm. Indiana; [*notulata* var. A Muls.].. **venusta** *Mel h.*
Similar in form to *venusta* but smaller and less strongly and less unequally punctured, black above, the elytra with violaceous reflection, the pronotum with a rather wide oblique pale border, becoming very narrow basally and extending very finely along the apex, with a small medial dilatation ; elytra each with a transverse pale fascia extending from inner third or two-fifths to and enveloping the margin, its posterior limit transverse and feebly sinuate, especially toward the margin, its anterior limit deeply sinuate, forming two acute points, one on the margin and one on the medial line, the inner flank of the inner point straight and oblique ; legs black ; epipleuræ with the outer edge black toward base. Length 5.0 mm.; width 4.0 mm. Louisiana ; [*notulata* var. B Muls.]........**notulata** *Muls.*

These two species seem to be amply distinct and not varietal forms of one—at least no intermediate forms are known. This may however be another case of dimorphism.

Cycloneda *Crotch.*

The type of this genus is the *Coccinella sanguinea* of Linné, described from Surinam. *Sanguinea* is therefore in all probability specifically different from any of our forms, and it is not included in the table given below. The species are all very closely allied ; they have the elytra pale red or yellow or black and immaculate, those with spotted elytra belonging to other genera. The metacoxal plates generally have no trace of the oblique dividing line, but in *hondurasica* there is a short but well developed line, which fails to attain the bounding curve by a long distance. The body is rounded or oval, very convex, minutely and obsoletely punctulate, with the side margins of the elytra quite broadly reflexed, the gutter extending around the outer and anterior parts of the humeral callus as in *Neoharmonia*, and the edge strongly and abruptly thickened ; the gutter is always more strongly, closely and subrugosely punctured toward base. The species before me may be outlined as follows :—

Pronotum black, with a narrow pale side margin extending with equal width posteriorly and internally along the base, terminating abruptly at about lateral sixth and sometimes extending more narrowly along the median parts of the apex, also with an isolated small pale spot at the middle of the length and lateral fourth. . 2

Pronotum as in the preceding section, except that the apical margin is always broadly pale, with a posterior medial spur in both sexes, and the sublateral pale spot is always united to the pale apex and sometimes also to the basal pale border, isolating a large black spot; body smaller............................6

Pronotum completely black....... 7

2—Body broadly oval but distinctly longer than wide, the marginal bead of the elytra not or only slightly darker. 3

Body extremely dilated, very nearly as wide as long, the marginal bead distinctly black. 5

3—Metacoxal plates with a distinct but short disconnected oblique line ; body rather more convex ; female without a white apical pronotal margin at the middle. Length 5.6 mm.; width 4.6 mm. Honduras..........**hondurasica**, sp. nov.

Metacoxal plates devoid of any trace of an oblique line 4

4—Elytra generally luteo-flavate, without distinct paler spaces at the sides of the scutellum ; pronotum of the female generally with the apical margin narrowly pale, usually subinterrupted at the middle and not posteriorly spurred. Length 5.7-6.0 mm.; width 4.7-5.0 mm. Florida............ .**immaculata** *Fabr.*

Elytra deep and bright scarlet, with a short transverse basal paler spot at each side of the scutellum which is black as usual ; pronotum relatively narrower than in *immaculata* and rather more strongly rounded at the base, the median length relatively greater when compared with the sides, having a narrow apical margin with narrow parallel posterior prolongation pale in the male, the female interruptedly margined with paler and without a medial spur. Length 4.4-5.8 mm.; width 3.7-4.8 mm. Texas (Brownsville), and California (San Diego and Los Angeles)....... .... **rubripennis**, sp. nov.

5—Body nearly as in the preceding but with the isolated pale spots of the pronotum smaller, the apical margin rather broadly pale in the male, with a parallel medial spur extending rather beyond the middle ; in the female the apical edge is wholly devoid of a pale margin, the pale border ending abruptly at the eyes. Length 4 5-5 5 mm.; width 4.2-4.9 mm. Bahama Islands (Egg Island)—Mr. Wickham........... .... **limbifer**, sp. nov.

6—Sublateral spur from the pale apical margin never joining the basal pale area, the medial spur of the pale apex short and triangular and not parallel as in the

preceding species ; body more elongate-oval, the elytra luteo-flavate, with more
narrowly reflexed margins, which are always paler.   Length 4.0–5.0 mm.; width
3.5–4.0 mm.   New York, Pennsylvania, Indiana and Iowa.......**munda** *Say*
Sublateral spur longer, frequently joining the basal pale border ; medial spur long
and narrow, gradually acuminate and extending to or beyond the middle of the
disk ; body smaller and rather more rounded, the elytra generally bright scarlet
in color and with almost completely obsolete punctures, occasionally yellow, and,
in a northern example, with more distinct punctuation.   Length 3.8–4.7 mm.;
width 2.9–3.4 mm.   California (Sta. Cruz to Siskiyou Co.), Washington State,
British Columbia and Idaho (Cœur d'Alène)...............**polita**, sp. nov.
7—Body broadly rounded and rather less convex, shining, deep black throughout above
and beneath, the sides of the prothorax rather less arcuate, and the basal angles
more narrowly rounded ; elytral punctures much larger and more distinct than
usual but sparse.   Length 4.3 mm.; width 3.5 mm.   Locality not recorded.

ater, sp. nov.

Ater is widely divergent, both in coloration and to some extent in
punctuation and form of the prothorax, but seems to be assignable to
Cycloneda.   The unique type was discovered in the Levette cabinet
but had no label attached.

Olla, gen. nov.

In this dimorphic genus the ventral plates are almost as completely
divided by an oblique line as in *Coccinella*, but the line does not quite
form a junction with the posterior bounding curve ; in view of the
close similarity of the body with *Cycloneda*, therefore, I have placed
the genus at this point of the series rather than near *Coccinella*, with
which it has little or no affinity.   The following species represents the
pale forms with spotted dorsal surface, more numerous in Mexico :—

Broadly oval and strongly convex, very finely and obsoletely punctulate, the side margins as in *Cycloneda* ; upper surface pale brownish-yellow, the head pale and immaculate ; pronotum with a basal black spot at two-fifths from the middle and a
short transverse spot before the scutellum, also with two posteriorly converging
black spots at the centre and a narrow elongate spot on the median line joining
the ante-scutellar spot, and, at lateral eighth and basal third, a small rounded
spot ; scutellum black in the male ; elytra each with a subbasal transverse series
of four small black spots, a medial series of three spots, the inner the largest and
transversely crescentiform and, at apical fourth near the margin, another small
rounded black spot; under surface and legs pale.   Length 4.25–5.25 mm.;
width 3.4–4.0 mm.   Indiana, Texas (Brownsville and El Paso), Arizona and
California (Sta. Cruz and San Francisco)................**abdominalis** *Say*

The large series before me exhibits an extremely small amount of
variation, which, considering its extended geographical range, is very
remarkable.   The male has the fifth ventral truncate, becoming very
feebly sinuate toward the middle, with the edge there slightly concave;

the female has the fifth segment a little longer and very broadly ogival at apex.

The following black species were said by Crotch to form a simple variety of *abdominalis*. My series of *abdominalis*, as before stated, and of two of the species given below, are quite extended, and I am unable to detect any noteworthy variation of any kind, even in the outline of the spots, the constancy of form and ornamentation being in fact one of the most remarkable instances of the kind known to me ; these series are each made up of males and females. Although I do not remember to have ever taken the black with the pale spotted form in California, where the latter is abundant, it should, however, be noted as a suspicious fact that several of the localities yielding *abdominalis* in my series are common also to the black species given below. It is, therefore, possible that we may have here a case of dimorphism, and the same may be true of *Adalia humeralis* and *bipunctata*, and of *Hippodamia divergens*, or allied species, and *mæsta*, but in the absence of intermediate forms any consanguinity in these very puzzling cases can only be proved by systematic biological observation. I might prefer rather to consider these perfectly constant and well-established aberrations of color—and, to some extent, of accompanying structure as well, such as the more finely reflexed side margins in the black forms,—more as protective adaptations to slightly changed environments. Of these black forms we have, at any rate, three quite well-defined variations, as follows :—

Body very broadly rounded, the head pale, sometimes more or less nubilate with piceous, the pronotum black, without a well-defined pale apical margin, but with a small central spur, the side margins obliquely pale, the pale area either curving narrowly around the basal angles or disappearing before reaching the base, its oblique inner boundary nearly straight and but slightly uneven ; elytra minutely punctulate, gradually rather more distinctly toward the sides, which are but narrowly though strongly reflexed, with the edge beaded, each with a large transverse and irregular spot before the middle, which is emarginate internally at apex and externally at base ; under surface, epipleuræ and legs black ; meso- and met-episterna, hypomera, tarsi and abdomen pale. Length 4.2–4.9 mm.; width 3.5–4.2 mm. Texas (Brownsville and El Paso), Arizona and California (Los Angeles and San Francisco)................ ....**plagiata**, sp. nov.

Body as in the preceding but larger and similarly colored, except that the apical margin of the pronotum is narrowly pale with the medial spur distinct and the oblique inner boundary of the lateral pale area more distinctly spurred at its middle point ; elytral pale spot before the middle smaller, triangular and feebly oblique, the outer side truncate, the inner angle narrowly rounded. Length 5.0–5.7 mm.; width 4.5–4.8 mm. Florida...............**sobrina**, sp. nov.

Body narrower and oval, more strongly convex, the pronotum similarly colored but
without a pale apical margin or medial spur, the oblique lateral spot not reaching the base in the type, but with a minute detached spot at the basal angles;
elytra with a large pale spot having a straight transverse base, from the extremities of which anteriorly the outline is evenly semi-circular. Length 4.8 mm.,
width 3.75 mm. New Mexico (Las Vegas)...........**fenestralis**, sp. nov.

One of these species was described by Mulsant under the name
binotata Say (=*affinis* Rand.), which belongs to *Hyperaspis*, and the
oculata of Fabricius, to which they were referred by Crotch, is described as having a large rounded pale spot at each side of the pronotum, and must therefore apply to some other species, possibly of
Neda.

Cleis *Muls.*

The species which I have ventured to assign to this genus are
rather small in size and have a distinctly oblong-oval form, with irregular elytral ornamentation. Those before me may be recognized by
the following characters :—

Pronotum with three spots forming a central posteriorly pointed triangle, the posterior
the smallest and elongate-oval, the anterior each with a small spot attached
antero-externally, also with a larger irregular basal spot at the middle of each
side and another subtriangular at the middle and lateral eighth, some or all of the
spots generally united, forming an irregular design with a large M-shaped central
figure ; scutellum black ; elytra a little longer than wide, somewhat broadly
ogival at apex, distinctly but not very unequally punctate, pale in color with a
piceous-black design, the most conspicuous feature of which is a longitudinal and
slightly oblique vitta from the callus to apical fifth, the two united transversely
across the suture behind and at basal two-fifths, and with a subcontiguous spot
externally at the posterior limit ; in the most developed form the entire elytra are
black, with a pale border dilated internally at the middle, a large discal spot behind the middle and a basal fascia irregularly dilated ; in the paler forms the
dark fascia at two-fifths is broken up and all the lines much reduced in width ; , ' ,,
under surface and legs pale reddish-brown, the prosternum, hypomera, median / ,
parts of the meso- and metasterna, epipleuræ and entire parapleuræ of the hind
body pale yellowish-white. Length 4.0–5.0 mm. ; width 3.0–3.5 mm. Massachusetts, New Jersey and Wisconsin (Bayfield) ; [*concinnata* Melsh., *contexta*
Muls.]..**picta** *Rand.*
Pronotum similar, except that the sublateral spot is feebler and usually disintegrated ;
body similar in form and with but slightly feebler punctures but smaller and with
the dark design of the elytra paler in color and less developed, the external spot near
the posterior extremity of the vitta frequently prolonged irregularly to the side
margin. Length 3.7–3.9 mm. ; width 2.6–2.9 mm. California (Alameda and
Siskiyou Cos.)..**minor,** sp. nov.
Pronotum similar but relatively smaller and with the black design more irregular,
with a few black points at the middle of each side of the apex in addition ;

elytral design less developed, consisting of a fine straight vitta from the callus to the middle at apical fourth, where it is slightly dilated internally, each also with a small elongate dark spot near the vitta internally at two-fifths, and another at three-fifths from the base at the lateral margin and remote from the vitta. Length 4.0 mm. ; width 3.0 mm. Hudson Bay........**hudsonica**, sp. nov.

The last of these is quite distinct from the other two in the more depressed form and in the displacement of the small postero-external spot with reference to the dark vitta ; it also has the suture finely black throughout.

Anisocalvia *Crotch*.

The type of this genus is the European *14-guttata*, which is erroneously referred to *Harmonia* in our lists ; it is more narrowly oblong than any of our species and has the upper surface brownish-orange in color, the elytra with fourteen small rounded paler yellow spots. The pronotum has a longitudinal impression along the sides, close to, but independent of, the concave margin caused by the reflexed edge. The body is evenly oval, moderately convex, with rather narrowly reflexed side-margins, becoming broader around the base of the callus ; the punctures are coarse and unequal and the mesosternum quite deeply sinuate, the prosternum bilineate. Our species may be distinguished as follows :—

Elytra black, or black with pale spots........................ 2
Elytra pale, with eleven large rounded or oval black spots, of which three are on the
    suture, the one at the apex transverse 3
2—Body in the female black, the pronotum but little more than twice as wide as long,
    black with a narrow apical and lateral margin and median line pale, the sides
    feebly convergent, rounding and more convergent at apex ; elytra black through-
    out, with the reflexed side-margin pale ; beneath black, the epipleuræ pale.
    Length 5.3 mm. ; width 4.3 mm. New York (Adirondack Mts.) ; [*similis*
    Rand., *obliqua* Rand.] .... **cardisce** *Rand.*
Body in the female less broadly oval, black, the pronotum with a narrow apical and
    lateral margin and a fine median line not attaining the base pale ; sides rather
    strongly convergent, evenly and broadly arcuate from base to apex, the disk dis-
    tinctly more than twice as wide as long, strongly, moderately closely punctured ;
    elytra black, with a fine side margin, toward apex only, paler, and a rounded
    discal pale spot near the suture and two-thirds from the base ; femora black ;
    epipleuræ piceous-black, the abdomen pale around the entire limb ; in the male
    the body is similar, the pronotum similarly colored, but the elytra have fourteen
    pale spots, that near the suture and apical third being the largest and with its
    postero-external margin nubilate, the legs and epipleuræ pale throughout, the
    latter slightly black opposite the back areas of the upper surface, the abdomen
    pale, clouded with blackish toward the middle and base. Length 5.2 mm. ;
    width 4.0 mm. British Columbia....................**victoriana**, sp. nov.

3—Body evenly elliptical, pale yellowish in color, the pronotum black with apical
    and lateral margins and entire median line pale, the black area joining the basal
    margin at the middle of each side ; punctures strong and quite close-set ; sides
    evenly convergent, broadly and evenly arcuate ; elytra longer than wide, rather
    narrowly rounded behind, the spots large, separated generally by about one-half
    their widths ; epipleuræ, limb of the abdomen, tibiæ and tarsi pale. Length
    5.2 mm ; width 3.9 mm.   British Columbia............ **12-maculata** *Gebl.*
Body similar but smaller and rather more broadly oval, with the prothorax relatively
    smaller and having the sides very much more strongly convergent, the basal
    angles more broadly rounded and the punctures finer and sparser ; coloration
    similar, except that the elytral spots are relatively much larger and only very nar-
    rowly separated, the two transversely placed at the middle, generally confluent.
    Length 4.0 mm.; width 3.2 mm.   Hudson Bay. .... **elliptica**, sp. nov.

The form named *hesperica* by Crotch, is not included above and
must be regarded as a manuscript name.   If any modification whatever
of a species is worthy of a distinctive name, it is worthy also of a description better than this : " Ventral segments and metasternum almost
smooth—Arizona," which is not even of comparative worth, as these
parts in the *similis*, described immediately above under the name of
14-guttata, are not alluded to at all in regard to their sculpture.

Anatis *Muls.*

These are large, broadly oval or rounded and convex species, with
rather coarse unequal punctuation and deeply sinuate mesosternum.
The prosternum is rather broad between the coxæ, and is transversely
convex along the median line throughout, terminating at apex in a
conspicuous prominence.   The antennæ are moderately developed in
proportion to the size of the body, and the prothorax is less transverse
than usual.   The American species are as follows, *ocellata* being introduced for comparison :—

Body oval or subrhomboidal, the pronotum black with broadly pale side-margins and
    a black marginal spot extending from the basal angles to about two-fifths, angu-
    larly oblique internally but never attaining the central black area, the sides of
    which are feebly convergent, rectilinear but emarginate at the middle, also with
    two approximate pale basal spots at the middle.................... 2
Body broadly subrhomboidal, the pronotum black with a broad yellow vitta extending
    from base to apex, parallel and slightly distant from the side margin, which it
    joins at the apical angles, also with two very minute pale points near the basal
    margin at the middle 3
2—Elytra evenly oval, distinctly longer than wide, the side-margins black, the sub-
    marginal spot at two-fifths, elongate-oval and not laterally extended, the subsutural
    spot of the same range elongate ; basal pale spots of the pronotum subquadrate,
    not united at base ; pale apical margin transverse, finely interrupted at the mid-
    dle.   Length 8.5 mm.; width 6.3 mm.   Europe........... ***ocellata** *Linn.*

Elytra rounded or feebly dilated at two-fifths, scarcely as long as wide, the sides generally evenly arcuate with pale margin, the spots not ocellated, the external at
two-fifths rounded, generally not or only narrowly prolonged laterally ; basal spots
of the pronotum slightly oblique, never united at base, the pale apical margin
bioblique, interrupted or very nearly so at the middle. Length 6.5–8.7 mm.;
width 5.5–7.0 mm. Rhode Island, New Jersey, Indiana, Iowa and Arizona
[*labiculata* Say]............................ .. 15-punctata *Oliv.*

Elytra oval, not or scarcely appreciably dilated at two-fifths, rather longer than wide ;
submarginal spot at two-fifths geminate, the outer part enveloping the margin,
which is pale elsewhere with the fine thickened edge slightly darker ; spots all
surrounded by a broad pale border, the ground tint red-brown ; basal spots of
the pronotum short but rather large, angulate antero-externally, united at base ;
pale apical margin transverse and entire, not interrupted but rather broader at the
middle. Length 8.7–10.0 mm.; width 6.8–7.4 mm. Indiana, Wisconsin
(Bayfield) and Idaho (Cœur d'Alène).................... .. mali *Say*

Elytra decidedly rhomboidal, scarcely as long as wide, strongly dilated at two-fifths,
where there is a small marginal spot ; remainder immaculate or with faint vestiges
of one or two of the spots of the preceding species, the punctures much smaller
and nearly equal ; basal spots of the pronotum large, much extended antero-externally, unit ng with the lateral pale area and broadly united at base ; pale
apical margin transverse, not interrupted but rather wider at the middle. Length
8.3 mm.; width 6.9 mm. California (Siskiyou Co.)......... rathvoni *Lec.*

3—Elytra very broadly rounded or subrhomboidal, slightly more dilated at two-fifths,
scarcely as long as wide, the punctures strong but rather less coarse and more
nearly equal than in *15-punctata*, bright brownish-red or ochre, without trace of
maculation but having the entire limb deep black, the border clearly defined and
scarcely occupying the entire reflexed portion, broadening a little at two-fifths ;
pronotum scarcely three-fourths wider than long, broadly, feebly convex, deeply
impressed just within the lateral margins, rather finely and not closely punctate ;
head black ; entire legs and under surface black, the epipleuræ black in external
and red in internal half of their width from base to apex. Length 8.7–10.0 mm.;
width 7.5–8.7 mm. New Mexico (Fort Wingate).........lecontei, sp. nov.

It can be readily observed that *15-punctata* is not even closely related to the European *ocellata*. *Signaticollis* of Mulsant, I have not
seen, but it may be the same as *mali* Say. *Lecontei* somewhat resembles the Mexican *Pelina hydropica*, but I cannot see that it differs
generically from our other species of *Anatis ;* the antennal club is obtriangular with the three joints rather loosely articulated, shorter than
wide and but little more developed internally than externally.

Neomysia, gen. nov.

In the shorter, more feebly emarginate prothorax, with more broadly
rounded apical angles, the present genus evidently approaches the
Psylloborini closer than any other of the Coccinellini, and this is also

confirmed somewhat by the antennæ, which are rather long, slender, with very feebly dilated 3-jointed club having somewhat elongate and loosely connected joints. The anterior coxæ are not unusually widely separated, and the prosternum is not prominent at the middle of the apex ; the mesosternum is broadly sinuate. The genus seems to differ from *Mysia*, the type of which is *oblongoguttata*, in the more narrowly reflexed margins, very fine punctuation and polished surface ; it has but little affinity with *Anatis*. Our species are the following :—

Pronotum without a well-defined discal darker area...........................2

Pronotum with a large trapezoidal median dark area, which is well defined externally..3

2—Pronotum pale yellow, with a feeble red-brown clouded basal spot at lateral fourth and a small nubilate V-shaped spot just before the middle on the median line, also with a feeble disintegrated discal cloud near each side ; elytra yellow, each with three fine incomplete and interrupted subequidistant longitudinal vittæ of pale red-brown Length 7.0 mm.; width 5.7 mm. New Mexico (Fort Wingate) . ...**interrupta**, sp. nov.

Pronotum pale yellowish-brown, without maculation, except a feeble trace of the two basal clouded spots of the preceding ; elytra similar in color, with three very feeble incomplete nubilate vittæ on each, the two inner uniting near the apex and broader, the outer narrow and almost completely obsolete. Length 6.7 mm.; with 5.4 mm. California.............. ....**horni** Cr.

3—Elytra uniformly pale yellow-brown, sometimes slightly paler along the base and externally, rarely with feeble trace of two brown vittæ uniting near the apex at the middle of the width ; pronotum in the male black, with broad yellow side margins, obliquely subrectilinear internally, inclosing a detached central black spot and with barely a trace of a small pale spot before the scutellum, the apex rather broadly yellow in a straight line slightly broader at the middle ; female similar but with the dark area pale brown with clouded blackish lateral edges, the pale apex not dilated at the middle. Length 6.4–7.2 mm.; width 4.9–5.5 mm. Canada, New Jersey, Indiana and Texas (Galveston); [*notans* Rand.].

pullata *Say*

Elytra pale, with broad irregular longitudinal markings....................... 4

4—Pronotum black, obliquely yellow in outer fifth, the pale margin inclosing a small internally angulate black spot just behind the middle and equidistant from the margin and central black area, which is bordered broadly with yellow at apex, the margin dilated posteriorly along the median line for a short distance, also with a small pale bifurcate spot before the scutellum ; elytra pale, with a broad black subsutural vitta from the base for three-fifths, uniting broadly at base with a short broad median vitta, which extends one-fifth, with a triangular black spot in the same line just before the middle and continued again as a broad vitta from three-fourths to seven-eighths, the posterior extremity being in line with the subsutural vitta, also with a narrow external vitta from one-third to three-fourths ; suture finely black throughout, a whitish basal spot at each side of the scutellum ; under surface and legs black. Length 6.6 mm.; width 5.3 mm. Lake Superior.

randalli, sp. nov.

Pronotum black in a broad trapezoidal median region, separated from the apical margin by a very fine nubilous pale border not prolonged posteriorly at the middle,
and having, at each side behind the middle, a small lateral spur not extending
more than half way to the side margin, without trace of a pale spot before the
scutellum ; elytra much longer than wide, with an inner broad black vitta to
nearly two-thirds from inner third of the base, its posterior extremity subunited
with a slight dilation of the fine black sutural margin, the latter dilated near the
base, also with a broad vitta along the median line not united with the inner vitta
basally, extending unbroken from the base at outer two-fifths nearly to the apex,
angularly dilated within at the middle, and a fine external vitta from basal to
apical third or more ; legs black.   Length 6.3 mm.; width 4.5 mm.   Colorado.

montana, sp. nov.

These species are all evenly oval and strongly convex, and vary
much less in size individually than is usual in this family.   *Subvittata*
of Mulsant, I have failed to recognize ; the description of the pronotal
ornamentation will not apply, even approximately, to any form described above.

PSYLLOBORINI.

In the structure of the front, the Psylloborini are evidently intermediate between the Coccinellini and Epilachnini.   The two following genera are very closely related to each other, and inhabit the
eastern and western hemispheres respectively.   The surface of the head
is pubescent in both.   The body is small in size, convex, the pronotum small, diaphanous at the edges and broadly reflexed at the sides;
body pale in color, spotted with a darker tint above ; mesosternum
truncate, the claws with a large quadrate tooth internally at base.   The
two genera before me may be characterized as follows :—

Elytra more broadly reflexed at the sides ; scutellum well developed.... ...* **Thea**
Elytra very narrowly reflexed at the sides, the scutellum minute....... **Psyllobora**

In almost every other character these two genera are so nearly
similar, that it might scarcely be conducive to taxonomic convenience
to maintain them distinct.   Still, there are certain peculiarities in the
types of ornamentation that render them easily separable at first sight.
The genera *Halysia* and *Neohalyzia* are composed of larger species,
which also belong to the Psylloborini.

Psyllobora *Chev.*

A large genus, of which but a small proportion of species have yet
been described.   As in many other genera, the same general scheme
of arrangement of the elytral spots is common to many species, and
the material of our fauna has never been critically examined.   The

species in my cabinet inhabiting the United States may be readily
identified as follows :—

Elytra without common sutural spots, the sutural margin pale............. 2
Elytra with two common sutural spots at one-third and two-thirds from the base, the
    sutural margin narrowly black throughout............. 10
2—Elytral spots uniform in color throughout..................................3
Elytral spots unequal in intensity of coloration among themselves................9
3—Middle of the three subbasal spots broadly confluent with the small spot on the
    callus, forming a single spot.   Atlantic regions......................... 4
Middle spot narrowly united with the external basal spot, the latter semi-detached or
    well defined by a deep strangulation ; elytral punctures minute and sparse.   Pacific
    coast regions 8
4—Each elytron with nine spots, some of which are more or less confluent among
    themselves, the outer basal considered as having disappeared by fusion ; punc-
    tures distinct 5
Each elytron with a large discal reniform spot, the punctures minute and sparse.....7
5—Form broadly oval, the elytral spots black................................6
Form narrowly oval, the elytral spots brown in color ; pronotum faintly punctulate,
    the ante-scutellar spot distinct ; elytra much longer than wide, quite strongly but
    not closely punctured, each with two large subequal and approximate basal spots,
    the inner more oblique, the outer rounded, also with two equal subsutural spots,
    slightly elongate-oval, at basal third and near apical fourth, three submarginal
    at two-fifths, three-fourths and subapical, increasing in size posteriorly, a large
    discal median spot fused with a smaller one in the same line at two-thirds, the
    central spot equal in size to the subapical.   Length 2.15 mm. ; width 1.4 mm.
    Iowa (Keokuk)............................... .....**obsoleta**, sp. nov.
6.—Pronotum finely but distinctly punctate, the ante-scutellar spot small but distinct ;
    elytra strongly and very closely punctured, the spots well developed and occupy-
    ing together as much area as the pale interspaces, arranged as in *obsoleta*, but
    with the outer basal much larger and more prolonged posteriorly, and the sub-
    apical much smaller, oblique and subdivided into two small equal spots, the two
    discal confluent spots similarly united to the subsutural and submarginal spots
    near two-thirds.   Length 2.1-2.7 mm. ; width 1.6-2.0 mm.   Rhode Island,
    New Jersey, Iowa and Wisconsin . ...**20=maculata** *Say*
Pronotum subimpunctate, the ante-scutellar spot obsolete ; elytra as in the preceding,
    barely as long as wide, distinctly but much less closely punctured, the spots oc-
    cupying nearly the same relative positions but very much smaller, the pale in
    excess, the spots all isolated, the submarginal at a much greater distance from
    the edge, the outer basal smaller and not prolonged posteriorly, the subapical
    quadrate.   Length 1.9 mm.; width 1.5 mm.   Florida (Palm Beach).
                                                        parvinotata, sp. nov.
7—Body very small, rounded, with very minute sparse punctures ; pronotum subim-
    punctate, the five spots present but pale brown in color ; elytra very pale yellow-
    ish-white, with brown markings consisting, on each, of two subbasal spots, the
    outer the larger and with a lobe on the callus, a small faint subsutural cloud at
    one-third, a large bilobed discal spot extending from basal third to apical fifth,

prolonged and acuminate antero-externally and a large bilobed and less well-defined subapical spot. Length 1.6 mm.; width 1.3 mm. Texas (Brownsville)—
Mr. Wickham...... ....**renifer**, sp. nov.

8—Similar in form and size to *20-maculata ;* pronotum impunctate, the five spots
smaller and somewhat clouded ; elytra as long as wide, narrowly rounded at
apex, the punctures extremely minute, sparse and not impressed, the spots black
and well defined, nearly coincident in position with those of *20-maculata*, but
with the outer basal subdetached, the submarginal at one-third very small, anterior subsutural much more elongate and the oblique subapical more nearly subdivided and in the form of a dumb-bell. Length 2.7 mm.; width 2.0 mm.
Idaho (Cœur d'Alène) ....**borealis**, sp. nov.
Similar to *20-maculata* but rather more broadly rounded, smaller than *borealis* and
with the elytral punctures impressed though minute and sparse ; elytral markings
as in *borealis* but pale brown in color, the subapical completely divided, forming
two small rounded and widely separated spots. Length 2.2 mm ; width 1.75
mm. California (Siskiyou Co.) ....**separata**, sp. nov.

9—Body broadly oval ; pronotum subimpunctate, the five spots more or less nubilate, the
two anterior transversely triangular ; elytra about as long as wide, with rather fine
but impressed sparse punctures, pale in color, with spots arranged as in *20-maculata*
but pale suffused brown in color, except the inner basal and the two submarginal
at one-third and two-thirds from the base, which are well developed, particularly
the anterior, which three spots are blackish in color ; subapical spot very faint
but usually completely divided   Length 1.9–2.6 mm.; width 1.4–1.9 mm.
California (coast regions from San Diego to Humboldt Co.)......**tædata** Lec.
Body broadly oval and similar in punctuation and ornamentation to *tædata*, except
that the pronotal spots are so faint as to be scarcely traceable and the anterior of
the two darker submarginal spots of the elytra almost completely obsolete, the
two subbasal nearly equal in depth of coloration, and that the outer—which is
perfectly simple and elongate-oval in *tædata*, uniting generally with the central
spots—is here abbreviated and isolated and united to a distinct semi-detached
spot on the callus. Length 2.6 mm.; width 1.9 mm. California.

deficiens, sp. nov.

10—Much more narrowly oval than *20-maculata* but similar in size, and with the five
pronotal spots similarly placed and large, except the ante-scutellar, which is very
small and punctiform ; elytra much longer than wide, white in color, very minutely,
sparsely punctulate, the punctures not impressed, the markings deep black and
abruptly defined, consisting, besides the sutural marks, of eight spots on each :
two basal, the outer irregular and obliquely prolonged postero-externally parallel
to the margin for a short distance, one large and triangular, nearer the suture
than the margin at two-fifths, one small submarginal at one-third, another larger
at three-fourths obliquely united to a small spot behind the discal, and two isolated subapical, the inner the larger. Length 2.7 mm.; width 1.8 mm. Florida
(Dry Tortugas) ....**nana** Muls.

The form of the outer basal spot of the elytra seems to be a valuable character, and the large series before me show that most of the

others employed in the table are sufficiently constant to afford specific criteria.

A very extensive tribe, especially in the tropics of the western hemisphere, but of which only two or three species occur within the United States. It is probable that the great genus *Epilachna* may be subdivided for convenience, as there is a remarkable variety in form, sculpturé and style of ornamentation among its species.

Epilachna *Chev.*

The two species known to me may be defined as follows :—

Body very broadly oval, shining, pale orange-yellow, the punctures rather coarse, deep, unequal and moderately close ; pubescence short, moderately abundant ; head immaculate, the pronotum pale, with an apical and basal black spot on the median line, the basal the larger, and one at each side just behind the middle near the margin ; elytra each with two elongate-oval sutural spots just behind the middle and at basal fifth, the posterior much the larger, also with two submarginal in range with the two subsutural, a median subbasal very small, a central subequal to the posterior submarginal, and a large subquadrate subapical spot ; metasternum blackish ; legs pale.   Length 7.2–8.0 mm.; width 6.0–6.6 mm.   Eastern United States.................................. .... **borealis** *Fabr.*

Body more narrowly oval and distinctly smaller, duller in lustre, densely pubescent and very closely, unequally punctured, pale yellowish-brown in color, the head and pronotum without spots ; each elytron with three very small subbasal spots, the median less basal, and three in a transverse range just before the middle, scarcely larger than the subbasal, the median a little larger, and two near apical fourth, as small as the subbasal, placed near inner fourth and outer third ; under surface and legs pale throughout.   Length 6.4–7.0 mm. ; width 5.0–5 5 mm. Sonoran regions...... ....**corrupta** *Muls.*

Mexicana Guér., is said to occur within the United States, but I have seen no examples from this country; the upper surface is black throughout, the elytra each with six large rounded pale spots in two equilateral triangles ; my specimens, from Guerrero, have the legs pale, the femora black except at apex, in fact colored exactly as in *defecta*, from Honduras.   *Defecta* is, however, a shorter and more broadly ovular species, with less pronounced dilatation at basal fourth of the elytra.   The metacoxal plates in *Epilachna*, are arcuate but not quite entire, and are always much shorter than the first segment.

This tribe includes the genera *Pentilia*, *Cryptognatha* and probably *Bura* of South America and the West Indies, *Lotis* and *Xestolotis* of

Africa and *Sticholotis* of Asia. They are rounded, subglobular insects of small or moderate size, recalling Chilocorini in general appearance but with the formation of the front nearly as in Coccinellini. The minute species of the United States, which we have heretofore designated by the name *Cryptognatha*, because of prosternal structure, together with *Œneis*, belong to another taxonomic division of the family characterized by a more compact body and narrow epipleuræ. *Nestolotis* will be characterized in an appendix to the present paper.

It is possible that *Menoscelis* and *Thalassa* may also form either a part of this tribe or a special tribe closely related, but I have seen no examples.

CHILOCORINI.

The genera of this tribe have quite a different general habitus from those of the Coccinellini, being still more strongly convex and even subcompressed, with the outer part of the epipleuræ still more steeply descending ; the prevailing type of ornamentation, also, is different, being black with pale spots, while in the latter it is usually pale with black spots. Besides the radically different structure of the epistoma, the antennæ diverge widely from those of the preceding tribes, except some of the Pentiliini, being very short, compact and narrowly clavate. The three American genera represented before me are the following : —

Tibiæ obtusely dentate externally near the base ; pronotum pubescent toward the sides, with a double marginal line laterally at the base ; posterior legs moderately retractile, the abdomen and epipleuræ concave for the femora ...... **Chilocorus**
Tibiæ not dentate externally ; pronotum not pubescent toward the side margins, with the double marginal line at the sides of the base not evident ; in *Axion*, however, with the edge impressed near the sides of the base, forming a closer junction with the edges of the elytra.................................2
2—Posterior legs strongly retractile, epipleuræ and base of the abdomen deeply concave for the femora ; body large, extremely convex or subcompressed and very minutely punctulate. .... **Axion**
Posterior legs not retractile, the abdomen and epipleuræ not concave behind the coxæ ; body small, usually with more distinct punctuation ; ornamentation variable.
Exochomus

In *Chilocorus* and *Axion* the upper surface is deep black, the combined elytra having two or three red spots ; the former occurs on both sides of the continent but *Axion* seems to be peculiarly characteristic of the Sonoran fauna.

Chilocorus *Leach.*

In this genus the species have a remarkable superficial community

of habitus, and are consequently difficult to define ; they are generally
larger than in *Exochomus*, but smaller than in *Axion*. Those before
me may be identified as follows :—

Sterna black, the abdomen red, generally black toward the middle of the base.....2
Sterna in great part red, the prosternum alone black ; abdomen red throughout, the
    legs black as usual ; prothorax more narrowly rounded at the sides ; body deep
    black above..... 4
2—Elytral spot small, rounded ; body black above, very broadly oval, the elytral
    punctures generally stronger and becoming quite coarse toward the margins ;
    head distinctly pubescent. Length 4.4–5.0 mm.; width 3.8–4.3 mm. Vermont,
    New York, Pennsylvania, Indiana and Iowa.............**bivulnerus** *Muls.*
Elytral spot more or less evidently larger and always transverse, the head less con-
    spicuously pubescent ; elytral punctures finer 3
3—Broadly oval and less compressed, black with distinct bluish reflection ; sides of
    the pronotum but little more than a third as long as the median line ; elytral spot
    large, transversely oval, extending from basal fifth to the middle and from inner
    fifth or sixth to outer fourth. Length 4.2–4.8 mm.; width 3.4–3.8 mm. Cali-
    fornia.................................. ....**orbus**, sp. nov.
Narrowly oval and more pointed behind, smaller and narrower than *bivulnerus*, com-
    presso-convex, deep black above without metallic reflection ; sides of the pro-
    notum fully two-fifths as long as the median line ; elytral spot distinctly variable
    in size, but as an average extending from rather more than basal fourth to a little
    before the middle and from inner to outer third or fourth. Length 3.7–4.75 mm. ;
    width 3.0–3.8 mm. California (San Francisco) to Washington State.
                                                 fraternus *Lec.*
4—More broadly oval ; pronotum deeply impressed apically near the angles in the
    male, with the edge there rufescent ; elytral spot more uneven in outline, gener-
    ally extending from basal fourth to the middle and from inner fifth or sixth to
    outer fourth or fifth. Length 5.0 mm.; width 4.5 mm. Honduras.
                                                     cacti *Linn.*
Narrowly oval and more compressed, the pronotum in the male not, or only very
    feebly and indefinitely, impressed apically near the angles, with the edges there
    not at all paler ; elytral spot more evenly outlined, generally extending from basal
    fifth to the middle and from inner fifth or sixth to outer fourth ; punctures very fine,
    becoming slightly larger toward the margins. Length 4.4–4.6 mm ; width
    3.75 mm. California (San Diego)....................**confusor**, sp. nov.

The longitudinal impression on the flanks of the elytra are analogous to those previously noted in *Olla*, of the Coccinellini.

Axion *Muls.*

These species are the largest of the tribe and are colored nearly as
in *Chilocorus*, but with a greater development of the red spot. The
surface of the elytra is almost completely impunctate ; the pronotum is
feebly punctate near the side margins, and the apical margin near the

angles is always more or less pale.   The four species in my cabinet
may be separated by the following characters :—

Elytra together with two large obliquely oval red spots, the side margins not at all
   thickened ; abdomen black. Sonoran regions...................... 2
Elytra with three smaller red spots one of which is sutural, the edges with a strongly
   thickened bead ; abdomen red throughout.   Atlantic regions...............4
2—Elytra quite broadly reflexo-explanate at the sides ; upper surface strongly shining ; body large, broadly rounded behind in both sexes, the male with the
   elytral spot rather small, but slightly oval, extending from basal fifth or sixth to
   the middle and from inner third or fourth to outer fourth or fifth, the spot in the
   female larger, extending from very near the base at outer two-thirds to the middle and from inner fourth to outer sixth or seventh near the humeri.   Length
   6.0–6.7 mm.; width 5.2–5.75 mm.   Arizona ; [*texanum* Lec.].
                                                               plagiatum *Oliv.*
Elytra very narrowly, and but slightly, less declivous toward the edges, the body
   smaller 3
3—Body broadly rounded behind, alutaceous in lustre, the elytral spot in the female
   rather small in size, rounded, with the anterior outline oblique and emarginate,
   extending from basal sixth to the middle and from inner two-fifths to outer fourth ;
   abdomen and legs black as in *plagiatum*.   Length 5.3 mm. ; width 4.6 mm.
   New Mexico (Las Vegas)........................ .**alutaceum,** sp. nov.
Body pointed and ogival behind, the elytra polished, the spot similar in the sexes and
   very large, obliquely and broadly oval, extending from the basal margin—which
   it very narrowly attains or virtually attains at outer two-fifths—to three-fifths of
   the length and from inner fifth or sixth to outer eighth, where the outline is
   parallel to the side margin for a considerable distance.   Length 5.3–5.7 mm. ;
   width 4.6–5.1 mm.   California (Los Angeles) and Arizona.. ...**pleurale** *Lec.*
4—Body very broadly oval and compresso-convex, the upper surface strongly shining,
   the pronotum more alutaceous, with the entire apical margin very finely and indefinitely paler ; elytra very broadly ogival at tip, each with a small parallel-sided
   red spot extending, parallel to the side-margin, from the base at outer two-thirds
   for one fifth the length, and also with a small oval red spot on the suture at
   apical third ; legs black.   Length 6.6 mm. ; width 5.6 mm.   Rhode Island.
                                                          tripustulatum *DeG.*

Tripustulatum does not seem to be at all abundant, and my cabinet
contains only the single specimen taken some twenty years ago.   *Pila-
tei* of Mulsant, because of its red abdomen, is almost surely specifically
different from *plagiatum ;* it is said to be from Texas but I have not
seen a representative.

Exochomus *Redt.*

The metacoxal plates are rounded as usual, but they are not complete as stated by Crotch, the bounding arc not quite attaining the
basal margin of the first segment.   The species are rather numerous,
and are much smaller and generally less convex than in the preceding

genera, only rarely exhibiting any trace of lateral compression. The
punctuation is very minute or subobsolete, but in *marginipennis* becomes
quite distinct though sparse. The species before me may be outlined
as follows :—

Body strongly compresso-convex as in *Chilocorus*, the anterior tibiæ more dilated and
arcuately sublaminate externally ; body rounded, deep black above, the under
surface and legs throughout testaceous ; head slightly rufescent at the apical margin ; pronotum with the edge slightly rufescent at the apical angles ; elytra minutely punctulate, more distinctly toward the margins, which are evenly declivous to the edge and not at all reflexed, with a very fine marginal bead, each with
an elongate-oval red spot on the median line, extending two-fifths from the base,
with its margins rather nubilate. Length 3.7-3.9 mm. ; width 3.0-3.2 mm.
Arizona. . . . .**arizonicus,** sp. nov.
Body evenly and less strongly convex, not at all compressed, the anterior tibiæ nearly
straight externally and not laminate, the elytral margins narrowly but abruptly
reflexed, and with a more distinct marginal bead. 2

2—Pronotum black throughout ; body oval ; elytra black, with a large humeral and
small discal posterior spot pale, the marginal bead black .3
Pronotum black throughout ; body rounded, more convex, the elytra pale with black
spots and marginal bead black 4
Pronotum black, nubilously pale at the sides or at the apical edges near the angles ;
body rounded or oval, moderately convex, the elytra with a black design, the
side margins always pale 5

3—Elytra polished or feebly alutaceous, obsoletely punctulate, the pale humeral spot
parallel with the side margin, about twice as long as wide, without tendency to
prolongations along the basal or lateral margins, the discal spot rounded, clearly
defined, situated at apical fourth and inner third ; under surface and legs black,
the epipleuræ pale except behind the middle. Length 2.8-3.8 mm. ; width
2.3-3.0 mm. California (San Francisco to Humboldt Co.).
californicus, sp. nov.
Var. A—Similar but with the elytra strongly alutaceous, and with the humeral
spot extending narrowly along the margin for a short distance posteriorly but
not along the base. San Francisco.
Elytra polished, very minutely punctulate, black, the pale humeral spot more sinuate
within and more angular internally at its posterior limit, continued along the
margin with broadly sinuate internal outline and gradually narrowing, becoming
extinct at a point opposite the discal spot, also extended narrowly along the basal
margin very nearly to the scutellum ; discal spot subtriangular, at posterior fourth
or fifth and much nearer the suture than the margin. Length 3.0-3.3 mm.;
width 2.2-2.7 mm. Indiana? . . . .**ovoideus,** sp. nov.
Elytra polished, minutely and sparsely but somewhat more distinctly punctulate, the
humeral spot oblong and about twice as long as wide, as in the two preceding
somewhat prominent within at its posterior limit, abruptly narrowed and continued along the lateral and basal margins as in *ovoideus* but more broadly at the
base, the discal spot subtriangular, at the same position but continued forward

narrowly becoming nubilously extinct two-fifths from the base ; under surface and legs as in the preceding. Length 3.15–3.3 mm.; width 2.4–2.6 mm. Nevada.

desertorum, sp. nov.

4—Body broadly rounded, polished, minutely, very obsoletely punctulate ; head and pronotum black throughout ; elytra pale orange, the sutural, basal and external margins extremely finely black, with a common transverse spot across the suture at the apex ; each also with two very small rounded black spots, the anterior on the callus, the posterior slightly larger and near apical third nearly on the median line ; under surface and legs black, the epipleuræ pale, edged externally and finely with black. Length 3.3–4.0 mm.; width 2.8–3.4 mm. Texas (El Paso).

högei *Gorh.*

5—Elytra very finely but evidently punctulate, entirely pale, each with a transversely oval black spot near the apex, approaching the suture rather nearer than the external margin ; head and pronotum pale in the male, the latter with a median dark cloud toward base, black in the female with the pronotum broadly and nubilously pale at the sides ; legs pale or so in great part. Length 2.6–2.9 mm.; width 2.0–2.4 mm. Texas (Austin); [*guexi* Lec.]..........**childreni** *Muls.*

Elytra pale, with two broad transverse fasciæ of black...... 6

Elytra black on the disk ; body in general more broadly rounded ; punctures very minute and sparse...9

6—Anterior fascia not attaining the base and always separated from the posterior ; elytral punctures extremely minute and subobsolete.......................7

Anterior fascia broadly attaining the base and broadly united with the posterior fascia at the median line of each elytron ; punctures sparse and fine but very distinct...8

7—Body broadly oval, almost rounded and larger, the sides of the pronotum broadly and nubilously pale in both sexes, the head blackish in the female ; thoracic margins very strongly convergent. Length 2.8–3.3 mm.; width 2.5–2.9 mm. Texas (Brownsville)—Mr. Wickham ....**latiusculus,** sp. nov.

Body more narrowly oval, the head and pronotum black, apparently in both sexes, the apical angles only nubilously and not very markedly pale, the thoracic sides much less convergent from base to apex. Length 2.4–2.9 mm.; width 1.8–2.1 mm. Southern California (Pasadena, Los Angeles and San Diego).

fasciatus, sp. nov.

8—Body not very broadly oval ; head and pronotum black, the apical angles of the latter distinctly pale in color; elytra black, with a rounded or oval pale spot at each side of the scutellum and a common, transversely rhombiform spot on the suture at three-fifths, extending laterally as if to narrowly unite with the median projection of the pale margin, which extends from the base very nearly to the apex and broadly bisinuate within, not tending to spread along the basal margin. Length 2.5–2.8 mm.; width 1.9–2.2 mm. Tennessee and Florida [*prætextatus* Muls.]................... .... **marginipennis** *Lec.*

9—Head and pronotum black, the apical angles of the latter nubilously paler; elytra black, with a broad pale margin extending, with its inner margin parallel, to nearly three-fifths, there obliquely and abruptly narrowed and continued narrowly almost to the apical angles ; body smaller and much more broadly rounded than in *marginipennis*, with less obvious punctuation. Length 2.2 mm.; width 1.8 mm. Texas (El Paso)................... ....**subrotundus,** sp. nov.

Head, pronotum and elytra deep black throughout ; under surface and legs also black, the tarsi picescent.   Length 2.9 mm.; width 2.4 mm.   New Mexico.

æthiops *Bland*

The Mexican *contristatus* is said to be distinct from *childreni* by Gorham, being larger, more compresso-convex and with the elytra immaculate. *Marginipennis* was described by the elder LeConte, and, to distinguish the two authors, I would suggest that the contracted name of the latter be printed " LeC." that of the younger LeConte remaining " Lec."

Ovoideus and *desertorum* of the table, are in all probability subspecies of *californicus*, but my material is not sufficient to decide at present, and the forms from *childreni* to *æthiops* may be regarded as derivatives of the *marginipennis* type, but in my opinion specifically distinct.

PLATYNASPINI.

The species of this tribe somewhat recall the Chilocorini in form, but are always pubescent. The body is oval, convex but not compressed, generally black with small pale spots above, the legs retractile within shallow depressions. The abdomen differs from that of the preceding tribe in having the sixth segment distinct, the fifth being as short as the fourth, and the metacoxal arcs also differ, being nearly as in the Coccinellini, the bounding curve extending rapidly to the apical margin. The antennæ are very short, and the fourth joint of the maxillary palpi strongly securiform. The species are all foreign to the American continents and are only moderately numerous.

TELSIMIINI.

This tribe is necessary for two very small species, having a structure of the epistoma and eyes similar to that of the Platynaspini, and with a convex, pubescent body, but having the maxillary palpi somewhat as in Pharini though stouter, the fourth joint being conical, with the apex obliquely truncate. The abdomen differs from that of the preceding tribe in being purely five-segmented, as in Pharini, the fifth longer and strongly rounded. The metacoxal arcs curve outward, becoming rectilinear and parallel to the apical margin at a point between the middle and apex of the segment, and attain the sides of the body. The epipleurae are rather wide and descend strongly externally, and the legs are moderately retractile. The scutellum is very small and the eyes are finely faceted and pointed antero-internally. The anterior margin of the prothorax is broadly angulate at the middle of the emargination. The types are African and will be described in an appendix to the present paper under the generic name *Telsimia*.

PHARINI.

In this remarkable tribe the abdomen consists of five segments, the fifth long and strongly rounded, and the metacoxal arcs curve rapidly to the apex of the first segment, which they follow externally. The legs are only feebly retractile, the impressions being very shallow and the tarsi are elongate and generally rather compressed, with the basal node of the third joint more or less free. The fourth joint of the maxillary palpi is slender, gradually drawn out to a finely acuminate point, and the antennæ are moderate in length, straight, with the club narrow. The epistoma is sinuato-truncate at apex and extends only to the eyes, which are not emarginated by it, but which have a very minute notch as in Scymnillini. The prosternum is flat, rather widely separates the coxæ and has two parallel entire and widely separated carinæ. The two genera before me belong to the old world fauna and are as follows:—

Body pubescent above, the epipleuræ descending externally. ....*Pharus.*
Body subglabrous, the epipleuræ wide but horizontal................*Pharopsis.*

Species of both these genera will be alluded to in the appendix. Although the palpal structure is remarkably aberrant in this tribe, there is no necessity at all for considering it a distinct section of the family, as is proposed in the catalogue of Heyden, Reitter and Weise, and the palpi of the preceding tribe are to some extent intermediate. In fact this character is no more unusual than the dilated clypeus of Chilocorini, and the peculiar form of the fourth palpal joint is evidently due to extreme obliquity of truncature, seen in a transition stage in *Xestolotis*. *Pharopsis* appears to be distinct from any of the African genera recently proposed by Weise.

ŒNEINI.

The genus *Œneis* of Mulsant, so far from being identical with *Cryptognatha*, in reality belongs to a different division of the family because of the narrow and subhorizontal epipleuræ. Our small species hitherto placed in *Œneis* by LeConte, and *Cryptognatha* by Crotch and Horn, really constitute a different genus because of the less convex median parts of the upper surface, sculpture and structure of the anterior legs. In fact the indications point to several genera among these small obscure forms. As a guess, the species from *auri-culata* to *æthiops* (Crotch—Rev. Cocc., p. 206), may be assigned to *Cryptognatha*, those from *reedi* to *nigrans* to *Œneis*, and *pusilla* and *puncticollis* to the new genus described below. The Ceylon species *flavescens*, *nigritula* and *lateralis* probably constitute another distinct genus. The species of Œneini are either wholly or in great part glabrous, and are all among the most minute members of the family.

Delphastus, gen. nov.

In some respects this genus is allied to *Smilia*, although so different in prosternal structure and retractility of the legs ; the upper surface, for example, has rather long, stiff and very remotely scattered erect setæ, corresponding to the very short and microscopic erect hairs of that genus ; the pronotum has an oblique line at the apical angles, closer to the margin than in *Smilia*, and finally the antennal foveæ are at the apex of very deep lateral emarginations of the front, rather remote from the eyes, which latter are entire. Were it not for the radically different structure of the abdomen, *Smilia* could therefore enter the present tribe quite as well as the Scymnini, the deflexion of the prosternum not being in general an essentially tribal character, any more than the crural impression of the epipleuræ. In the Coccinellidæ tribal characters must be determined from the general structure of the body, rather than from any special modifications, and, considering all points, it seems to me that *Smilia* should either constitute a distinct tribe just before Hyperaspini, or else enter the Scymnini.

In *Delphastus* the body is very broadly oblong-oval and only moderately convex, highly polished, subglabrous and subimpunctate, the antennæ well developed, with a compressed elliptical club, the coxæ all very remotely separated, the epipleuræ narrow, horizontal and feebly concave, the anterior femora greatly dilated, so that in repose the under side of the prothorax may present an almost unbroken surface from side to side, the anterior tibiæ and tarsi being completely concealed beneath the expanded femora lying deeply within the prosternal depressions ; the meso-crural excavation is very deep and abruptly limited, and extends to the outer margin of the epipleuræ. The tarsi are long and slender, and may be flexed upon the tibiæ in repose, but are not received in grooves ; the posterior tibiæ are, however, slightly expanded and broadly subangular externally. The claws are small, slender and abruptly bent behind the middle with an internal swelling at base. The abdomen appears to be similar in structure in the sexes, the fifth segment ogival and longer than the three preceding combined. The prothorax is as wide as the elytra or very nearly, short and transverse, with the fine intromarginal line receding from the edge at the apical, as well as the basal, angles, and the scutellum is well developed and a little longer than wide. Our species may be defined as follows :—

Elytra black throughout...2
Elytra castaneous 3
Elytra and entire body pale testaceous.................................4
2—Head and sides of the pronotum pale in the male, entirely black in the female ; legs red, the femora sometimes picescent ; head and pronotum finely, sparsely punctate. Length 1.3–1.4 mm.; width 1.0–1.1 mm. Pennsylvania, North Carolina (Ashville) and Texas (Austin)....................**pusillus** *Lec.*
    Var. A—Similar but slightly larger and with the punctures of the pronotum more distinct ; body and legs black. Southern States....**puncticollis** *Lec.*
Head pale in the male, the pronotum black throughout, with distinct but sparse punctuation ; legs red. Length 1.15–1.3 mm.; width 0.85–1.0 mm. California (southern) and Arizona (Tucson)....................**sonoricus**, sp. nov.
3—Castaneous ; middle of the prothorax and a narrow space at the base of the elytra piceous ; head and legs yellow ; pronotum with a few scattered punctures near the middle. Length [1.5 mm]. Sta. Catalina Island, coast of Southern California..**catalinæ** *Horn*
4—Similar to *pusillus* in form but very small and entirely testaceous. Length [0.8 mm]. Florida (Sand Point).........................**pallidus** *Lec.*

I have seen no representative of *puncticollis, catalinæ* or *pallidus*. It is quite possible that the first may be a perfectly distinct species, as the length is given .07 inch by LeConte.

SCYMNILLINI.

In abdominal structure this tribe, which in some respects may be allied to the Ortaliini, resembles the preceding and departs widely from the Hyperaspini or Scymnini ; the ogival fifth segment is, however, shorter than in Œneini, and is generally but little longer than the two preceding together, perfectly similar in the sexes, except that the fifth segment is more broadly rounded and a little shorter in the male. The body is oval, small to very minute in size, more or less pubescent or setulose, with the head strongly deflexed and deeply inserted in the prothorax, the latter obviously narrower than the elytra, abruptly so in *Zagloba*, deeply emarginate at apex, with narrowly reflexed side margins, the base feebly lobed before the scutellum, which is moderate in devel-

opment and subequilateral. The eyes are well developed, with their
inner sides nearly straight and parallel, and having a narrow deep anterior emargination, the antennæ very short but apparently of eleven
joints, inserted very close to the eyes, exposed at base, the clypeus
narrowed and feebly sinuato-truncate. The fourth joint of the maxillary palpi is securiform throughout. The anterior coxæ are remotely
separated, with the prosternum flat and devoid of carinæ, the apex
feebly deflexed in some species of *Zagloba*, but not enough to afford
protection to the trophi. The legs are perfectly free, the epipleuræ
narrow and flat and devoid of any trace of impression, even the basal
pit of *Scymnus* being rudimentary. The tibiæ are slender and can be
folded back into a feeble femoral depression, the tarsi well developed
and free, and the claws slender and apparently simple. The genera
and species are few in number as thus far discovered. The genera may
be defined as follows :—

Metacoxal arcs small and short, semi-circular and either entire or failing to attain
the base externally ; body coarsely pubescent......... .. ..**Zagloba**
Metacoxal arcs curving outward at a slight distance from the suture, and almost attaining the sides of the body, nearly as in the subgenus *Scymnobius;* body
smaller in size and subglabrous.. ....**Scymnillus**

These genera are both represented in the more southern parts of
the United States from the Atlantic to the Pacific.

Zagloba, gen. nov.

The body is broadly rounded or oval, and clothed rather plentifully
above with moderately long erect or semi-erect bristling whitish hairs,
which, on the elytra, stream irregularly, forming partial vortex-like
arrangements of the pubescence. The species are rather few in number
and are invariably mixed up in cabinets with *Scymnus*, from which they
differ radically in abdominal structure. Their departure from *Scymnus*
was recognized by Dr. Horn, but that author, neglecting to observe
the abdomen, placed the only species thus far described in *Cephaloscymnus*, with which it has no real affinity, and no resemblance, except
a slight similarity in the form of the eyes and prothorax. Our species
known to me are as follows :—

Metacoxal arcs entire, joining the base of the first segment 2
Metacoxal arcs not attaining the base of the first segment externally ; body broadly
rounded. Atlantic regions...6
2—Body very broadly rounded ; prosternum slightly deflexed at apex. California....3
Body narrowly oval or oblong-oval, the prosternum perfectly flat and less remotely
separating the coxæ. Atlantic regions 5

3—Metacoxal arcs extending to the middle of the segment, black, the elytra slightly piceous, each with two large nubilous pale areas; prothorax short and transverse, abruptly and distinctly narrower than the elytra, the sides feebly convergent and nearly straight to beyond the middle; elytra finely, sparsely, somewhat unequally punctulate, one-half wider than the prothorax.   Length 1.7 mm.; width 1.25 mm.   California (Sta. Cruz Mts)..................................**ornata** *Horn*
Metacoxal arcs extending distinctly beyond the middle ; elytral sculpture similar...4

4—Prothorax only slightly and not very abruptly narrower than the elytra, the sides strongly convergent and evenly, strongly arcuate from base to apex; elytra oval, but little more than a third wider than the prothorax and rather longer than wide in the female, shorter in the male, piceous, each with two large nubilous pale areas.   Length 1.6-1.8 mm.; width 1.25 mm.   California (exact locality not recorded) ....**laticollis,** sp. nov.
Prothorax abruptly very much narrower than the base of the elytra, much smaller than in *laticolis*, the sides rather strongly convergent and nearly straight to the middle, then broadly rounded ; elytra pale testaceous, with a large basal subquadrate brown area on the suture which is emarginate at each side, at least two-thirds wider than the prothorax.   Length 1.7 mm.; width 1.3 mm.   California (Sonoma Co.)...**orbipennis,** sp. nov.

5—Elongate-oval, moderately convex, the stiff whitish pubescence of the upper surface very conspicuous ; head and prothorax throughout pale rufo-testaceous, the elytra black ; legs and abdomen testaceous, the latter blackish toward base ; eyes narrow, not at all covered by the pronotum ; prothorax only slightly and not very abruptly narrower than the elytra, the sides feebly convergent and arcuate, the apex much less deeply emarginate than in the preceding species ; elytra longer than wide, rather narrowly rounded behind, finely, not densely and somewhat unequally punctate ; metacoxal plates extending far beyond the middle.   Length 1.5 mm.; width 0.9 mm.   Florida (near Palm Beach).**bicolor,** sp. nov.

6—Very broadly rounded, moderately convex ; head and prothorax testaceous, the latter broadly and nubilously blackish toward the middle, abruptly and conspicuously narrower than the elytra, with the sides feebly convergent, rounded at apex, the latter deeply emarginate ; elytra black, shining, scarcely visibly punctulate, not as long as wide, rather narrowly rounded at tip ; under surface piceous or paler, the legs testaceous ; metacoxal arcs extending but slightly beyond the middle.   Length 1.35-1.6 mm.; width 1.05-1.2 mm.   Texas (Brownsville)—Mr. Wickham. ....**hystrix,** sp. nov.

Scymnillus *Horn.*

The members of this genus are all small, and number among them some of the most minute of the Coccinellidæ. The surface is apparently glabrous, but minute hairs can generally be discovered on the head or pronotum, and the elytra usually have some very small, erect and widely scattered setæ. The epistoma is very short before the antennæ. The three species before me may be thus outlined :—

Body oval, black, the abdomen piceous toward the edges, the legs blackish ; head and
pronotum quite strongly and closely punctured throughout, each puncture bearing a very short but distinct subdecumbent hair, short, transverse, the sides
almost continuous, strongly convergent, evenly and moderately arcuate, the
apical emargination moderately deep ; elytra fully as long as wide, polished,
glabrous, ogival at apex, minutely but distinctly, sparsely punctate, the humeral
callus quite pronounced.   Length 1.0–1.45 mm.; width 0.75–1.0 mm.   California.
                                                                    aterrimus *Horn*
Body very broadly rounded, minute and subglobular. 2
2—Piceous-brown, the median parts of the pronotum and sterna of the hind body more
    darkly shaded ; legs pale ; head minutely, sparsely punctate, each puncture with
    a short and inconspicuous hair ; pronotum minutely, sparsely punctulate, sub-
    glabrous except near the abruptly reflexed lateral edges, where the hairs are erect,
    stiff and bristling, very much narrower than the elytra but with the sides almost
    continuous ; elytra almost circular, glabrous and subimpunctate, about as long as
    wide.   Length 0.85 mm.; width 0.72 mm.   Bahama Islands (Eleuthera).
                                                                    lateralis, sp. nov.
Black throughout, the legs not paler ; body very broadly rounded, the head and pro-
    notum finely but rather strongly, moderately closely punctulate, the former very
    feebly pubescent, the latter subglabrous, with a very few microscopic hairs, es-
    pecially toward the sides, the latter nearly continuous, very strongly convergent,
    with the margin very minutely reflexed ; elytra minutely but distinctly, sparsely
    punctulate, not as long as wide, extremely obtusely ogival at tip, glabrous.
    Length 0.78 mm.; width 0.65 mm.   Bahama Islands (Eleuthera)—Mr. Wick-
    ham. . . . . **eleutheræ**, sp. nov.

HYPERASPINI.

Besides the genera defined below, it is probable that *Tiphysa* and
Hinda, distinguished by the elongate scutellum, can also legitimately
enter this tribe, which is closely related to the Scymnini, but recog-
nizable at a glance by the perfectly glabrous upper surface. The
scutellum in all the genera mentioned below is well developed and
equilatero-triangular. As a special peculiarity of this tribe, although
evident to a generally less degree in Scymnini, it should be stated that
the genital segment is greatly developed in both sexes, assuming almost
perfectly the appearance of a true sixth segment in form and sculpture,
and is more conspicuously developed than in any other tribe of the
family—in the genus *Smilia*, however, which is somewhat aberrant
among the Scymnini, forming a connecting link with the present tribe
in some respects, the genital segment is equally well developed, and it
is also very strongly developed in the South African *Cranophorus*. In
the males of *Hyperaspis* and probably *Helesius*, there is no visible
segment beyond the sixth, but in *Brachyacantha* and *Hyperaspidius*,

there is a second supplementary segment in that sex. Although seven segments can thus be counted in the males and six in the females, there is no difficulty whatever in perceiving that the true abdominal segments terminate, as in all other tribes, with the fifth, and that the one or two additional are parts of the genital armiture, and what might be termed pseudo-segments.

In the Hyperaspini, the fourth joint of the maxillary palpi is always strongly securiform, the eyes well developed and very finely faceted, the antennæ short and 11-jointed, and the legs rather short and stout, with the anterior tibiæ modified according to the genus. The anterior coxæ are narrowly or moderately separated, and the prosternum flat. The metacoxal plates are largely developed, attaining the first suture or very nearly, and frequently extend along the latter for some distance, then curved strongly forward but apparently never quite attaining the base externally. The genera are few in number, and those before me may be readily separated as follows:—

Epipleuræ foveate for the tips of the hind femora.............................2
Epipleuræ completely devoid of foveæ, narrow and flat........................4
2—Eyes with a small anterior emargination; anterior tibiæ not dilated beyond the middle but with an acute external edge, spinose externally at about basal two-fifths, without external apical plate but with an oblique double edge from the spine to the apex; tarsal claws with a large internal, pointed or subquadrate tooth at base; body very convex; epipleuræ very narrow, more or less horizontal; ornamentation well defined; prosternum not bicarinate.........**Brachyacantha**
Eyes entire; anterior tibiæ with an external plate delimited by an oblique cariniform line at apex; epipleuræ narrow but generally slightly descending externally; prosternum bicarinate between the coxæ; anterior tibiæ not spinose...............3
3—Anterior tibiæ slender, the apical plate never more than feebly oblique toward the tip; claws with an internal quadrate tooth at base which is slightly variable in size; upper surface with clearly defined ornamentation [*Cleothera* Muls., *Oxyny. chus* Lec.] ....**Hyperaspis**
Anterior tibiæ thickened externally, especially beyond the middle, the apical plate very oblique toward the tip; claws simple, arcuate and slender; body with suffused coloration ....**Helesius**
4—Anterior femora slender, without an apical external plate; tarsal claws simple, arcuate and slender; prosternum feebly bicarinate; ornamentation generally well defined ....· ....**Hyperaspidius**

All of these genera, except *Helesius*, which is Sonoran, are widely distributed throughout the United States.

Brachyacantha *Chev.*

Next to *Hyperaspis*, this is the most abundant genus of the tribe, and presents the same difficulties in regard to discrimination of the species. The male sexual characters of the abdomen are, however, much more pronounced and are frequently very valuable in defining closely related forms. The forms which seem to merit distinctive names may be defined as follows :—

Elytra pale at base, or each with a pale spot near the middle of the basal margin. . 2
Elytra never conspicuously pale or maculate at base, except sometimes at the humeral
    angles. 12
2—Elytra each with five clearly defined and isolated pale spots, two basal, two in a
    transverse line very near the middle and one subapical, the humeral constant in
    both sexes. 3
Elytra with the basal and lateral margin pale, and each with a discal pale spot. 8
Elytra black, with a basal and subapical pale spot but without a spot near the
    centre. 9
Elytra pale, each with two black spots, one anterior and one posterior. 11
3—Spots generally separated from each other longitudinally by more than their own
    dimensions. 4
Spots relatively larger, whitish, separated by their own diameter or less. 7
4—Basal spot almost fully circular, only slightly truncated by the basal margin ; body
    small, elongate-oval, piceous-brown in color ; head and subquadrate sides of the
    pronotum flavate in the female ; elytra finely but strongly, sparsely punctate,
    polished, the spots nearly equal, moderately large, the subapical largest and the
    humeral smallest ; under surface piceous ; legs pale throughout. Length 1 9–2.1
    mm. ; width 1.3–1.45 mm. Indiana. . . . . **stellata**, sp. nov.
Basal spot never much more than semi-circular, broadly truncated by the basal
    margin. 5
5—Male with the two median lobes of the basal black area of the pronotum narrowly
    rounded 6
Male with the two median lobes broadly and rectilinearly truncate, the dividing spur
    of the apical pale margin short and very minute or obsolete, body more broadly
    oval than in *ursina* and more variable in size, finely punctulate ; spots small,
    variable in size and form among themselves, the subapical usually the most con-
    spicuous. Length 2.1–3.6 mm. ; width 1.6–2.75 mm. North Carolina (Ashe-
    ville). . . . . **congruens**, sp. nov.
6—Black area of the pronotum in the male more extended, its two approximate
    median lobes approaching rather close to the apical margin ; elytral spots, except
    the humeral, well developed and subequal in size ; body elongate-oval. Length
    2.75–3.75 mm. ; width 2.0–2.7 mm. Massachusetts, New York, Pennsylvania
    and Indiana. . . . . **ursina** *Fabr.*
Black area less developed, the apex broadly pale even before the median lobes ;
    elytral spots smaller and very unequal, the two median much smaller than the
    basal or apical ; body smaller, with the punctures much less fine and notably
    sparser. Length 1.8–2.3 mm. ; width 1.2–1.6 mm. Pennsylvania and Mary-
    land. . . . . **10-pustulata** *Melsh.*

7—Form elliptical, the spots, excepting the humeral, subequal in width and isolated
at about their own diameters or a little less; head and subquadrate side spot of
the pronotum pale in the female, the surface finely, rather closely punctate
and fully three-fourths longer along the median line than at the sides; elytra
finely, sparsely, punctate, piceous-black; legs pale. Length 2.7 mm.; width
2.0 mm. Texas (Brownsville)—Mr. Wickham...........**testudo**, sp. nov.
Form very broadly rounded, the basal spot of the elytra broader than the discal and
with a tendency to join the latter; spots all very large and relatively narrowly
separated; pronotum in the male one-half longer at the middle than at the sides,
broadly pale anteriorly, the two median lobes of the black area not much advanced and broadly rounded; punctures fine and sparse; under surface black,
the abdomen piceous; legs pale throughout. Length 2.5 mm.; width 2.1 mm.
Texas (Dallas) ....**bolli** Cr.
8—Oblong-oval, black, the head and oblique sides of the pronotum angularly lobed
within at the middle, pale in the female, the apical margin also very narrowly
pale, the prothorax two-thirds longer at the middle than at the sides, finely,
sparsely punctate; elytra with a narrow rufo-flavate margin from the scutellum to
the sides, narrowest at outer two-fifths, the lateral margin obliquely pale at the
humeri, the pale side margin extremely narrow at basal fourth, then dilated to
outer fourth at the middle opposite the discal pale spot, then narrowed at apical
fourth, thence gradually expanded and extending transversely to within two-
thirds of its apical width of the suture, receding somewhat from the side margin
as in the *fimbriolata* group of *Hyperaspis*; punctures fine but strong, moderately
sparse, closer near the base; under surface black throughout, the femora blackish,
paler at apex. Length 4.1 mm.; width 3.0 mm. Colorado (Beaver Brook—
6000 feet elevation)..................................**illustris**, sp. nov.
9—Elytra with the median marginal spot; basal, marginal and subapical very nearly
equal in size, rufo-flavate, the humeral spot wanting at least in the female; head
and pronotal sides broadly pale in that sex, the apical margin very narrowly;
upper surface black, polished, finely, sparsely punctate. Length 2.8 mm.;
width 2.0 mm. Georgia................................**flavifrons** *Muls.*
Elytra without either of the median spots.................................10
10—Broadly rounded, polished, finely, rather sparsely punctate; male with the head
yellow, the prothorax pale with a basal black area extending to lateral eighth or
ninth of the base, approximately bilobed in the middle, extending to apical
fourth or fifth; basal spot of the elytra more than semi-circular, the subapical
slightly larger, the humeral spot oblique; female with the head black, the front
nubilously paler in V-shaped design, the pronotum black throughout, except a
very narrow margin about the apical angles; elytra similarly maculate, except that
the humeral spot is wanting. Length 3.2–3.6 mm.; width 2.5–2.75 mm.
Massachusetts and Indiana; [*confusa* Muls. ♂, *quadripunctata* Melsh. ♀ and
diversa Muls. ♀].................................**basalis** *Melsh.*
Less broadly rounded and much smaller; head and tips of the apical pronotal angles
piceous in the female, the remainder black, finely not closely punctate; elytra
with a rufo-flavate pale area at base, extending rather beyond basal third from the
lateral margin nearly to the scutellum, truncate behind, rounded and receding

somewhat from the basal margin internally, the subapical spot oval, adjoining the limb and distant nearly half its width from the suture. Length 2.4 mm.; width 1.8 mm. Florida . . . . **quercetI** *Schz.*

11—Narrowly oval, finely, not closely punctulate, with flavate pale areas ; male with the head pale, the pronotum black, with a very narrow pale apical margin perfectly even in width but gradually wider from the eyes and extending to the basal angles, the black area with a feeble angular extension at apical third ; scutellum black ; elytra pale, the suture more or less broadly black from a short distance behind the scutellum, narrowest just behind the middle, extending at apex anteriorly along the sides very narrowly to about apical third, also with a triangular black spot involving the callus and another, rounded but posteriorly sinuate, behind apical third at outer third of the width ; under surface black throughout. Length 3.2 mm.; width 2.2 mm.   Kansas. . . . . **albIfrons** *Say*

Narrowly oval, with flavate pale areas anteriorly, reddish on the elytra, finely, not closely punctulate ; male with the head pale, the pronotum black in a basal area between the basal angles, the anterior margin of which curves evenly from the basal angles to anterior third at lateral two-fifths, then feebly sinuate and then extending forward in two rounded lobes separated by a narrow deep fissure to apical fifth or sixth ; scutellum black ; elytra pale, the suture more or less broadly black from the scutellum to the apex, narrowly at the scutellum and for a short distance just behind the middle, the external marginal bead also black, becoming broader at the apex and joining the sutural black area ; anterior black spot obliquely oval, sending off a nubilous connecting isthmus to the sutural black area, the posterior spot smaller, at apical and outer third ; under surface black throughout, the posterior femora in great part black, the intermediate less so.   Length 3.7 mm.; width 2.4 mm.   California (Sta. Monica).

pacifica, sp. nov.

12—Elytra each with two transversely confluent pale spots before the middle and a subapical spot. 13

Elytra each with three small, widely isolated pale spots 15

13—The confluent pale spots very nearly separated ; male with a pale oblique humeral spot which is absent in the female ; head pale in the male, the pronotum pale, with a large and abruptly defined median black area from the base to apical fifth or sixth, deeply emarginate at each side; female with the head pale, sometimes narrowly darker at the edges, the pronotum similarly colored, the margins of the black area less sharply defined as usual and extending nearly to the apical margin. Length 2.9–3.7 mm.; width 2.1–2.8 mm.   Texas (Brownsville)—Mr. Wickham . . . . **decora,** sp. nov.

The spots very broadly confluent, forming a fascia slightly emarginate on both sides.14

14—The fascia a third as wide as the length of the elytra, broadening within, broadly truncate opposite the suture ; subapical spot very large, extending along the limb, the marginal bead black ; head in the female pale throughout, the pronotum very broadly and intero-angulately pale at the sides.   Length 5 2 mm.; width 3.6 mm.   Kansas. . . . .**socialis,** sp. nov.

The fascia not more than a fourth or fifth as wide as the length of the elytra, the sexes perfectly similar throughout in coloration ; head black, with a very large pale

area, the pronotum black, with an angulate lateral pale spot ; elytral fascia parallel-sided and slightly oblique externally, the subapical spot oval and slightly
distant from the limb ; male with the abdomen impressed along the middle toward tip and with the third segment medially bicuspid as usual in this group.
Length 4.2–4.75 mm.; width 2.8–3.2 mm.   Arizona.........**dentipes** *Fabr.*

15—Body very much smaller ; male with the head and pronotum pale yellowish-
white, the latter with a basal black area extending to lateral fifth or sixth, the
median part feebly bilobed and extending to apical fourth or fifth ; elytral spots
small, at the margin slightly behind basal third, near the apex and further from
the suture than limb, and at basal third and inner two-fifths ; under surface
black, the legs rather slender and pale ; sexual characters feeble. Length
2.5 mm.; width 1.75 mm.   Rhode Island...............**indubitabilis** *Cr.*

Lepida is not represented in the material before me and *bistripustu-
lata* (= *erythrocephala*) is represented by *decora* of the table ; the second is allied to *dentipes* but in the typical form has the two ante-median
spots separated, the inner the larger.   The species from *stellata* to
bolli are more or less close derivatives of the *ursina* type and those
from *socialis* to *dentipes*, probably including *tau* and *quadrillum*, which
I have not examined, may be considered as subspecies of the *dentipes*
type, but in each case the peculiarities of form, size or ornamentation
hold good through extended series.   In fact, as in many other parts
of the Coccinellidæ, we may have a succession of what can only be regarded as distinct forms, with all the fixed characteristics of species,
having an identical general scheme of ornamentation.   This is evident
also in many other parts of the Coleoptera as in *Cicindela, Omophron*
and *Heterocerus*.   Ornamentation may become in other words as important a generic structural character as any other special modification.
In the present tribe there is even an intergeneric similarity or parallel-
ism of ornamentation, as shown in *B. decempustulata* and *Hyperaspis
troglodytes*, which can scarcely be mutually distinguished superficially,
and the same is well known in *Chilocorus* and *Exochomus*, showing
that ornamentation in the Coccinellidæ has been evolved for a useful
purpose and that it should form a correspondingly important criterion
in classification.

Hyperaspis *Chev.*

The tarsal claws seem to vary gradually and between somewhat
narrow limits in this genus, being occasionally almost simple, but I do
not find this character to be of much importance in classification and
have therefore not employed it at all.   The comparative definition of
the species is difficult, as there is little or no structural variety and the

two sexes frequently differ in coloration. In adopting type of coloration as a primary taxonomic character however, this is restricted below to the patterns of the elytra, as sexual divergencies in ornamentation are almost exclusively confined to the head and prothorax, which are very often in part pale in the male and entirely black in the female. In fact this seems to be the only possible means of distinguishing the males from the females, as the external structure of the abdominal apex is very nearly similar in the two sexes. The species are numerous and those known to me may be distinguished as follows :—

Body very broadly rounded and strongly convex 2
Body elongate-oval or oblong-oval and frequently more depressed...............15
2—Elytra black, with a pale red margin not attaining the sutural angles and with
    which a rounded discal spot is broadly confluent a little behind the middle ;
    elytra strongly and moderately closely punctured. Length 3.0 mm. Illinois.
                                                          bolteri *Lec.*
Elytra black, with three marginal or submarginal pale spots.....................3
Elytra black, with a short marginal vitta from the humeral angle, a submarginal oval
    or rounded spot near the apex and another at or near the middle and near inner
    third of the width, the latter obsolete in var. *omissa*.........................4
Elytra black, with a single marginal or submarginal spot far behind the middle or
    near the apex..7
Elytra black, with two marginal or submarginal pale spots, the anterior of which is
    not basal..8
Elytra black, without marginal or submarginal spots but with a single spot near or be-
    hind the centre of the disk...10
3—Black, shining, finely but distinctly punctate, the pronotum closely, the elytra
    rather sparsely ; head black, the pronotum with a quadrate lateral spot almost as
    wide as long ; elytra with a humeral marginal vitta between two and three times
    as long as wide in less than basal third, an internally rounded marginal spot just
    behind the middle, a rounded subapical spot equidistant from the margin and
    suture and a spot on the disk at basal two-fifths and inner third or fourth, which
    is rounded but with its anterior edge broadly sinuate ; legs black, the tibiæ and
    tarsi pale, the posterior tibiæ blackish ; sides of the abdomen narrowly reddish.
    Length 2.8 mm. ; width 2.2 mm. Arizona.............**8-notata,** sp. nov.
Black, shining, the pronotum finely and not very closely punctulate, the elytra more
    strongly and quite sparsely punctate ; head pale, the basal third black ; pronotum
    with a very narrow parallel pale side margin ; elytra with a very small narrow
    humeral, and a slightly larger but narrow and parallel post-median, yellow spot,
    and a large subapical spot nearer the margin than the suture, also with a small
    elongate-oval spot just before the middle and at inner two-fifths ; legs pale pice-
    ous, the hind thighs darker ; abdomen not visibly pale at the sides. Length 2.15
    mm. ; width 1.55 mm. Nevada (Reno)..............**notatula,** sp. nov.
4—Humeral vitta very narrow and inwardly prolonged along the base for a short dis-
    tance, terminating posteriorly just before the middle ; male with the head and a

narrow parallel side margin and very fine apical margin of the pronotum yellow,
the female with the head and pronotum black throughout, the latter very finely
but rather closely punctate ; elytra very finely and rather sparsely punctured ; abdomen black throughout ; legs black, the anterior tibiæ and tarsi pale. Length
2.7–3.2 mm. ; width 2.0–2.35 mm. Montana (western)—Mr. Wickham.

montanica, sp. nov.

Humeral vitta broad, not inflexed along the basal margin, the pale spots generally
deep red. 5

5—Male with the head and a triangular marginal spot and very fine apical margin of
the pronotum pale, the female with the head and pronotum black, the latter with a
narrower triangular marginal spot pale, the apical margin not at all paler ; humeral
vitta terminating at basal two-fifths ; abdomen margined with testaceous throughout ; legs in great part pale. Length 3.2–3.4 mm.; width 2.35–2.6 mm. Arizona.

pinguis, sp. nov.

Male with the head and a narrow parallel lateral margin and more or less fine apical
margin of the pronotum pale, the female with the head and pronotum throughout.
black ; humeral vitta extending to the middle. 6

6—Form very short and broad, the apical pale pronotal margin of the male very distinct, about half as wide as the lateral pale border ; elytral punctures sparse and
fine but distinct, the discal pale spot at the middle of the length ; eyes blackish.
Length 2.6–3.0 mm.; width 2.15–2.4 mm. Texas (El Paso).**lateralis,** *Muls.*

Var. A—Similar to *lateralis* in form, size and disposition of markings, except
that the discal spot near the middle is wholly obsolete. Arizona (Grand
Cañon of the Colorado)—Dr. Prudden. . . . .**omissa,** n. var.

Form less dilated, the pronotal apex exceeding finely pale in the male, the elytral
punctures extremely minute and still sparser, the discal spot slightly in front of
the middle ; eyes bright green. Length 2.8 mm.; width 2.2 mm. California (San
Diego). . . . .**lævipennis,** sp. nov.

7—Elytra each with a rounded subapical spot three times further from the suture than
from the margin, the spot rather large, circular and reddish ; pronotum finely
but distinctly, rather sparsely punctured, having a wide internally rounded marginal spot in both sexes, the head pale in the male and black in the female ;
elytra quite strongly and not closely punctured ; legs black. Length 2.7–2.9
mm.; width 2.1–2.25 mm. New York (Adirondacks) and Indiana [*guexi*
Muls.]. . . . .**bigeminata** *Rand.*

Elytra each with a smaller and generally yellowish irregularly rounded pale spot near
the apex, twice as far from the suture as from the margin, and also a larger deep
red rounded spot just before the middle and a little nearer the margin than the
suture ; head and prothorax black throughout in the female ; punctures smaller
than in *binotata* ; the prothorax more transverse. Length 2.7 mm.; width 2.1
mm. Northern Atlantic States . . . . **signata** *Olv.*

Elytra each with a very small rounded disco-marginal spot at posterior third of the
edge, and another similar in the same transverse line at inner third at posterior
fifth viewed vertically, also with a much larger rounded spot just before the
middle and just visibly nearer the margin than the suture, the spots deep red and
the exterior of the two posterior frequently almost obsolete ; pronotum with a

moderately wide internally rounded yellow marginal spot, the head apparently black in both sexes.   Length 2.0–2.7 mm.; width 1.6–2.0 mm.   Rhode Island, Pennsylvania, North Carolina (Asheville) and Indiana—also a specimen from Las Vegas, New Mexico, which represents a slight variety (*trinifer* n. v.), still more broadly oval, with yellow elytral spots, the two posterior but little smaller than the anterior...**proba** *Say*

Elytra each with a large elongate pale spot along the margin, exten-ling from a little behind the middle to apical fifth or sixth of the edge, internally rounded in out-line, the apical edge narrowly picescent thence to the sutural angle, also with a large oval spot, slightly longer than wide, just before the middle at inner two-fifths ; pronotum black, with a very broad internally rounded marginal pale spot and narrow pale apical margin ; head entirely pale ; pale areas very pale straw color throughout ; legs throughout and posterior half of the abdomen pale, the latter dusky toward the middle.   Length 2.15 mm.; width 1.7 mm.   Texas (Brownsville) ....**rotunda**, sp. nov.

8—Elytra with a parallel marginal vitta extending from basal fifth to three-fifths of the length and more than twice as long as wide, also with a large and rounded but antero-laterally sinuato-truncate spot near the apex, equidistant from margin and suture, and a large oval and feebly oblique spot at basal third, less than half as far from the suture as from the margin ; head pale ; pronotum with a very broad and internally angulate pale margin and a narrow pale apical margin joining the lateral pale areas, the latter yellowish-white throughout ; punctures moderately distinct ; under surface black throughout, the legs in great part pale.   Length 2.2 mm.; width 1 6 mm.   Texas (Brownsville)—Mr. Wickham.

gemma, sp. nov.

Elytra without a marginal vitta but with a rounded pale spot at or near the middle..9

9—Marginal pale spot just before the middle ; each elytron also with another similar in size near the apex and very near the edge, less than half as far therefrom as from the suture, also with a slightly larger rounded spot a little before the middle and half its width from the suture ; head pale ; pronotum black, with a broad marginal spot as wide as long, broadly rounded internally, the apex not at all pale ; punctures quite deep and strong but only moderately close-set ; under sur-face of the hind body black, the abdomen pale at the limb throughout ; legs very pale throughout ; ornamentation yellowish-white.   Length 1.8–2.1 mm.; width 1.4–1.6 mm.   Texas (Brownsville)—Mr. Wickham. ......**medialis**, sp. nov.

Marginal pale spot slightly behind the middle small, each elytron also with a still smaller transversely oval subapical spot, almost as far from the margin as from the suture, and a small rounded discal spot, distinctly before the middle and slightly nearer the suture than the margin ; head pale ; pronotum with a narrow pale lateral margin slightly narrowed to the base, the inner margin straight ; apex not pale, the pale areas reddish-yellow ; punctures fine ; under surface black, the abdomen paler at the edges, broadly behind ; legs in great part piceous.   Length 2.3 mm. ; width 1.8 mm.   Arizona (Benson)—Mr. Dunn.

triangulum, sp. nov.

10—Discal spot of each elytron irregular in form, red, extending from basal fourth to apical fifth and from inner to outer fourth of the width, obliquely truncate anteriorly, subparallel for less than half its length, then rapidly and rectilinearly narrowed to a blunt point ; head black ; pronotum black, with a moderately wide yellow side margin longer than wide and broadly rounded internally, the apex not pale ; punctures very fine but rather close-set; under surface black ; anterior tibiæ and tarsi pale.   Length 3.0 mm ; width 2.2 mm.   Florida (Jacksonville).

regalis, sp. nov.

Discal spot circular or oval in form..11

11—The spot situated slightly before the middle of the length 12

The spot circular, moderate in size and situated more or less distinctly behind the middle 14

12—The spot obliquely oval from the base outwardly........................13

The spot rather small, circular or very nearly ; head and pronotum black throughout in the female, the latter margined at tip and sides with yellow in the male, finely but strongly, rather closely punctured, the elytral punctures strong and sparser, the spot before the middle and rather nearer the margin than the suture, red in color.   Length 2.0–3.8 mm. ; width 1.7–3.1 mm.   New Hampshire, Pennsylvania, Maryland, Indiana and Wisconsin ; [*signata* Lec. nec Oliv., *normata* Say, *affinis* Rand. and *leucopsis* Mels.]................**binotata** *Say*

13—The spot extending from basal two-fifths and inner two-fifths to apical three-fifths and outer five-sevenths ; pronotum of the female with a subparallel yellow margin.   Length "3.3 mm. ; width 2.6 mm."   L'Amérique septentrionale—Dejean..**inedita** *Muls.*

The spot extending from basal fourth and inner third to three-fifths of the length and outer third, red in color; head and pronotum of the female entirely black throughout ; punctures very fine and inconspicuous, moderately sparse ; under surface black.   Length 2.3 mm. ; width 1.9 mm.   Texas (Austin).

bicentralis, sp. nov.

14—The spot just visibly behind the middle and equidistant from the suture and margin ; male with the head pale, the pronotum black, with a narrow apical and broad lateral margin pale, the latter feebly arcuate internally, the female with the head and pronotum black, the latter having a pale, internally rounded side margin, as wide as that of the male ; punctures fine but strong and close-set ; legs pale, the femora blackish ; ornamentation yellowish-white in color.   Length 1.9 mm. ; width 1.45 mm.   Texas (Brownsville)—Mr. Wickham.

globula, sp. nov.

The spot just before apical third and distinctly nearer the margin than the suture ; head pale, the pronotum black, with narrow apex and broad side margin pale, the latter rather wider than long and internally rounded ; punctures rather fine but strong, moderately sparse ; legs red throughout ; ornamentation dark yellow in color.   Length 2.5–3.2 mm. ; width 1.9–2.5 mm.   Texas (Brownsville)—Mr. Wickham ....**wickhami,** sp. nov.

15—Elytra without a discal spot near the middle..........................16

Elytra with a discal spot at or near the middle..........................33

Elytra with a discal vitta which is occasionally more or less obsolete, and, in *simulans* altogether wanting, the elytra being black without indication of subapical pale spot ; sides of the pronotum narrowly pale..........................38

16—Elytra with a pale marginal vitta which is sometimes abbreviated or resolved into three spots, of which only the middle one remains in several instances..... ..17

Elytra without a marginal vitta or median marginal spot but with a subapical pale spot..27

17—Elytra without ornamentation, other than a circular spot very slightly behind the middle and adjoining the side margin....................................18

Elytra without ornamentation, other than a basal marginal vitta extending to slightly behind the middle...................... 19

Elytra each with three widely separated marginal or submarginal spots 20

Elytra with an internally sinuate marginal vitta, extending from the base to distinctly behind the middle, and, in addition, with a laige transversely oval subapical spot...... 21

Elytra with a continuous or subcontinuous marginal vitta, bisinuate within and not attaining the sutural angle..22

18—Lateral spot larger, yellow, nearly two-fifths as wide as the elytron ; pronotum of the female black, with a very narrow faint pale streak at the margin anteriorly, finely, sparsely punctate, the sides moderately convergent ; head nearly as wide as an elytron, black. Length 2.5 mm.; width 1.8 mm. California (Siskiyou Co.).
osculans *Lec.*

Lateral spot very small and reddish, scarcely more than a sixth as wide as the elytron ; pronotum of the female black throughout, strongly and closely punctate, the sides strongly convergent ; head black, very much narrower than an elytron ; head of the male pale, the side margin of the pronotum also narrowly pale from the apex to basal third. Length 1.75-2.3 mm.; width 1.5-1.7 mm. Texas (El Paso).
pleuralis, sp. nov.

19—Marginal vitta extending from very near the basal margin for two-thirds the length, much dilated internally and with rounded outline in its posterior two-thirds, the dilated part emitting a slender transverse spur extending to inner third of the width ; elytral punctures fine and sparse but rather strong ; head and pronotum black throughout in the female. Length 2.6 mm.; width 1.8 mm. California (San Diego).....................................**tæniata** *Lec.*

Marginal vitta beginning at about its own width from the basal margin and continuing to apical two-fifths, only feebly dilated internally with rounded outline posteriorly ; elytral punctures minute and sparse ; head and pronotum black throughout in the female ; body more narrowly oval than in *tæniata*. Length 2.4 mm.; width 1.65 mm. Nevada (Reno).....................**nevadica**, sp. nov.

20—Basal spot rounded, not quite enveloping the basal margin, prolonged posteriorly for a short distance by a rapidly and acutely acuminate spur which is medial with reference to the spot and not marginal ; second spot at the middle larger and semi-circular internally ; subapical spot smaller than the medial, transversely oval, slightly nearer the limb than the suture but quite distant from both ; punctures sparse and fine ; head and pronotum black throughout in the female. Length 2.2-2.35 mm.; width 1.65 mm. California (Alameda).
psyche, sp. nov.

21—Marginal vitta extending from very near the base to apical third, gradually narrowed from its base for two-thirds its length and then expanded with rounded internal outline; subapical spot rather large, transversely oval, very close to the limb and about twice as far from the suture; punctures fine and sparse but rather strong; head and pronotum black throughout in the female. Length 2.6 mm.; width 1.8 mm.   California (Siskiyou Co.)..............**dissoluta** *Cr.*

22—Apical extremity of the marginal vitta not anteriorly extended; head and narrow apical and lateral margin of the pronotum pale in the male..................23

Apical extremity of the vitta greatly expanded, truncate along the suture and prolonged anteriorly for some distance......... 26

23—Posterior of the two internal sinuosities rounded and forming an angle which is more than right, the vitta varying but little in width throughout its length.....24

Posterior internal sinus angulate and right or less in extent, the vitta rather broad and more irregular in width.................. 25

24—Marginal vitta wide, deflecting but very narrowly from edge posteriorly.   Length 2.0–2.7 mm.; width 1.4–1.9 mm.   Colorado, Texas, Arizona and California [*rufomarginata* Muls.].................... ......**fimbriolata** *Melsh.*

Marginal vitta narrow, deflecting widely from the edge posteriorly; body smaller and more narrowly oval.   Length 1.8 mm.; width 1.25 mm   California (San Diego). **limbalis**, sp. nov.

25—Median part of the vitta moderately arcuate internally, the apical part generally not tending to separate as a spot, but in one male the apical part is wholly detached as a subapical spot, and, in another male, the median part emits a broad angulate spur extending transversely to inner two-fifths, nearly as in *tæniata;* body more broadly oval than in *fimbriolata* or *tæniata,* and with a smaller, more rapidly narrowed prothorax.   Length 2.0–2.5 mm.; width 1.4–1.9 mm.   Arizona (Grand Cañon of the Colorado)—Dr. Prudden.............**cincta** *Lec.*

Median part of the vitta strongly but evenly rounded internally, the apical part much narrower, departing more from the edge than in *cincta* and always semidetached; body smaller and more narrowly oval than in *cincta*.   Length 1.9–2.0 mm.; width 1.3 mm.   California (Humboldt Co.)...........**nupta**, sp. nov.

26—Larger, evenly elliptical, the marginal vitta reddish, rather broad, only feebly dilated internally at the middle but strongly at its sutural termination, the internal sinuosities rounded; posterior part deflecting but narrowly from the edge; punctures strong; head and pronotum black throughout in the female.   Length 2.6–2.75 mm.; width 1.8–1.95 mm.   Dakota—Mr. Wickham.

inflexa, sp. nov.

27—Upper surface moderately convex, the elytral punctures more or less fine and sparse...28

Upper surface depressed, the elytral punctures strong and close-set..............32

28—Subapical spot bright yellow and sharply defined.......................29

Subapical spot very small, darker or obscure yellow and with nubilate outline....31

29—Body elongate-subelliptical, the prothorax more transverse and less narrowed from base to apex, the sides narrowly yellow in the female with rounded inner outline; subapical spot of the elytra large, triangular and outwardly pointed, its margin parallel and close to the limb.   Length 2.9 mm.; width 1.8 mm.   California (locality not indicated).......................**elliptica**, sp. nov.

Var. A—Body equally or even more distinctly elongate-elliptical, the narrow yellow margin at the sides of the pronotum in the female narrower, parallel,

not quite attaining the base and with its inner outline nearly straight ; subapical spot small, transversely and evenly oval, remote from the limb and
nearly twice as far from the suture. Length 2.65 mm.; width 1.65 mm.
California (Mendocino Co.)..................... ......**angustula**, var. nov.
Body more briefly oval, with more arcuate sides ; subapical spot transversely oval; size
smaller..................... 30
30—Subapical spot large, its antero-lateral outline irregular, approaching close to the
limb anteriorly ; head yellow in the male as usual, the pronotum narrowly yellow
at the sides in both sexes ; elytral punctures fine and sparse. Length 2.3–2.75
mm. ; width 1.6–2.0 mm. California (Siskiyou Co.)...........**postica** *Lec.*
Subapical spot small, evenly and transversely oval, parallel to the limb and but
slightly less distant therefrom than from the suture ; size smaller ; coloration of
the head and prothorax similar ; elytral punctures fine but much stronger and a
little closer. Length 1.8- 2.0 mm. ; width 1.2–1.35 mm. California (Humboldt
and Siskiyou Cos.),............................**oculaticauda**, sp. nov.
31—Obtusely oval, the head pale in the male but sinuately black at base, the pronotum
black with a narrow parallel pale side-margin ; elytra sparsely and very finely
punctate, the subapical spot small, transversely oval, twice as wide as long, remote from the limb and one-half further from the suture ; legs piceous-brown.
Length 2.0 mm. ; width 1.4 mm. California (Placer Co.)...**efteta**, sp. nov.
32—Evenly elliptical, subdepressed ; sides of the pronotum in the female narrowly
yellow, with somewhat irregular and nubilate inner outline ; elytra black, strongly
punctate, with feeble nubilous marginal pale streak at the humeral angles and a
very small, transversely oval, obscure yellowish and nubilous subapical spot remote from the limb and still more distant from the suture ; under surface
piceous. Length 2.3 mm. ; width 1.6 mm. California (Alameda).
                                         subdepressa, sp. nov.
33—Elytra with a pale spot very near the basal margin and inner third..........34
Elytra without a subbasal pale spot, the subcentral spot generally more or less elongate-oval,..35
34—Punctures of the elytra fine ; head and a narrow lateral and apical margin of the
pronotum pale in the male ; elytra with a humeral and a median elongate marginal spot and another, transversely oval and subapical, also with an elongate spot
just behind the middle and nearer the suture than the margin. Length 2.5 mm.;
width 1.7 mm. Massachusetts and Lake Superior........**disconotata** *Muls.*
Punctures of the elytra rather coarse and deeply impressed, somewhat sparser ; ornamentation somewhat similar to the preceding, except the spots are less elongate
and the subcentral one rounded ; size smaller. Length 2.2 mm. ; width 1.55
mm. Rhode Island ; [*discreta* Lec.]...................**troglodytes** *Muls.*
35—Elytra with a narrow, internally sinuate marginal pale vitta extending two-thirds
to three-fourths from the base, the vitta frequently wholly wanting..........36
Elytra with an entire marginal vitta, internally bisinuate, not extending quite to the
suture and which is never wanting but sometimes resolved by individual variation into three separate spots...37
36—Subapical spot smaller and slightly elongate-subquadrate, less distant from the
suture than from the limb ; discal spot at basal third almost equidistant from
suture and margin, the punctures fine and rather close-set ; head and pronotal

apex narrowly, and sides more broadly with angulate inner outline at the middle, pale in the male.   Length 2.65 mm. ; width 1.8 mm.   Massachusetts ; [*venustula* Muls., *jucunda* || Lec. and *lecontei* Cr.] ....**lugubris** *Rand.*

Subapical spot larger, slightly transverse, much nearer the limb than the suture ; discal spot but slightly before the middle and somewhat nearer the suture than the margin ; punctures not coarse but strongly impressed, moderately sparse ; head and a narrow parallel pronotal side-margin pale in the male, the female having the pronotal sides similar to the male but with the head black.   Length 1.9–2.7 mm. ; width 1.3–1.8 mm.   California (Los Angeles to Sonoma Co.) ; [*horni* Crotch and *elegans* Gorh. nec Muls.] ....**4-oculata** *Mots.*

37—Marginal pale vitta broader, deeply bisinuate within ; size larger.   Length 2.0–2.7 mm. ; width 1.4–1.8 mm.   Rhode Island, New York, Pennsylvania, Indiana, Iowa and Wisconsin ; [*elegans* Muls., *maculifera* Melsh. and *guttifera* Weise] ....**undulata** *Say*

Marginal pale vitta narrow and very feebly bisinuate within ; size much smaller, the pronotum more alutaceous, with the apex and side margin similarly pale in the male.   Length 1.6 mm. ; width 1.0 mm.   Florida.........**paludicola** *Schz.*

38—Elytra with a well-marked and constant, internally and feebly bisinuate pale margin, not quite extending to the suture, and a generally constant discal vitta, extending from very near the basal margin near the middle obliquely toward the sutural angle... 39

Elytra without a well-defined and continuous marginal pale vitta, the discal vitta wholly obsolete or only distinct posteriorly................................40

39—Discal vitta joining the marginal near the sutural angles ; body larger and less narrowly oval.   Length 2 3–2.6 mm. ; width 1.6–1.75 mm.   California (San Francisco)...................................... ....**annexa** *Lec.*

Discal vitta not joining the marginal but separated therefrom near the sutural angles by a space not as wide as its own width ; elytra more obtusely subtruncate at tip.   Length 2.2 mm. ; width 1.35 mm.   Kansas. ....**4-vittata** *Lec.*

40—Elytra with remnants of the discal vitta behind the middle, sometimes with three narrow and feeble marginal spots, the margin frequently black throughout ; body more depressed.   Length 2.1 mm. ; width 1.4 mm.   Lake Superior ; [*consimilis* Lec.—*Oxynychus*]..**mœrens** *Lec.*

Elytra wholly without pale markings of any kind, except a narrow suffused humeral streak at the margin ; prothorax relatively more elongate along the median line.   Length 2.1 mm. ; width 1.5 mm.   Arizona.............. ....**simulans,** sp. nov.

I have been unable to examine any exponents of *cruentata, lewisi, tædata, gemina, pratensis, punctata, tristis* or *floridana,* following the order of the most recent table of LeConte (Trans. Am. Ent. Soc., VIII, 1880, p. 186), but in assigning them to places in the above table would venture to place the first immediately after *tæniata, lewisi* and *pratensis* after *pleuralis, tædata* after *regalis, gemina* after *proba, punctata* after *paludicola* and *tristis* after *effeta,* with which it is evidently very closely related.   *Floridana* cannot be identified.

In the above arrangement it is evident that the species from *montanica* to *lævipennis* are close derivatives of the *lateralis* type, that those from *tæniata* to *inflexa* are close, and those from *elliptica* to *subdepressa*, but slightly less close, derivatives of the *fimbriolata* type, and further, that those from *disconotata* to *paludicola*, and then from *annexa* to *simulans*, are also more or less closely related to the same type. Most of the others are rather isolated in relationship, except, perhaps, *signata* and *binotata*, which may possibly be varietal forms of one type, but I have no evidence to prove this and have never seen a series from any one locality which contained the two forms intermingled. There is before me a large series of *binotata* collected in Indiana, not one of which has a vestige of the subapical spot, and my only representative of *signata* was taken in a wholly different region.

Although it is possible that many of the forms in the table above may prove to be more properly subspecies of a few type forms, which could only be definitely determined by future collecting and careful investigation, they are at least apparently worthy of distinctive names for future reference, and that is all that can be positively affirmed at present ; anything else would, in the absence of evidence, be mere speculation and individual opinion. The genus is an extremely difficult one so far as the differentiation of species is concerned.

Helesius, gen. nov.

The two species for which this generic group seems to be desirable, differ from *Hyperaspis* in having the anterior tibiæ thickened externally, and in having a suffused coloration, devoid of any trace of the abruptly defined pale areas of that genus. They may be defined as follows :—

Form oval, strongly convex, moderately shining, the head and prothorax rufo-piceous, the latter gradually black toward the middle, finely but distinctly, rather closely punctulate, more closely so toward the sides, the length at the middle nearly one-half greater than at the sides, the base evenly rounded in circular arc ; elytra barely as long as wide, the sides continuous with those of the prothorax, evenly rounded behind, very minutely, sparsely and obsoletely punctulate, black throughout ; under surface piceous, the legs rufo-piceous. Length 2.7 mm. ; width 1.8 mm. Texas (Brownsville)—Mr. Wickham...................**nubilans**, sp. nov.

Form oblong-oval and less convex, shining, the pronotum alutaceous in one sex, the head and prothorax rufous, the latter minutely punctulate, more strongly and closely toward the sides, the latter nearly three-fourths as long as the median length, the base broadly rounded or subparabolic ; elytra longer than wide, the sides feebly arcuate, the base not quite as wide as the base of the prothorax, the apex very obtusely rounded, black or paler, minutely and sparsely but distinctly punctulate ; under surface and legs pale. Length 2.3 mm. ; width 1.4 mm. Colorado (Florissant).....................................**nigripennis** *Lec.*

The latter of these species was described as a *Scymnus* by its author, under the supposition probably that the pubescence had been accidentally removed ; the example before me is slightly smaller than the type.

Hyperaspidius *Crotch.*

This is an aberrant genus in the present tribe, in having the elytral epipleuræ devoid of depressions for the posterior femora, although in every other feature it is perfectly normal. The type of ornamentation differs from anything observed in *Hyperaspis* or *Brachyacantha*, and the species are much smaller as a rule. The absence of epipleural foveæ shows that the presence or absence of this character is not so important in itself as it has been assumed to be, and that it is not necessarily a tribal character at all ; this is shown also in the Chilocorini, and the same statement can be made regarding the structure of the tarsal claws. The epipleural depression in the Hyperaspini never assumes the form of an abruptly excavated pit, as it does in some ptinids and to some extent in *Delphastus*. The species are few in number and may be defined as follows :—

Elytra with a pale discal vitta not joining the marginal pale area.................2
Elytra with the discal vitta entirely wanting 3
Elytra with the discal and marginal vitta largely united, leaving merely an elongate black spot from the callus...4
Elytra completely pale...5
Elytra black, with suffused humeral and subapical pale markings 6
2—Head and pronotum pale in the male, the latter with the basal margin to lateral fourth and two median dashes, converging posteriorly and united with the dark basal area, piceous-black, black in the female, with a narrow lateral margin of the pronotum pale ; elytra in both sexes fully as long as wide, oblong, subtruncate, finely, rather sparsely punctate, with a pale yellowish-white basal margin, continued along the sides and apex very nearly to the sutural angle, receding slightly from the edge at apex, and also with its inner basal limit continued posteriorly as a rather broad, sharply defined vitta to or very slightly beyond apical third, receding visibly from the suture from a point slightly behind the scutellum. Length 1.9–2.2 mm.; width 1.25–1.4 mm. Texas (El Paso) ; [*vittigera* Lec.].

trimaculatus *Linn.*

Head and pronotum piceo-rufous in the female, the side margin of the latter very narrowly pale and rather nebulously so toward base ; elytra scarcely as long as wide, subquadrate, with arcuate sides and subtruncate apex, blackish, finely but strongly, sparsely punctate, with a basal and marginal pale area nearly as in the preceding species but with the subsutural vitta nearly straight, almost parallel throughout to the suture and extending posteriorly to apical fifth ; legs pale. Length 1.8 mm. ; width 1.2 mm. California (Alameda Co.).

comparatus, sp nov.

Head and pronotum of the male pale yellowish-white, the latter with a black anteriorly sinuate basal margin extending to lateral sixth ; elytra oblong, longer than wide, obtusely subtruncate at tip, with pale vitta and lateral and basal margins nearly as in the two preceding species, except that the basal margin is almost interrupted at outer third by the black vitta, the pale vitta parallel and very close to the suture, as wide as the black vitta or wider, obtusely acuminate at apical fifth ; punctures very fine and rather close-set ; body much narrower than in the preceding. Length 1.6 mm.; width 0.8 mm. New Mexico (Las Cruces)—Mr. Cockerell...**ingenitus,** sp. nov.

Head and pronotum of the male pale flavate throughout, the latter suffused with reddish toward base, of the female dark rufo-piceous, the pronotum blackish toward base and with a narrow side-margin, extending inward slightly at apex, pale ; elytra in both sexes black, finely, rather sparsely punctate, oblong, very broadly rounded at tip, with a narrow pale basal and lateral margin terminating near the sutural angles, as in the preceding species but narrower and more deeply sinuate around the base of the callus, and also sinuate at apical fourth, the pale vitta forming merely an elongate discal spot just behind the middle in the female, and extending from basal two-fifths to apical third in the male, the elytra in the latter barely as long as wide, rather longer in the female ; body much larger. Length 2.65–2.8 mm.; width 1.4–1.75 mm. Colorado (Colorado Springs)—Mr. Wickham................................   ...   .**insignis,** sp. nov.

3—Elytra entirely black, with a narrow pale margin along the base and down the sides as far as the middle ; front of the head and apical margin of the pronotum irregularly yellow in the male. Length [2.0 mm.]. California. .**arcuatus** *Lec.*

4—Broadly oblong-oval, broadly rounded behind, finely punctate, the head and pronotum black in the female, the latter with a narrow parallel pale side-margin ; elytra about as long as wide, pale flavate, the suture broadly black, the vitta constricted slightly just behind the scutellum and strongly at the apical angles, each with an elongate triangular black dash involving the callus and extending from the basal margin at outer third for two-fifths. Length 2.1 mm.; width 1.4 mm. Florida.......................................**militaris** *Lec.*

5—Body oblong, subtruncate behind, pale luteo-flavate throughout above and beneath, except the head, which is piceous ; prothorax only slightly wider than the head, very feebly sinuate at apex, the latter only very slightly narrower than the base, the sides feebly arcuate, the punctures very fine ; elytra slightly longer than wide, finely but rather strongly, moderately sparsely punctate. Length 1.9 mm.; width 1.3 mm. Massachusetts (Mt. Tom)...............**transfugatus,** sp. nov.

6—Body almost evenly oval, only slightly obtuse at apex, the sides strongly arcuate, dark piceo-castaneous throughout, the legs scarcely paler, the head nebulously paler toward the apex; the pronotum very narrowly flavate at the sides toward apex, finely punctulate ; elytra but slightly wider than the prothorax, about as long as wide, finely and sparsely but distinctly punctate, the humeral angles, extending more or less briefly along the edge posteriorly, and two subapical spots arranged subtransversely and frequently coalescent, pale flavate. Length 1.4–1.6 mm.; width 0.85–1.0 mm. California (Monterey)......... **conspiratus,** sp. nov.

CRANOPHORINI.

The remarkable development of the pronotum over the head, with total or partial obliteration of the anterior thoracic emargination, so universal in the family, is probably due to environments essentially similar to those of *Sacium*, where the structure is similar and points apparently to a true affinity between these genera, confirming the relationship between the Coccinellidæ and Corylophidæ, which is well known to exist. The body is narrowly oval, usually rather pointed behind, the abdomen with the genital segment large and well developed, generally with a terminal seventh segment in the males, the metacoxal arcs entire in *Cranophorus* but extremely short. The middle coxæ are widely separated, the anterior very narrowly for the present family, the scutellum rather small, the palpi securiform, the antennæ only moderately short, with the joints of the club well defined though not very loose, and the legs are perfectly free. The three genera known to me may be thus defined :—

Antennal club narrow, parallel or fusiform and 3-jointed ; body small or minute..2
Antennal club gradually inflated and 5-jointed.................................3
2—Pronotum evenly rounded in circular arc at apex, the cephalic opening beneath
   horizontal, the prosternum convex in the middle, not at all deflexed; fifth ventral
   segment not much longer than the fourth ; epipleuræ gradually dilated toward
   base but relatively narrow even at the widest part and horizontal or feebly
   descending externally ; tarsal claws cleft within beyond the middle and also with
   an internal basal enlargement ; body distinctly pubescent......* **Cranophorus**
Pronotum truncate at the middle of the apex, the cephalic opening inclined upward
   posteriorly, the mouth protected in part by the prosternum, which is very strongly
   deflexed, flat, with strongly arcuate apex ; fifth ventral very much longer than the
   fourth, the sixth less developed ; epipleuræ not dilated, very narrow, flat, but
   little wider than the met-episternum ; tarsal claws small, apparently merely enlarged within at base ; body minute and subglabrous.................**Nipus**
3—Body larger ; metacoxal arcs complete and about two-thirds as long as the segment.
   * **Oryssomus**

Cranophorus is South African and several new forms will be described in the appendix to the present paper. *Oryssomus* is South American, and *Nipus* is Californian and perhaps Sonoran.

Nipus, gen. nov.

The two species of this genus at present known may be defined as follows :—

Body elongate-oval, the elytra gradually obtusely pointed behind, black, the pronotum
   nebulously pale and broadly impresso-explanate at the sides, especially toward

apex, one-half wider than long, the sides continuing the curvature of the elytra, impunctate, alutaceous, the pubescence more visible toward the sides ; elytra one-half longer than wide, rapidly narrowed from slightly behind the middle, finely but rather strongly, somewhat unequally and very sparsely punctate, each with a large oval central red spot, which is nebulously defined ; pubescence very inconspicuous. Length 1.2 mm.; width 0.7 mm. California (Los Angeles).

                                                        biplagiatus, sp. nov.

Body narrowly and evenly elliptical, rounded behind, black or piceous-black, the under surface and legs rufescent, the elytra not maculate ; pronotum slightly pubescent and paler laterally in the impressed area, two-thirds wider than long, impunctate, alutaceous, the sides rather more arcuate than the contiguous limb of the elytra, the latter nearly one-half longer than wide, rather coarsely, deeply and not very sparsely punctate, the hairs erect and microscopic. Length 0.85 mm.; width 0.55 mm. California (Sonoma Co.)........... **niger,** sp. nov.

Both the above species have the elytral suture finely margined, except toward base.

SCYMNINI.

The numerous small species of this tribe may be distinguished at once by the distinct pubescence, there being but one genus in which the body becomes virtually glabrous throughout the dorsal surface. The antennæ are short and the eyes entire or subentire, and the posterior legs are always free. The genera may be defined as follows :—

Antennæ free, rapidly descending along the sides of the head before the eyes in repose, the front not dilated ; head and maxillary palpi moderate in size, the metasternum not foveate ; anterior coxæ moderately separated ; antennæ apparently 11-jointed...2

Antennæ resting in repose upon wide dilatations of the front under the antennal foveæ and before the eyes, apparently 9-jointed, with a narrow solid 4-jointed club ; head large ; anterior coxæ very widely separated, the prosternum flat and not carinate...6

2—Pronotum with a fine oblique line at the apical angles, the body apparently glabrous, each puncture of the upper surface bearing an extremely minute erect hair, only visible under considerable enlargement with oblique illumination ; antennæ inserted in very deep narrow emarginations at the sides of the front, strongly bent, the last three joints forming a compressed club ; last joint of the maxillary palpi narrow, the apex narrowly oblique ; clypeus narrow and rather long ; prosternum short, flat, with two abbreviated parallel carinæ, the apical margin abruptly deflexed for a short distance, but not enough to afford protection to the mouth ; metacoxal arcs joining the apical margin of the first segment near the sides of the abdomen ; tarsal claws simple ; body oval, convex, the elytra more or less pointed at apex, the eyes small......................**Smilia**

Pronotum without an oblique line near the apical angles ; body always distinctly pubescent ; tarsal claws cleft within....................................3

3—Clypeus extremely short before the rather large and well-developed eyes, truncate with rounded angles, the antennæ inserted under its sides adjoining the eyes, straight, the club small, with the three joints equal in length ; prosternum transversely convex, not carinate, broadly and gradually deflexed, forming a protection to the mouth in repose ; metacoxal arcs complete, the first suture nearly obliterated at the middle ; tarsal claws cleft at the middle.................**Stethorus**

Clypeus prolonged for a considerable distance before the eyes, the sides converging, the antennæ inserted in very small shallow emarginations just before the eyes..4

4 - Last joint of the maxillary palpi narrow, obliquely pointed at tip ; antennæ bent, with the club well developed, the head very small, with well-developed eyes ; prothorax much narrower than the elytra ; prosternum rather narrowly separating the coxæ, with two short feeble carinæ, gradually and feebly deflexed toward apex but not affording protection to the mouth ; metacoxal plates entire..**Didion**

Last joint of the palpi large and normally securiform ; antennæ with the club moderate ; prosternum flat, not at all deflexed toward tip, generally bicarinate............5

5—Head exserted and moderately deflexed, the eyes small and not attaining the anterior margin of the prothorax, which is parallel at the sides ; antennæ well-developed and straight ; prosternum flat, with two distant parallel and entire carinæ ; metacoxal arcs short, widely incomplete, not extending to the first suture ; body parallel and straight at the sides................. **Selvadius**

Head inserted within the prothorax, the eyes well developed and partially covered ; antennæ shorter and bent ; body more or less oval, the prothorax narrowed in front, the prosternum and metacoxal arcs varying subgenerically......**Scymnus**

6—Metasternum with a large circular and densely pubescent pit near each lateral margin ; body oblong-oval, rather depressed, pubescent ; metacoxal plates entire ; eyes oval, entire or virtually so, somewhat prominent and barely attaining the prothorax ; maxillary palpi very long ; first two tarsal joints short, the last long, the claws well developed and simple..................∴**Cephaloscymnus**

These genera are all very widely distributed, except *Didion* and *Selvadius*, which are founded upon local types. *Cephaloscymnus* is a remarkably aberrant and specialized form, but its general affinity with *Scymnus* is sufficiently evident.

.Smilia *Weise*.

These are small, apparently glabrous species, formally assigned to *Pentilia*; they inhabit the entire territory of the United States. Those thus far discovered may be identified as follows :—

Elytra uniform in coloration throughout, the suture finely and distinctly margined...2

Elytra bicolored, each having a large oval red spot; sutural margin " not distinct "..6

2—Head transverse, the clypeus broader and shorter, rapidly expanding before the antennal sinus.......................................3

Head but little wider than long and perfectly flat throughout, the clypeus narrower and more prolonged, only very feebly expanding before the antennæ ; species very minute, black, the pronotum minutely reticulate and alutaceous but not distinctly punctured.......................................5

By looking carefully in April and May, the beetle (fig. 5) can easily be distinguished on the trunks of the trees, with its larva, which is flattened and of a uniform brownish color.

Although so small and insignificant, it accomplishes a great deal in the destruction of the scale insects. The larvæ, hatching in the spring, at the same time with the young scale insects, immediately begin their warfare upon them, which they continue even after they have transformed into beetles.

It would be difficult to estimate the benefit derived. from these little beetles. After May and June, the majority of them suddenly disappear, and we do not see them again until the great fall brood of scales hatch.

Through the kindness of Dr. GEO. H. HORN, I am enabled to give the following description :

HYPERASPIDIUS COCCIDIVORA, N. S.—Broadly oval, convex, piceous, shining, each elytron with a large, badly-defined rufous space, which sometimes reaches the side margin and suture. Thorax sparsely, finely punctate. Elytra more coarsely punctured. Body beneath, and legs piceous, shining. Length, .04 inch.

This insect resembles some of our smaller Scymnus. but it is entirely without pubescence. It is not larger than PENTILIA PUSILLA, and from its resemblance to that insect, except in color, would have been referred to that genus, but there are six abdominal segments.

3—Body castaneous in color throughout, rather more broadly oval, shining, the elytra
    rather strongly, moderately closely punctate. Length 1.1 mm. ; width 0.75
    mm. Michigan (Marquette)............................ **marginata** *Lec.*
Body black throughout, the under surface and legs more piceous in *atronitens*......4
4—Pronotum minutely but strongly reticulate and alutaceous ; elytra finely but rather
    strongly, sparsely and somewhat unevenly punctate. Length 0.8–1.0 mm. ;
    width 0.6–0.7 mm. Pennsylvania to Texas (Brownsville)........**misella** *Lec.*
Pronotum perfectly devoid of minute reticulation and highly polished throughout like
    the elytra, finely punctulate, the sides almost continuous ; elytra distinctly longer
    than wide, gradually narrowed behind, the punctures extremely minute but deep,
    even and relatively sparse ; size much larger. Length 1.15–1.3 mm. ; width
    0.8–0.9 mm. California (Siskiyou Co.)............... **atronitens**, sp. nov.
5—Narrowly oval, the prothorax as wide as the base of the elytra, the sides nearly
    continuous ; elytra narrowed behind from far before the middle, finely and not
    very conspicuously punctured. Length 0.85 mm. ; width 0.55 mm. Texas.
                                                             minuta, sp. nov.
More broadly oval, the prothorax much narrower than the elytra, with the sides dis-
    tinctly discontinuous ; elytra but little longer than wide, more rapidly narrowed
    behind from a point which is but little before the middle, the punctures strong,
    deep and rather close-set, much more conspicuous than in *minuta*. Length 0.8
    mm. ; width 0.65 mm. California..................... **planiceps**, sp. nov.
6—Spots oval, narrowly separated at the suture ; pronotum not distinctly punctate.
                                                       coccidivora *Ashm.*

Ovalis, said by Dr. Horn to be the same as *felschei* Weise, is
omitted from the table, as I have not seen a specimen ; the suture is
said by Dr. Horn to have the marginal stria not evident, but this is
not borne out by the description of LeConte, or by the other species ;
it is brown in color, 0.8 mm. in length and inhabits Florida. It is
quite possible that *coccidivora* may differ generically, but not having
an example before me I am unable to decide.

Stethorus *Weise.*

The species of this genus are as small as in *Smilia*, but differ very
greatly structurally ; they differ from *Scymnus* in the fc mation of the
front of the head and prosternum. The genital segment is as large
and conspicuous as in the Hyperaspini. *Stethorus* is probably cosmopolitan and the species are rather difficult to distinguish among themselves. The following table contains all that are known to me at
present, those from Europe and Africa being introduced for the sake
of completeness : —

Metacoxal plates short and very broad, never extending much beyond the middle of
    the segment......... 2
Metacoxal plates narrow and strongly rounded, much longer, approaching closely to
    the hind margin of the segment ; species more minute as a rule..............6

2—Legs pale and bright flavo-testaceous, the femora black with the apex distinctly and conspicuously pale; sides of the prothorax nearly continuous and strongly converging..3
Legs piceo-fuscous, the femora blackish; body similar in form to the preceding....5
3—Metacoxal plates shorter, frequently extending much less than half way to the suture; elytral punctures moderately close-set, quite strong and distinct, the pubescence short. Length 1.2 mm.; width 0.8 mm. Europe; [*minimus* Payk.].
* punctillum *Weise*
Metacoxal plates less transverse, extending to or beyond the middle, varying somewhat according to the sex of the individual; body somewhat smaller, with shorter and more transverse prothorax and less conspicuous elytral punctures but longer pubescence..4
4—Elytral punctures fine but strong and quite sparse. Length 1.0–1.2 mm.; width 0.75–0.9 mm. Lake Superior to Pennsylvania and Delaware, North Carolina and Kansas...**punctum** *Lec.*
Elytral punctures finer, feeble and less sparse; prothorax slightly more transverse, the body a little more oblong-oval but otherwise extremely similar. Length 1.15 mm.; width 0.8 mm. Cape of Good Hope (Cape Town).
* jejunus, sp. nov. *)*
5—Elytra very distinctly longer than wide, finely but strongly, sparsely punctured, the pubescence moderately long, recurved as usual. Length 1.0–1.3 mm.; width 0.75–0.9 mm. California (Humboldt, Sonoma and Sta. Cruz Cos.).
picipes, sp. nov.
Elytra not obviously longer than wide; body smaller and more broadly oval, the prothorax shorter and more transverse; elytral punctures stronger and more close-set. Length 0.9 mm.; width 0.75 mm. California (Siskiyou Co.).
brevis, sp. nov.
6—Legs pale rufo-testaceous throughout; body evenly oval; elytral punctures very small and sparse, the pubescence moderately long, recurved as usual. Length 0.9 mm.; width 0.72 mm. Florida (Haulover near Jupiter Inlet)..**utilis** *Horn*
Legs pale testaceous, the femora black except at apex; body narrower, more elongate and more oblong, with much less arcuate sides; prothorax transverse, with the sides continuous as in *utilis* but rather less arcuate; elytral punctures stronger and less sparse, the pubescence nearly similar but somewhat fuscous in color. Length 0.95 mm.; width 0.7 mm. Texas (Columbus).....**atomus,** sp. nov.

Punctum and *picipes* are both abundant, and the difference in the coloration of the legs, noted in the table, appears to be constant; in *picipes* the form is a trifle more elongate and more narrowly oval; *jejunus*, five specimens of which were taken by the writer about sixteen years ago, resembles *punctum* so closely that the two could scarcely be distinguished unless examined in series. *Gilvifrons* Müll., which is associated with *punctillum* in the European catalogues, I have not seen, but the genus *Stethorus*, which is there considered a subgenus of *Scymnus*, is in no wise to be so regarded; it is a perfectly valid genus.

) Probably = St. aethiops Weise. See Sicard (Ann. Soc. Ent. France, 1909, p 147

Didion, gen. nov.

This genus resembles *Scymnus* in most of its structural features, but differs in its narrow prothorax with rapidly converging sides, small, deeply inserted but feebly inclined head, with narrowly oval eyes and flat surface, in the feebly deflexed prosternum, and especially the narrow and obliquely pointed last joint of the maxillary palpi. The pubescence is rather abundant but very short and decumbent. Individuals appear to be very rare, and the genus is confined as far as known to the Upper California *Sequoia* belts. The two species represented before me may be defined as follows from the female only :—

Metacoxal plates more narrowly rounded, extending extremely near to the suture, black throughout, the legs piceous ; body oval, convex, moderately shining, the head finely, sparsely punctate, the eyes widely separated ; prothorax not quite twice as wide as long, about two-thirds as wide as the elytra, the sides strongly convergent and evenly but very feebly arcuate throughout, the punctures very minute and inconspicuous ; elytra much longer than wide, evenly rounded at apex, finely but strongly, rather closely punctate. Length 1.6 mm. ; width 0.95 mm. California (north of San Francisco) .... **longulum**, sp. nov.

Metacoxal plates distant from the suture by a third or fourth of their own length ; in coloration and sculpture nearly similar to *longulum*, the pubescence slightly longer, the body much smaller, the prothorax rather more than twice as wide as long, with the sides much less convergent but more strongly arcuate and notably more discontinuous with those of the elytra, the latter more broadly and obtusely rounded behind, much longer than wide. Length 1.25 mm. ; width 0.8 mm. California (Sonoma Co.) .... **parviceps**, sp. nov.

These species are both represented by single examples thus far, but very recently Dr. Blaisdell has sent me a male from Calaveras Co., which appears to be identical with *parviceps*.

Selvadius, gen. nov.

Differs remarkably from *Scymnus* in its narrow parallel body, exserted, feebly inclined and transversely orbicular head, small eyes and longer straight antennæ. The maxillary palpi are larger than usual in *Scymnus*, thick, with the last joint strongly securiform. The single type may be described as follows :—

Body narrowly oblong, rather feebly convex, moderately shining, piceous-brown in color, with the legs, palpi and antennæ yellow ; punctures fine but strong and close-set throughout, those of the elytra larger but shallower than those of the pronotum ; pubescence short, fine and decumbent ; head relatively well developed in size, feebly convex, the eyes small, convex, oval, entirely exposed before the prothorax and entire, the vertex very broad between them ; antennæ nearly as long as the head, 11-jointed, the second joint subglobular, three to five subequal, narrower, elongate and cylindric, six and seven shorter, the latter a little broader toward tip, eight to eleven forming the usual narrowly oval compact club, the eleventh joint short and somewhat spongy-pubescent; prothorax but little more than twice as wide as long, the sides parallel and straight, rounding and slightly convergent at apex; elytra but little wider than the prothorax, much longer than wide, obtusely and broadly subtruncate at tip; mesocoxal arc not attaining the episternal suture, the metasternal curving outward and very short, attaining apical fourth of the segment; genital segment distinct and well developed. Length 1.4 mm.; width 0.65 mm. Arizona (Tuçson)........**rectus**, sp. nov.

The type was taken by the writer some years ago, but no note relative to habits can be found; if my memory serves however, it was taken while sorting riparial detritus.

Scymnus *Kug.*

This is one of the largest genera of American Coleoptera. The species possess a remarkable uniformity of appearance, the body being oval or oblong-oval and always pubescent throughout, with the legs almost completely free, the anterior alone being somewhat contractile, with an attendant depression or well-defined pit at the base of the epipleuræ for the tip of the femur. The prosternal ridges are important, on some occasions, in discriminating species which may be closely allied otherwise. The postcoxal plates or arcs of the first ventral segment serve as sharply defined criteria in grouping the species, but the several sections can scarcely be regarded as distinct genera.

The species have been almost completely neglected in the United States, as far as systematic work is concerned, and the recent revision of Dr. Horn (Tr. Am. Ent. Soc., XXII, p. 81) had no further aim than an exposition of the groups, into which the genus can be advantageously divided, together with the publication of a few of the more strikingly distinct species. The latter are very difficult to discriminate in many parts of the series, and especially in the small and obscure forms of the Pacific coast and Arizona. I am not at all confident that my interpretations may be entirely correct, but it can be said at least that the total number of species here recorded will be increased rather than diminished in the future. I have been accumulating a large material during many years, with the object of monographing the genus, and all localities are tolerably well represented. The following table may assist in identification, but actual comparison will be necessary in many cases :—

Abdominal lines arcuate throughout, curving forward externally..................2
Abdominal lines extending outward externally parallel to the edge of the segment
   and at a slight distance therefrom ; prosternum relatively slightly wider between
   the coxæ, flat and wholly devoid of carinæ ; genital or "sixth" ventral segment
   unusually developed. (*Scymnobius* sg. n.)............................ .71
Abdominal lines gradually curving into the first suture externally and forming a part
   thereof ; prosternum scarcely as wide between the coxæ as in *Scymnobius*, but
   always rather flat and finely but strongly bicarinate, the carinæ straight, widely
   separated and gradually converging ; eyes occasionally with a very small and
   feeble emargination. (*Diomus* Muls.)...............................79
2—Abdominal plates entire, the bounding arc extending to the basal margin of the
   first segment ; prosternum rather narrow and convex between the coxæ, with two
   strong and well-developed carinæ, which are but rarely abbreviated in front.
   (*Pullus* Muls.)..3
Abdominal plates incomplete externally, the bounding arc not attaining the basal
   margin ; prosternum somewhat variable between the coxæ, the carinæ always
   present but frequently abbreviated in front and more feebly developed than in
   Pullus. (*Scymnus* in sp.)...64
3—The carinæ entire or subentire...4
The carinæ greatly abbreviated, attaining about the middle of the prosternum ; abdominal plates very small, broader in *nanus ;* prothorax varying in form..... 63
4—Abdominal plates large and long, attaining the apical margin of the first segment ;
   prosternal carinæ arcuate, most narrowly separated well behind the apex ; body
   oblong-oval, about one-half longer than wide, evenly pale flavo-testaceous above,
   the head and under surface piceous-black ; last ventral segment and legs pale ;
   head and pronotum very finely and remotely punctulate, the latter less remotely
   and rather more visibly toward the sides, strongly transverse, with the sides
   strongly convergent, broadly and evenly arcuate, almost continuous with the outline of the elytra, the latter finely, evenly and sparsely punctate, the hairs laid
   longitudinally and evenly almost throughout ; under surface strongly and closely
   punctured, the abdomen more finely and less closely, the plates polished and almost impunctate throughout. Length 2.3 mm. ; width 1.5 mm. Colorado.
                                                       flavescens, sp. nov.
Abdominal plates normal, always shorter than the segment ; prosternal carinæ generally straight but sometimes bent outward through a short distance from the
   apex..5
5—Elytra uniform in coloration on the disk, not considering the apex............6
Elytra bicolored on the disk, the pale areas either clearly defined and constant spots
   or nubilate and variable..57
6—Elytra entirely pale in color ; prosternal carinæ entire, feebly converging throughout...7
Elytra black, with the common apex more or less broadly pale, the anterior margin of
   the pale area biarcuate and generally very well defined....................8
Elytra black, with the apex not paler or more or less finely so, in which case the
   anterior limiting line of the pale area is usually quite well defined but transverse
   or not biarcuate 16

7—Somewhat narrowly and evenly oval, very pale throughout, except beneath, where the body is black, the hypomera, epipleuræ, tip of the abdomen and legs pale; prothorax but little more than twice as wide as long, the sides rather strongly convergent and evenly arcuate, the punctures scarcely visible; elytra quite closely punctate, the punctures very minute, with larger punctures which tend to lineal arrangement intermingled toward base. Length 1.3–1.6 mm.; width 0.9–1.1 mm. California and Arizona..................**pallens** *Lec.*

Broadly oval and more convex, shining, dark rufo-testaceous, the pronotum slightly clouded toward the middle and base; elytral suture very finely piceous; body beneath and legs pale, except the post-sterna, parapleuræ and abdomen which are black, the latter pale at tip; head and broad sides of the pronotum paler than the elytra, the prothorax short, nearly three times as wide as long, with moderately converging and feebly arcuate sides, which are not continuous with those of the elytra, the punctures sparse and very minute, closer and distinct toward the apical angles; elytra finely but distinctly, evenly and rather sparsely punctured; under surface closely punctate, the ventral plates distant from the hind margin of the segment by two-fifths of their own length; legs rather slender. Length 1.6 mm.; width 1.2 mm. Colorado......**nugator**, sp. nov.

8—Form very broadly oval, the elytra not at all longer than wide............... 9

Form more narrowly oval, the elytra longer than wide.......................11

9—Prothorax entirely orange yellow, a little more than twice as wide as long, the sides evenly arcuate and subcontinuous with those of the elytra, the latter finely but distinctly, not very densely punctate, the pale apical area advancing to apical two-fifths at the sides and beyond apical fourth on the suture; abdomen and legs throughout pale. Length 1.6 mm.; width 1.1 mm. Florida (Indian River).

semiruber *Horn*

Prothorax partly or complete'y black in color and more transverse...............10

10—Apex of the fifth ventra' segment very broadly and feebly sinuato-truncate in the male; pale area of the elytra very constant, extending to about apical sixth or seventh. Length 2.1–2.3 mm.; width 1.6 mm. North Carolina, Texas and Arizona; [*astutus* Muls.]...............................**creperus** *Muls.*

Var. A—Nearly similar to *creperus* but more broadly oval, with a larger thoracic black area and with the apical red area of the elytra smaller, less transverse and more strongly biarcuate at its anterior margin. Length 2.4 mm.; width 1.8 mm. Middle States.......................**fraternus** *Lec.*

Apex of the fifth ventral deeply and evenly sinuate in circu'ar arc with well-defined and somewhat prominent, though obtuse, lateral limits; red area of the elytra extending through apical third to two-fifths; punctuation distinct and rather sparse; pronotum red at the sides, more narrowly toward base. Length 2.2–2.4 mm.; width 1.6–1.7 mm. Florida and Indiana.........**hemorrhous** *Lec.*

Var. A—Similar to *hemorrhous* but with the red of the elytra extending well beyond the middle. Length 2 2 mm.; width 1.5 mm. Kansas.

divisus, var. nov.

Var. B—Similar to *hemorrhous* but larger, the pronotum completely black and more densely and distinctly punctured toward the sides; apical red area similar; female only observed. Length 2.7 mm.; width 1.8 mm. Canada.

laurenticus, var. nov.

Var. C—Similar in form, punctuation and sexual characters to *hemorrhous*, the upper surface entirely black with feeble æneous lustre, the sides of the pronotum and apex of the elytra appearing very faintly red in a strong light in areas similar in position and extent to those of *hemorrhous*, the pubescence rather finer and less conspicuous ; legs almost black throughout. Length 2.2 mm. ; width 1.6 mm.   Texas (Columbus)......**subæneus,** var. nov.

11—Red area of the elytra extending forward to about the middle, its bounding line somewhat feebly defined ; body oval, shining, black, the abdomen black throughout ; legs pale, with the femora infuscate ; prothorax but little more than twice as wide as long, the sides almost continuous, evenly and moderately arcuate and strongly convergent ; punctures fine and equal throughout, very sparse and but slightly more close-set toward the sides, the latter broadly and indefinitely pale ; elytra finely but strongly and closely punctured.   Length 2.25 mm. ; width 1.5 mm. Wyoming ....**postpinctus**, sp. nov.

Red area of the elytra not extending much beyond apical third ; species much smaller..12

12—Red area sharply defined ; elytra oval, finely but distinctly and closely punctate..13

Red area not well defined, its boundaries nubilate ; elytra more coarsely and sparsely punctate, gradually narrowed behind from near the humeri................15

13—Pronotum rufous, with a parabolic median black area extending from the base almost to the apex, two and one-half times as wide as long, the sides rather strongly convergent, feebly and evenly arcuate and almost continuous ; elytra closely punctured, the red area not extending further at the sides than at the suture, ending, along the median line of each, slightly beyond apical third ; legs and abdomen red, the latter black toward base, the male having a transverse fovea in the apical margin of the fifth segment, the first not modified in the middle and punctured throughout.   Length 1.75 mm. ; width 1.25 mm.   Texas (Columbus) ....**texanus**, sp. nov.

Pronotum black, rufous at the sides or apical angles 14

14—More broadly oval, the pale area at the sides of the pronotum abruptly defined and not extending to the basal angles ; pale area of the elytra extending to apical third, its most anterior point at outer third ; abdomen black, with the last three segments pale ; legs pale throughout ; male sexual characters as in *texanus*, the fifth segment less truncate, with the fovea smaller.   Length 1.7 mm.; width 1.1 mm. Kansas.......................... ....**rubricauda**, sp nov.

Very narrowly oval and more pointed behind, the lateral pale area of the pronotum more extended and indefinitely limited internally ; pale area of the elytra nearly similar in form to that of *rubricauda* but smaller, not extending quite to apical third.   Length 1.7 mm.; width 1.0 mm. Pennsylvania (near Philadelphia).
chromopyga, sp. nov.

15—Small and very narrowly suboval, shining, black, the sides of the pronotum abruptly but very narrowly pale, the pale area scarcely extending to the basal angles ; prothorax distinctly narrower than the elytra, the sides not continuous, moderately convergent and moderately though evenly arcuate, the punctures very fine and somewhat close-set toward the sides ; elytra rather prominently rounded at the humeri, the pubescence fine and rather sparse, the pale apex extending scarcely beyond apical fourth ; under surface black, the abdomen pale at tip ; legs pale, the hind femora slightly infuscate toward base.   Length 1.5 mm.; width 0.9 mm., Rhode Island (Boston Neck)..........**canterius,** sp. nov.

16—Pronotum entirely pale in color......................................17

Pronotum pale, with a median parabolic black spot at base, which is normal in the males throughout but much extended in the female of *marginicollis,* where it involves all the disk except the apical angles and a fine apical margin........18

Pronotum black, with pale side-margins or apical angles......................28

Pronotum black throughout ; elytra with the fine marginal bead at apex paler, becoming wider in *renoicus.*...43

17—Prothorax subequal in width to the base of the elytra, the latter about as long as wide, finely and quite closely punctured, the apical margin extremely narrowly reddish.   Length 1.5–1.9 mm.; width 1.0–1.3 mm.   North Carolina (Asheville) and Alabama....................................**cervicalis** *Muls.*

Prothorax at base abruptly narrower than the elytra, the sides discontinuous in curvature, strongly convergent, rather strongly and almost evenly arcuate ; disk minutely, sparsely punctulate scarcely more distinctly toward the sides, twice as wide as long ; elytra distinctly longer than wide, rather strongly and somewhat sparsely punctured, the apical margin very narrowly red ; legs red throughout.   Length 1.8–1.9 mm.; width 1.15 mm.   Kansas ....**kansanus,** sp. nov.

18—Surface polished, the pronotum evidently punctate, the punctures of the elytra more or less coarse and distinct 19

Surface alutaceous and minutely granulato-reticulate, the pronotum impunctate except the scars of fallen hairs ; elytral punctures very minute ; pronotal black spot small and basal...27

19—Pronotal punctures equal in size throughout the disk ; male modifications at the middle of the first ventral segment generally pronounced..................20

Pronotal punctures unequal, coarser, more close-set and more conspicuous toward the middle of the disk—contrary to the general rule—and finer and sparser laterally ; male modifications of the first ventral less pronounced ; legs red throughout..22

20—Pronotal punctures very small and sparse throughout ; male with a tubercle in the middle near the apical margin of the first ventral, the coloration of the pronotum different in the two sexes, the male having a small transverse black spot at the middle of the basal margin, the female having that somite black, with pale apical angles and fine apical margin ; legs black or blackish throughout.   Length 1.6–2.0 mm.; width 1.1–1.4 mm., California (coast regions from Humboldt to San Diego and Calaveras Co.); [*californicus* Boh.].....**marginicollis** *Mann.*

Pronotal punctures fine but distinct, more close-set toward the sides, sparser toward the middle ; larger species from the Mississippi Valley and Great Lakes, broadly oval in form...21

21—Male with a small shallow rounded pit at the middle of the apical margin of the first segment, the fifth with a small deep and rounded median sinuation ; sides of the prothorax very strongly convergent and broadly arcuate, continuous with those of the elytra, the latter rather coarsely and sparsely punctured, with the apical margin very narrowly and feebly rufous ; legs red throughout.   Length 1.9–2.3 mm.; width 1.5–1.7 mm.   Lake Superior....................**consobrinus** *Lec.*

Male with a large, elongate, acutely triangular, feebly impressed, polished and glabrous median area of the first ventral, defined by finer and denser pubescence, the
fifth with a larger but rather more broadly rounded median sinus ; prothorax two
and one-half times as wide as long, the sides continuous in curvature, strongly convergent, broadly and evenly arcuate, the punctures very distinct and less sparse
throughout than in *consobrinus ;* elytra quite coarsely and somewhat closely punctured, the apical margin only extremely narrowly rufescent ; legs rather short,
red throughout.   Length 2.3 mm.;   width   1.75   mm.,   Iowa (Keokuk).

                                                        iowensis, sp. nov.

Male with a shorter, broader and entirely unimpressed median area at the apex of the
first segment, the adjoining punctuation finer and closer, the fifth segment very
short and transverse, truncate at apex but not at all sinuate, the surface with a
strong transverse and anteriorly rounded impression-bevel toward apex in median
third or fourth ; sides of the prothorax almost continuous, strongly convergent and
broadly, evenly arcuate, the punctures very small, sparse and inconspicuous,
scarcely closer toward the sides; elytra not very coarsely but strongly, evenly
and quite sparsely punctate, the apical margin very finely testaceous ; legs red
throughout.   Length 2.2 mm.; width 1.7 mm.   Mississippi (Natchez).

                                                        natchezianus, sp. nov.

22—Elytra with a wider red apical margin, this being equal in width to a fourth or
    fifth of the length of the prothorax............................ 23

Elytra only very narrowly rufous at the apical edge......................... 25

23—First ventral of the male with a large median area at apex which is glabrous and
    impunctate, but not impressed or defined by particularly dense pubescence ; fifth
    segment very broadly and just visibly sinuato-truncate at apex, the sinuous por-
    tion with a very feebly convex bevel ; prothorax but little more than twice as
    wide as long, sparsely punctulate, the sides continuous, moderately convergent
    and evenly arcuate ; elytra quite coarsely and closely, but not densely punctate.
    Length 1.85 mm.; width 1.35 mm.   Alabama............... **caudalis** *Lec.*

First ventral of the male virtually unmodified and punctured almost throughout, the
    fifth broadly and much more decidedly sinuate and beveled at apex ; black spot
    of the pronotum small, not extending much beyond the middle............. 24

24—Sides of the prothorax continuous, strongly convergent, broadly and evenly
    arcuate, the elytral punctures rather small but strong and quite sparse ; body
    smaller and more narrowly oval, highly polished, the pubescence rather coarse
    and sparse, easily denuded.   Length 1.8 mm.; width 1.25 mm.   Texas (Co-
    lumbus)....................................**medionotans,** sp. nov.

Sides of the prothorax not quite continuous with those of the elytra, very much less
    convergent, broadly and evenly arcuate ; elytral punctures coarser and more close-
    set but not dense ; body more broadly oval.   Length 1.9–2.0 mm.;   width
    1.35–1.6 mm.   Texas (Brownsville and Galveston)... **subtropicus,** sp. nov.

25—Black discal spot of the pronotum very small, not extending beyond the middle,
    the punctures very sparse ; species very small, oblong-oval, the sides of the pro-
    thorax continuous but only moderately convergent, feebly arcuate ; elytral punc-
    tures strong, rather coarse and sparse, the pubescence whitish and coarse.   Length
    1.5 mm.; width 1.0 mm.   Florida (Palm Beach)—Mr. F. Kinzel.

                                                        kinzeli, sp. nov.

Black discal spot large, extending to the apical margin or very nearly, the elytral punctures distinct but not coarse and rather close-set ; pubescence coarse......26

26—Form more elongate-oval, the prothorax very conspicuously punctured toward the middle, the sides not quite continuous with those of the elytra, only moderately convergent but more distinctly and evenly arcuate ; elytra fully as long as wide ; legs red.   Length 2.2 mm.; width 1.5 mm.   Indiana ; [*socer* Lec].

collaris Melsh.

Form short and very broadly oval, the prothorax rather sparsely punctate even toward the middle, the sides almost continuous with those of the elytra, strongly convergent but only feebly, evenly arcuate ; elytra not quite as long as wide ; legs red. Length 1.9–2.2 mm.; width 1.5–1.65 mm. Arizona (Pinal Mts.)—Mr. Wickham, (Grand Cañon of the Colorado)—Dr. T. Mitchell Prudden, (near the southern boundary)—Mr. Morrison........................ **horni** *Gorh.*

Form short and broadly oval, the size smaller ; prothorax smaller and more transverse, the sides not quite continuous with those of the elytra, only moderately convergent, evenly and moderately arcuate ; punctures sparse throughout, very small, feeble and inconspicuous in the middle and gradually almost wholly obsolete toward the sides ; elytra distinctly shorter than wide ; legs red, the hind femora infuscate toward base.   Length 1.8 mm.; width 1.4 mm.   New Mexico (Albuquerque)—Mr. Cockerell........................**cockerelli,** sp. nov.

27—Rather narrowly oval and moderately convex, black, the prosternum, tip of abdomen, legs throughout, head and pronotum pale testaceous, the latter with the sides almost continuous, strongly convergent, evenly and feebly arcuate, the disk with a small parabolic basal spot occupying median third of the base and extending to the middle and varying but slightly in excess ; elytra with the extreme apical edge paler.   Length 1.8–2.0 mm.; width 1.1–1.3 mm.   Utah (southwestern)—Mr. Weidt............................... **uteanus,** sp. nov.

Somewhat narrowly oval, larger than *uteanus* but almost similar throughout in coloration and sculpture, the prothorax equally short and transverse but with the sides less convergent, feebly, evenly arcuate and not quite continuous ; elytra more elongate, the metacoxal plate more broadly rounded ; legs darker rufous throughout.   Length 2.0 mm.; width 1.3 mm.   Indiana........... **rhesus,** sp. nov.

28—Species of the Atlantic regions...29

Species of the Sonoran and Pacific regions..... 32

29—Elytra with a rather broad and well-defined red apex extending to apical seventh or eighth, its anterior margin transverse, tending to slight prolongation along the lateral edges, black, the abdomen red, blackish toward base ; head red, blackish basally ; pronotum broadly and rather abruptly at the sides and legs throughout, testaceous ; prothorax two and two-fifths times as wide as long, the sides very discontinuous with those of the elytra, only feebly convergent, arcuate at apex, becoming straight posteriorly, the disk finely but strongly punctate, the punctures sparse and inconspicuous at the middle, becoming coarser and unusually close-set toward the sides ; elytra rather strongly, somewhat coarsely but not densely punctured.   Length 2.6 mm.; width 1.8 mm.   Indiana ; [*chatchas* Muls.].

fastigiatus Muls.

Elytra black, with the extreme apical margin or beaded edge alone paler.........30

30—Pronotum black or blackish, broadly but gradually and indefinitely paler toward
the sides; head and legs uniform in color throughout but testaceous to blackish;
tip of abdomen narrowly red; prothorax moderately transverse, the sides strongly
convergent, almost evenly and moderately arcuate throughout and almost perfectly continuous with those of the elytra, the punctures fine and rather sparse,
closer and quite conspicuous toward the sides; elytra quite coarsely but evenly
and rather sparsely punctured, the pubescence moderately coarse and conspicuous.
Length 2.2-2.4 mm.; width 1.6-1.8 mm. Pennsylvania (near Philadelphia);
[*puncticollis* Horn nec Lec.]..........................**indutus,** sp. nov.
Pronotum black, with the apical angles alone pale, the pale color but very seldom extending to the basal angles; head red; sides of the prothorax distinctly discontinuous, more or less feebly convergent and evenly and moderately arcuate
thoughout; body broadly oval.....................................31
31—Larger species; legs red, the femora all more or less blackish toward base; pronotal punctures very fine, close toward the sides but not conspicuous; elytral
punctures not coarse but strong and quite sparse, the pubescence rather fine.
Length 2.0 mm.; width 1.55 mm. Rhode Island (Boston Neck).
agricola, sp. nov.
Small species; legs red throughout; pronotum shorter and more transverse, very
minutely, sparsely and scarcely visibly punctate, the punctures still sparse and
scarcely larger toward the sides; elytra barely as long as wide, polished, rather
finely but strongly and still more sparsely punctate, the pubescence sparser and
coarser. Length 1.5 mm.; width 1.1 mm. North Carolina (Asheville).
innocens, sp. nov.
32—Legs black throughout, the tarsi alone sometimes pale....................33
Legs red, the hind femora black throughout or with the tip alone paler..........34
Legs red, the hind femora slightly blackish at the base; body broadly oval......35
Legs uniform and clear red throughout............. 36
33—Sides of the pronotum obliquely and abruptly pale, the pale area scarcely extending to the basal angles; body large, convex and very broadly oval, shining,
black; head pale in the male, black in the female; tarsi not evidently paler;
prothorax relatively small, two and three-fifths times as wide as long, the
sides strongly discontinuous, very moderately convergent, evenly and rather
strongly arcuate; punctures fine, sparse in the middle, broadly close-set toward
the sides; elytra not quite as long as wide, not very coarsely but strongly, evenly
and not very closely punctured, the apical reflexed edge alone pale; male with
the first ventral obliquely bituberose at the middle near the apex, the fifth very
broadly and feebly sinuate. Length 2.4 mm.; width 1.7 mm. California
(Truckee)...**solidus,** sp. nov.
Pronotum almost black throughout, the apical angles alone feebly and gradually picescent; body smaller and much more narrowly oval; legs black, the tarsi red;
sides of the prothorax evidently discontinuous, rather strongly convergent, evenly
and somewhat feebly arcuate, the punctures very minute, sparse, becoming very
close toward the sides; elytra distinctly longer than wide, the apical edge
scarcely at all paler, the punctures not very coarse but strong and unusually
dense. Length 1.9 mm.; width 1.3 mm. Nevada (Reno).
desertorum, sp. nov.

34—Prosternal carinæ widely separated at base, straight and strongly convergent to
apical third, thence parallel and well separated to the apical margin; body
broadly oval, shining, black, the pronotum gradually pale testaceous toward the
apical angles, short, the sides almost perfectly continuous, strongly convergent,
evenly and moderately arcuate; punctures minute and inconspicuous, slightly
closer toward the sides; elytra scarcely as long as wide, the apical margin very
finely testaceous; punctures fine but strong, not very close-set. Length 1.75–
1.9 mm.; width 1 3–1.4 mm. Arizona (Yuma)......**apacheanus,** sp. nov.
Prosternal carinæ straight and feebly convergent throughout, becoming almost obliterated in basal half; body narrowly oval, the pronotum gradually testaceous toward
the apical angles, less transverse and relatively larger than in *apacheanus,* the
sides of the body being less arcuate; sides almost continuous, rather strongly
convergent and evenly, somewhat feebly arcuate; punctures minute and inconspicuous; elytra nearly a fourth longer than wide, pale at the apical margin,
quite coarsely and conspicuously, but not very closely punctured, the pubescence
coarse. Length 1.9 mm.; width 1.25 mm. Colorado....**monticola,** sp. nov.
Prosternal carinæ very strong, gradually convergent and feebly arcuate throughout,
moderately separated at the apical margin; body broadly oval, the pronotum
gradually testaceous toward the apical angles, only moderately transverse, the
sides evidently discontinuous, moderately convergent, evenly, moderately arcuate,
the punctures equal in size throughout, rather fine and sparse, but little closer toward
the sides; elytra slightly longer than wide, evenly rounded in semicircle behind;
punctures not very coarse but strong and somewhat close-set; pubescence coarse,
pale, somewhat abundant and conspicuous; legs pale rufo-testaceous, the middle
femora at base and the posterior to far beyond the middle black   Length 2.2
mm.; width 1.55 mm. Utah (southwestern)—Mr. Weidt....**aridus,** sp. nov.
35—Pronotum almost entirely black, becoming testaceous only at the extreme apical
angles, the surface almost completely impunctate, the base broadly angulate, the
sides almost continuous, strongly convergent and feebly, evenly arcuate; elytra
scarcely as long as wide, the sides feebly arcuate, the apex very broadly obtuse,
with the reflexed bead pale, the punctures sparse, very fine toward the suture,
fine but much stronger and more close-set externally. Length 1.8–1.95 mm.;
width 1.3–1.45 mm. California (Monterey and Sonoma Cos.).
                                    luctuosus, sp. nov.
Pronotum black, not very abruptly and obliquely pale at the sides, broadly in front,
very narrowly at the basal angles, the base almost transverse, lobed in the middle,
the sides strongly discontinuous, moderately convergent, evenly and distinctly
arcuate, the punctures fine but distinct, sparse, becoming close-set at the sides;
elytra scarcely as long as wide, evenly oval, with the apical bead alone pale, the
punctures somewhat coarse, deep, even and sparse throughout.   Length 1.9–2.2
mm.; width 1.4–1.7 mm. California (Siskiyou, Humboldt and Sta. Cruz Cos.).
                                    humboldti, sp. nov.
Pronotum black, abruptly and moderately broadly pale at the sides in a parallel area
almost equally wide at apex and base, the basal margin feebly bioblique, lobed
at the middle, the sides continuous but with a slight reëntering angle, strongly
convergent, evenly and distinctly arcuate; punctures minute, sparse and inconspicuous; elytra short, very obtusely rounded, somewhat alutaceous, the punc-

tures fine, feeble, moderately close-set, even and slightly asperulate, the pubescence rather short and closely laid ; male with a small glabrous subdepressed and
narrowly triangular area at the apex of the first ventral, surrounded by denser
vestiture. Length 1.9–2.1 mm.; width 1.45–1.7 mm. California (Sonoma
Co.)...**sonomæ**, sp. nov.
36—Pronotum broadly but obliquely red at the sides, the pale area very narrow at
    the basal angles; form broadly oval.. 37
Pronotum feebly and almost invisibly piceous to pale testaceous at the extreme apical
    angles only ; elytra with the apical reflexed bead paler, slightly wider in *jacinto*..40
37—Elytra with a narrow but distinct band of testaceous at the apical margin.....38
Elytra with the mere apical reflexed bead red, the paler tint scarcely extending further 39
38—Male with the fifth ventral segment broadly truncate toward the middle, the surface
    only feebly convex-beveled for a short distance at the middle ; first segment un-
    modified and punctured throughout ; prosternal carinæ widely separated at the
    apical margin ; prothorax rather small, short and transverse, very finely though
    distinctly, almost evenly punctured, the sides not quite continuous, rather feebly
    convergent and evenly, moderately arcuate ; elytra finely though distinctly,
    moderately closely punctured, polished and smooth. Length 1.6–1.8 mm.;
    width 1.2–1.3. Arizona (Benson and the Gila Valley)—One specimen, from
    San Diego, is much smaller and has the fifth ventral shorter and more broadly
    rounded in the female but does not otherwise differ............**gilæ**, sp. nov.
Male with the fifth ventral broadly, feebly sinuate, the surface strongly beveled in the
    middle, the first segment with an elongate impunctate area at the middle of the
    apex ; prosternal carinæ narrowly separated at the apical margin ; body similar
    to *gilæ* in form and sculpture, the sides of the prothorax more nearly continuous
    with those of the elytra and more convergent, and the base more oblique at each
    side. Length 1.8 mm.; width 1.3 mm. Utah (southwestern)—Mr. Weidt.
                                              decipiens, sp. nov.
39—Form less broadly oval, the prothorax relatively smaller, with the sides evidently
    discontinuous, moderately convergent, evenly and rather feebly arcuate, finely
    punctured, rather closely toward the sides ; elytra distinctly longer than wide,
    quite coarsely but not very closely punctured, the pubescence coarse, ashy and
    conspicuous. Length 2.25 mm.; width 1.6 mm. Colorado (Garland)—Mr.
    Schwarz...**garlandicus**, sp. nov.
Form very broadly oval, the prothorax relatively larger, the sides almost continuous
    with those of the elytra, strongly convergent, evenly but feebly arcuate, the
    punctures fine but strong, sparse, becoming notably close-set and distinct broadly
    toward the sides ; elytra not longer than wide, rather coarsely and strongly but
    not very closely punctate, the pubescence rather short, fine, more decumbent and
    not very conspicuous. Length 2.2 mm.; width 1.7 mm. California (Mokelumne
    Hill, Calaveras Co.)—Dr. Blaisdell...................**blaisdelli**, sp. nov.
40—Form rather narrowly oval, the elytra opaque and finely rugulose, finely, closely
    and asperulately punctate, the pubescence rather short and decumbent ; prothorax
    strongly transverse, as wide at base as the base of the elytra but with a feeble re-
    entering angle, smooth, polished, extremely minutely and sparsely punctulate, the
    sides rather strongly convergent, evenly and distinctly arcuate. Length 1.6 mm.;
    width 1.15 mm. California (Sonoma Co.)...............**advena**, sp. nov.

Form broadly oval, the elytra smooth and polished......................................41
41—Prothorax short, about two and one-half times as wide as long ; head wholly or
    partly red...42
Prothorax about twice as wide as the median length, the base strongly oblique at each
    side, the sides evidently discontinuous, only moderately convergent, evenly and
    feebly arcuate, the punctures minute and sparse ; elytra strongly and closely
    punctate.   Length 2.0 mm.; width 1.5 mm.   California (Sonoma Co.).

                                                              extricatus, sp. nov.
42—Sides of the prothorax evidently discontinuous, feebly convergent, evenly and
    feebly arcuate, the punctures strong and close-set in the middle, becoming finer
    and sparser toward the sides ; elytra evenly, finely but strongly, moderately
    closely punctured, the pubescence fine, infuscate and only moderately con-
    spicuous   Length 1.6–1.9 mm.; width 1.15–1.4 mm.   California (Monterey to
    Sonoma)..**ardelio** *Horn*
Sides of the prothorax nearly continuous, strongly convergent, evenly and distinctly
    arcuate, the punctures nearly as in *ardelio* but sparser throughout ; elytra finely
    but strongly, sparsely punctured, the pubescence rather coarse and distinct ; male
    with a feebly impressed, elongate-oval area at the middle of the apex of the first
    ventral, the fifth broadly sinuato-truncate and impressed, the characters nearly as
    in *extricatus* throughout.   Length 1.75–1.8 mm.; width 1.3–1.4 mm.   Cali-
    fornia (San Diego)...**jacobianus,** sp. nov.
Sides of the prothorax strongly discontinuous, very feebly convergent, evenly and
    feebly arcuate, the surface punctured nearly as in *jacobianus ;* elytra notably
    wider than the prothorax, rounded, finely but strongly, rather sparsely punctate,
    the apical margin red for a distance equal to about a fifth the length of the pro-
    thorax ; male with a very small, wholly unimpressed and feebly defined glabrous
    area at the middle of the apex of the first ventral, the fifth broadly sinuato-truncate
    and impressed ; pubescence of the upper surface coarse and conspicuous.   Length
    1.6 mm.; width 1.2 mm.   California (San Diego).........**jacinto,** sp. nov.
43—Species of the Atlantic regions...44
Species of the Pacific and Sonoran regions, *lacustris* and *abbreviatus* extending to the
    eastward as far as Lake Superior....................................... 45
44—Broadly oval, strongly convex, shining, black throughout, the legs uniformly
    colored but varying from pale testaceous to blackish ; pubescence rather coarse ;
    prothorax relatively rather small, finely but strongly, sparsely punctured, very
    closely near the sides, the sides discontinuous, strongly convergent, evenly and
    strongly arcuate ; elytra quite coarsely, strongly and sparsely punctured.   Length
    1.6–2.3 mm.; width 1.15–1.7 mm.   Atlantic States (from Massachusetts to North
    Carolina and Alabama)...................................**tenebrosus** *Muls.*
Narrowly oval, shining, black, the legs bright red ; prothorax relatively larger, the
    punctures extremely minute, sparse and subobsolete, becoming quite large but
    only moderately close-set near the sides, the latter almost continuous with those
    of the elytra, strongly convergent and rather feebly, evenly arcuate ; elytra dis-
    tinctly longer than wide, the punctures quite coarse, strong and somewhat sparse,
    the pubescence coarse and conspicuous.   Length 1.9 mm.; width 1.3 mm.   In-
    diana..**compar,** sp. nov.

45—Narrowly oval, small, black throughout, the apical angles of the prothorax perhaps
    becoming paler in some examples ; legs pale testaceous throughout ; prothorax
    small, much narrower than the elytra, the sides very discontinuous, only moder-
    ately convergent and straight, becoming feebly arcuate at the apex ; punctures
    sparse and scarcely visible throughout, really larger toward the middle but ex-
    cessively feeble and shallow and variolate as usual ; elytra somewhat strongly nar-
    rowed behind and evenly rounded from near the humeri, the apex rather nar-
    rowly rounded ; punctures fine, only moderately close, the pubescence rather
    short but coarse, ashy and distinct. Length 1.5 mm.; width 1.0 mm. Arizona—
    Mr. Wickham...**infans,** sp. nov.
Broadly oval and much larger, strongly convex, smooth and shining.............46
46—Legs black throughout, the tarsi pale ; body oval, convex, the sides of the pro-
    thorax almost continuous, strongly convergent, evenly and distinctly arcuate, the
    punctures quite coarse, not very close-set, as large as those of the elytra or larger,
    becoming gradually very fine, sparse and obsolescent toward the sides ; elytra a
    little longer than wide, moderately obtuse behind, not very coarsely but strongly,
    moderately sparsely punctured, more minutely toward the suture, the pubescence
    rather long and coarse ; under surface deep black throughout. Length 2.1 mm.;
    width 1.5 mm. Utah (southwestern)—Mr. Weidt...........**weidti,** sp. nov.
Legs red, the hind femora black, testaceous toward apex.......... 47
Legs red, the hind femora black at the extreme base...........................55
Legs bright and uniform rufo-testaceous throughout.........................56
47—Hind femora testaceous only well beyond the middle......................48
Hind femora becoming testaceous in about apical half......................53
48—Tip of the elytra pale testaceous in a border which is about a fifth or sixth as wide
    as the length of the prothorax, the latter relatively small, short and strongly trans-
    verse, the sides evidently discontinuous, rather feebly convergent, evenly and
    somewhat strongly arcuate, the punctures minute and inconspicuous ; elytra
    scarcely longer than wide, rather narrowly rounded behind, finely but strongly,
    evenly and not very closely punctured, the pubescence ashy-white, rather short
    and somewhat abundant ; tip of abdomen pale, the hind femora very gradually
    pale apically. Length 1.9 mm. ; width 1.35 mm. Nevada (Reno).
                                                     renoicus, sp. nov.
Tip of the elytra only paler along the fine reflexed marginal bead ; hind femora pale
    at apex only...49
49—Head pale toward the clypeal margin in both sexes but more broadly in the
    male..50
Head deep black throughout to the margin of the clypeus, at least in the female...51
50—Sides of the prothorax nearly continuous, strongly convergent, evenly and strongly
    arcuate, the punctures slightly closer and more evident toward the sides, fine but
    distinct throughout ; elytra rather coarsely, strongly, evenly and sparsely punctured;
    abdomen not pale at apex, the fifth segment of the male broadly sinuato-truncate,
    the surface deeply impressed in a transverse, posteriorly arcuate and well-defined
    concave bevel, the first with an elongate triangular glabrous area at the middle,
    defined by fine dense punctures. Length 2.2 mm. ; width 1.6 mm. Lake
    Superior ; [var. ? *nigrivestis* Muls. New Orleans, La.].......**lacustris** *Lec.*

Sides of the prothorax not quite continuous, less convergent, evenly and feebly arcuate ; punctuation similar ; elytra shorter, not quite as long as wide, very obtuse
at apex, the punctures even, rather fine but strong and quite sparse ; pubescence
coarse, yellowish-cinereous in color and conspicuous ; male characters nearly as
in *lacustris*, the fifth segment more feebly impresso-beveled and the glabrous area
of the first less defined.   Length 2.0–2.15 mm. ; width 1.5–1.65 mm.   California (Lake Tahoe)..................................**tahoensis**, sp. nov.
51—Pronotum impunctate at any part, the sides continuous with those of the elytra,
strongly convergent, evenly and rather strongly arcuate ; elytra distinctly longer
than wide, rather strongly but not closely punctate, the vestiture somewhat
whitish, coarse, not very abundant but rather conspicuous.   Length 2.15 mm. ;
width 1.5 mm.   Utah (southwestern)—Mr. Weidt .....**subsimilis**, sp. nov.
Pronotum distinctly but finely punctate, the punctures somewhat larger and more or
less close-set toward the sides, the latter not quite continuous with those of the
elytra, less strongly convergent, subevenly and moderately arcuate.........52
52—Elytra scarcely as long as wide, strongly, evenly but unusual y sparsely punctured, the pubescence long, coarse, not dense but very conspicuous, yellowish-
white in color.   Length 2.1 mm. ; width 1.6 mm.   Utah (southwestern)—Mr.
Weidt..........................   ....**mormon**, sp. nov.
Elytra fully as long as wide, somewhat less obtuse behind, rather strongly, evenly
punctate, the punctures moderately close-set, the pubescence shorter, finer, darker
in color, more decumbent and rather less conspicuous though more abundant.
Length 2.1–2.2 mm. ; width 1.6 mm.   California (Mokelumne Hill, Calaveras
Co.)—Dr. Blaisdell ; (Dunsmuir, Siskiyou Co.)—Mr. Wickham.
                                                 calaveras, sp. nov.
53—Head pale at the clypeal margin, more broadly in the male................54
Head black throughout ; body large, very broadly oval, the prothorax much narrower
than the elytra, the sides strongly discontinuous, feebly convergent, evenly and
strongly arcuate, the punctures fine and not conspicuous ; elytra about as long as
wide, rather coarsely, strongly and closely punctured, the pubescence fine,
short, decumbent, rather abundant but dark in color and inconspicuous ; under
surface densely punctate.   Length 2.5 mm. ; width 1.8 mm.   California (Siskiyou Co.)..**saginatus**, sp. nov.
54—Large, strongly convex, polished, the prothorax moderate in size, much narrower
than the elytra, the sides strongly discontinuous, moderately convergent, evenly and
moderately arcuate, the punctures fine and sparse, closer toward the sides ; elytra
as long as wide, not very coarsely but deeply, evenly and rather sparsely punctured, the pubescence moderately long and coarse but dark in color and rather
sparse ; male with a concave median glabrous area, defined by fine dense piliferous punctures, at the apex of the first ventral, the fifth broadly sinuato-truncate
and medially impressed as usual.   Length 2 4 mm. ; width 1.8 mm.   California
(probably near San Francisco)......................**strenuus**, sp. nov.
Smaller, equally convex and polished, less broadly oval, the prothorax shorter and more
transverse, the sides strongly discontinuous, rather feebly convergent, evenly and
somewhat strongly arcuate, the punctures fine, rather sparse, even, more close-set
toward the sides ; elytra a little longer than wide, evenly, almost semicircularly
rounded behind, not very coarsely but deeply, evenly and rather sparsely punc-

tured ; male with a small feebly concave glabrous area on the first segment bordered by finer denser piliferous punctures, the fifth sinuato-truncate and impressed as usual. Length 1.7–1.9 mm.; width 1.3–1.45 mm.   California (Humboldt and Siskiyou Cos.). . . . . **mendocino**, sp. nov.

55—Prothorax large, nearly as wide as the elytra, about two and one-half times as wide as long, the sides slightly discontinuous, feebly convergent, evenly and moderately arcuate, the punctures fine, sparse, but slightly larger and less sparse toward the sides ; elytra about as long as wide, finely, rather feebly and sparsely punctured, the pubescence moderately long and coarse, sparse and slightly dark in color ; head of the male red in apical third ; middle and hind femora black at base.   Length 2.0–2.25 mm.; width 1.5–1.7 mm.   California (Sonoma Co.).
                                                                                stygicus, sp. nov.

Prothorax relatively smaller, much narrower than the elytra, shorter and more transverse, the sides strongly convergent, evenly and strongly arcuate and very markedly discontinuous with those of the elytra, the punctures nearly similar ; elytra barely as long as wide, more coarsely, quite strongly, very evenly and not so sparsely punctured, the pubescence very fine, even, decumbent, dark in color and inconspicuous ; head black, the extreme apical margin of the clypeus pale in the female, probably more in the male ; hind femora black at base, the trochanter pale.   Length 2.1 mm.; width 1.6 mm.   California (Siskiyou Co.).
                                                                                tenuivestis, sp. nov.

56—Rather broadly oval, smooth, black, the clypeal apex, tip of abdomen feebly, and legs throughout, pale testaceous ; prothorax relatively rather small, much narrower than the elytra but only moderately transverse, the sides discontinuous, moderately convergent, evenly and feebly arcuate, the punctures rather large and close-set in the middle, shallow, variolate and feebly umbilicate, becoming fine and sparser toward the sides ; elytra evenly, finely but strongly, moderately closely punctured, the pubescence rather coarse but unusually short, subdecumbent, ashy and distinct.   Length 1.8 mm.; width 1.4 mm.   Arizona (Grand Cañon of the Colorado)—Dr. Prudden. . . . . **papago**, sp. nov.

57—Prosternal carinæ entire as usual ; elytral pale areas more or less nubilate and variable in extent. 58
Prosternal carinæ not quite attaining the apical margin 62
58—Body broadly oval. 59
Body narrowly oblong-oval 61
59—Body depressed, the pale areas of the elytra predominating 60
Body normally convex, testaceous, the pronotum with a large parabolic black spot nearly attaining the apex, the elytra black, each with a short, narrow, longitudinal and slightly oblique median vitta at the middle of the disk, which sometimes almost disappears, the under surface black ; legs dusky, the femora black except at tip ; prothorax closely and strongly punctulate, the sides slightly discontinuous, rather strongly convergent and evenly arcuate ; elytra fully as long as wide, finely but strongly, evenly and very closely punctate.   Length 1.8–2.0 mm.; width 1.25–1.45 mm.   Arizona (Yuma). . . . . **nubes**, sp. nov.

60—Elytral punctures moderately large and not very close-set ; upper surface testaceous, the pronotum with a broad parabolic black spot not attaining the apex, the elytra with a large triangular black common spot extending nearly to the sides at the base and having its apex on the suture at apical seventh or eighth, sometimes enveloping the entire base and extending posteriorly along the sides behind the middle ; under surface black, the legs dusky-testaceous, with the femora darker.   Length 1.8–2.0 mm.; width 1.3–1.4 mm.   Louisiana, Texas and Arizona (Tuçson)...**cinctus** *Lec.*

Elytral punctures fine, strong and very close-set ; upper surface testaceous, the pronotum with a broad parabolic and rather poorly defined black spot, extending to about apical fourth ; sides slightly discontinuous, the punctures minute and inconspicuous, very close broadly toward the sides ; elytra with a black sutural vitta, sinuously expanding toward base, where it does not extend laterally much beyond the middle of each, finely attenuate posteriorly and not attaining the apex ; under surface and legs as in *cinctus*.   Length 1.8–2.0 mm.; width 1.2–1.4 mm. California (Sta. Barbara); [*suturalis* ǁ Lec.]....................**lecontei** *Cr.*

61—Head testaceous, the pronotum black, with the apical margin narrowly, and apical angles more broadly, indefinitely pale, short and transverse, the sides strongly discontinuous, feebly convergent, evenly and moderately arcuate, the punctures fine but strong and very close-set throughout ; elytra dark rufo-testaceous, with sutural black vitta gradually expanding to the base and a nubilate lateral area not attaining base or apex, varying thence to entirely black, with a narrow oblique red discal streak on each closely approaching the suture posteriorly ; punctures fine, strong, even and extremely close-set ; pubescence rather long, coarse and conspicuous ; under surface black, the legs slender, testaceous, the femora black with the extreme tip red ; male with the fifth ventral less feebly sinuate at apex than in *lecontei*.   Length 1.8–1.9 mm.; width 1.1–1.25 mm.   California (Sonoma Co.) .. ..**sarpedon**, sp. nov.

62—Body broadly oval, each elytron with a distinctly defined discal red spot just before the middle, the spot obliquely subrhomboidal in form ; pronotum pale at the apical angles, the sides continuous and strongly convergent, arcuate ; elytra sparsely punctate, the pubescence coarse and distinct.   Length 2.5 mm. ; width 1.7 mm.   California..................................... ... .**paciiicus** *Cr.*

Body broadly oval, each elytron with a more elongate oblique red spot before the middle of the disk, the spot nearly attaining the suture ; pronotum entirely black. Length [2.5 mm].   New Mexico..........................**strabus** *Horn*

Body narrowly oblong-oval and much smaller, the elytra pale testaceous, with the suture narrowly blackish, the dark tint extending nubilously along the basal margin to the sides and sometimes prolonged backward along the latter for some distance, the punctures not very close ; prothorax much smaller, distinctly narrower than the elytra, the sides strongly discontinuous, feebly convergent and feebly arcuate, black, gradually paler toward the apical angles ; under surface and legs black, the ventral plate distant from the segmental apex by half of its own length.   Length 1.6–1.9 mm. ; width 0.8–1.1 mm.   California (Lake Tahoe, Truckee and Monterey) ....**coniferarum** *Cr.*

63—Very narrow and elongate-oval, polished, black, each elytron with a large triangular red spot at the centre of the disk ; under surface and legs black, the trochanters and tarsi paler ; prothorax unusually feebly transverse, scarcely twice as wide as long, the sides obviously discontinuous, feebly convergent and nearly straight, becoming feebly arcuate at apex ; punctures remote and almost obsolete ; elytra fully a third longer than wide, rather narrowly· obtuse behind, the punctures sparse and rather strong ; pubescence coarse ; male with the fifth ventral evenly sinuate at tip, the surface narrowly beveled along the sinus, the first gradually glabrous toward the middle.   Length 1.5 mm. ; width o.8 mm.   Pennsylvania.

punctatus *Say*

Much more broadly oval, the body smaller, less polished, black throughout, the pronotum feebly picescent at the apical angles, not more than three-fifths as wide as the elytra, scarcely more than twice as wide as long, the sides quite strongly convergent, very discontinuous and almost straight, the punctures very minute and inconspicuous ; elytra but little longer than wide, obtusely rounded at apex, very finely, rather feebly, evenly but not closely punctured, the pubescence rather short and fine, not very conspicuous ; legs rufo-piceous ; ventral plates approaching the hind margin of the segment by a third or fourth of their own length but rather narrow and strongly rounded.   Length 1.25–1.35 mm. ; width 0.8–0.95 mm.   Nevada (Reno).. ....**occiduus**, sp. nov.

Evenly and not very broadly oval, black, the frontal margin, mouth, apical angles of the pronotum and legs throughout pale ; marginal bead of the elytra at apex also testaceous ; prothorax nearly as wide as the elytra, finely, not densely and evenly punctate, the sides almost perfectly continuous, strongly convergent and feebly arcuate ; elytra distinctly longer than wide, rather finely but strongly, evenly and not closely punctate, the pubescence moderately long, cinereous and distinct ; metacoxal plate approaching extremely close to the suture, broadly rounded ; male with the fifth ventral broadly trapezoidal and sinuato-truncate, the edge narrowly beaded and the surface just anteriorly more convex.   Length 1.5 mm. ; width 0.95 mm.   New Mexico ....**nanus** *Lec.*

64—Upper surface black, each elytron with a large oval red spot on the median line of the disk just before the middle ; form very broadly oval, the head and pronotum black throughout, the latter finely, strongly and closely punctate ; elytra rather coarsely, evenly and moderately closely punctate, not pale at apex, the pubescence coarse ; under surface black throughout, the legs fusco-testaceous, the femora black.   Length 2.5 mm. ; width 1.85 mm.   Texas ; [Tennessee and Louisiana—Horn.].............................**circumspectus** *Horn*

Upper surface black, the prothorax bicolored ; elytra without discal pale spot.....65
Upper surface pale flavo-testaceous to piceous-black ; sides of the prothorax discontinuous 69

65—Elytra distinctly and somewhat broadly margined with red at the apical margin, oval, polished, the head and prothorax generally pale, the latter broadly black toward the middle and base, sometimes black throughout, the sides not quite continuous, rather strongly convergent and evenly, moderately arcuate, the punctures generally distinct but not very dense ; elytra coarsely and sparsely punctate, the pubescence coarse, rather long, ashy and conspicuous ; legs pale to blackish in color.   Length 1.8–2.7 mm. ; width 1.3-1.9 mm.   New York, Delaware, Mississippi, Texas, Indiana, Illinois and Iowa.............**americanus** *Muls.*

Elytra not red at tip, or only extremely finely so, the punctures very much smaller and more close-set, the body more narrowly oval ; vestiture rather short and decumbent, cinereous...66

66—Pronotum black, sometimes with the sides paler, minutely, sparsely and inconspicuously punctulate, the sides not quite continuous, moderately convergent,
evenly and rather strongly arcuate; elytra black, finely but very clearly, evenly
and not very closely punctate; legs red. Length 1.9 mm.; width 1.4 mm.
Oregon—Mr. Wickham......... ....**caurinus** *Horn*
Pronotum pale testaceous, with a parabolic and frequently somewhat ill-defined median
black spot at the base, extending almost to the apical margin, the sides subcontinuous with those of the elytra, convergent and rather strongly arcuate, the surface minutely reticulate and distinctly alutaceous, the punctures extremely small,
sparse and inconspicuous; elytral punctures very small, rather feeble; legs red
throughout 67
67—Abdominal lines only partially interrupted externally, approaching very close to
the hind margin of the segment, as apparently in the two following also; male
with the fifth ventral segment broadly sinuato-truncate, feebly impressed and very
inconspicuously pubescent along the sinuous portion. Length 1.7 mm.; width
1.15 mm. Nevada (Reno)............. ....**innocuus**, sp. nov.
Abdominal lines distinctly interrupted externally, as usual in the present group...68
68—Male with the fifth ventral segment very feebly sinuate at apex but conspicuously
clothed with coarse, dense, erect and subflavous pubescence. Length 1.9-2.1
mm.; width 1.3-1.55 mm. Indiana—Cab. Levette.......**rusticus**, sp. nov.
Male with the fifth ventral short and broadly truncate but scarcely at all sinuate, the
edge with a short and steep bevel and clothed with fine inconspicuous pubescence.
Length 1.8-2.0 mm.; width 1.25-1.5 mm. California (Sonoma Co.).
aluticollis, sp. nov.
69—Elytral punctures rather coarse and sparse though only moderately deep....70
Elytral punctures fine and close-set; body smaller and more narrowly oval uniform
piceous-black above or paler, with the pronotum still paler toward the apical
angles; prothorax relatively small, moderately transverse, much narrower than
the elytra, the sides moderately convergent and feebly arcuate, the punctuation
close but very fine; elytra longer than wide, not very obtuse at tip, feebly blackish toward the suture and side-margin in some of the paler forms, the pubescence
short, abundant and rather coarse. Length 1.75-1.9 mm.; width 1.0-1.2 mm.
California (Monterey to Humboldt Co.)............ ....**difficilis**, sp. nov.
70—Upper surface pale rufo-flavate, polished, immaculate, the pubescence rather short,
sparse, moderately coarse; prothorax much narrower than the elytra, minutely,
not very closely punctate, the sides only moderately convergent and more or less
feebly arcuate; elytra about as long as wide. Length 1.7-2.5 mm.; width
1.0-1.75 mm. British Columbia to northern California......**phelpsi** *Cr.*
Upper surface pale luteo-flavate, the elytra with small irregular blotches or dashes of
black, the pronotum frequently blackish except at the sides, strongly transverse;
elytral punctures binary, as in *phelpsi*, the larger sometimes tending to linear arrangement toward the suture and base; post-mesocoxal line generally entire but
sometimes more or less abbreviated, in one specimen only extending two-thirds the
distance to the episternal suture. Length 1.8-2.25 mm.; width 1.15-1.6 mm.
California (Humboldt to Los Angeles)............... .....**nebulosus** *Lec.*
71—Elytra black, each with a single sharply defined rounded discal pale spot....72

Elytra black, each with two sharply defined oval spots, or a design formed by an
amalgamation of such spots..73
Elytra black or piceous, with irregular paler design or maculation...............75
Elytra pale, or sometimes pale with the suture or margins dusky...... 77
72—Prothorax entirely testaceous, each elytron with a very large circular red spot just
behind the middle, the apex not paler ; pubescence rather coarse, cinereous
and conspicuous, the punctures very fine and not very dense ; legs flavo-testaceous,
Length 1.25 mm.; width 0.85 mm. Florida (Dry Tortugas)..**bivulnerus** *Horn*
Prothorax entirely black, the head red or black ; legs testaceous, the femora black,
especially the posterior ; elytra each with a smaller spot near apical third ; body
moderately large and stout, the sides of the prothorax nearly continuous with
those of the elytra ; elytral punctures rather small, the pubescence coarse, rather
abundant and conspicuous.   Length 1.9 mm ; width 1.3.   Pennsylvania.
                                                flavifrons *Melsh.*
    Var. A—Much smaller and generally somewhat more narrowly oval, the
    elytral punctures relatively rather larger, the pubescence not quite so con-
    spicuous.   Length 1.4–1.6 mm.; width 0.95–1.1 mm.   Pennsylvania, New
    Jersey, Delaware and Georgia.......................bioculatus *Muls.*
73—Spots of the elytra narrowly but clearly separated, oval.   Length 2.0 mm.;
width 1.2 mm.   Lake Superior ....**ornatus** *Lec.*
Spots of the elytra broadly coalescent, forming an elongate, bilaterally sinuate discal
maculation...74
74—Larger species and more broadly oval, the abdomen strongly and rather closely
punctured at the sides of the first segment, the epipleuræ scarcely attaining the
middle of the side-margin of the second segment, the arrangement of the punctures at the sides of the first segment indicating derivation from a form having
complete ventral plates, with the bounding line bending abruptly to the front
very near them argin ; prothorax black throughout, minutely and rather closely
punctate, the sides not quite continuous with those of the elytra, strongly convergent, evenly and strongly arcuate ; elytra much longer than wide, rather
strongly rounded at apex, finely but deeply, moderately closely and somewhat irregularly punctate ; legs red, the femora blackish.   Length 2.15 mm.; width
1.35 mm.   Massachusetts ....**sanguinifer,** sp. nov.
Small and narrowly oval but similar to the preceding in form, the abdomen finely and
sparsely punctate over the post-coxal areas, the lines curved forward at their extreme limit but not much prolonged, the epipleuræ attaining the apex of the
second segment, black, the elytral spot less defined than in *sanguinifer* ; the
punctures rather sparser and the apex more narrowly rounded.   Length 1.65
mm.; width 0.8–0.9 mm.   Colorado (Rocky Mts.).....**naviculatus,** sp. nov.
75—Black throughout, broadly oval, the legs piceous, each elytron with two transverse discal spots which are almost, or completely, divided each into two very
small pale spots, the outer of which are the more linear and oblique ; punctures
fine and very close-set, the pubescence rather coarse, cinereous and conspicuous
but easily denuded.   Length 1.8 mm.; width 1.2 mm.   California.
                                                guttulatus *Lec.*
Piceous-black, narrowly oval, the legs dark testaceous throughout, each elytron with a
transverse reniform pale spot just behind apical third, and also paler toward the

apical angles, the suture, however, dark to the apex; prothorax very minutely
punctulate, the sides not quite continuous, feebly convergent and rather strongly
arcuate; elytral punctures fine and moderately close, the pubescence coarse.
Length 1.4 mm.; width 0.8 mm.   California...............**scitus,** sp. nov.
Piceous-black, the head and prothorax dark testaceous, sometimes feebly infuscate
toward the middle of the base; form broadly oval, each elytron with a large pale
area, evidently consisting of two transverse spots longitudinally prolonged backward and forward, so as to unite and wholly or partially enclose an oval dark
spot at the centre; elytral punctures minute, the pubescence rather coarse and
distinct, cinereous; legs pale throughout......................... 76
76—Sides of the prothorax only slightly discontinuous, strongly convergent, evenly
and moderately arcuate; posterior transverse spot short, not extending to the
apex. Length 1.75 mm.; width 1.2 mm. California (Humboldt Co.).
                                                       suavis, sp. nov.
Sides of the prothorax strongly discontinuous, feebly convergent and very feebly
arcuate; posterior spot extending nearly to the apex, the elytra being pale with
the base broadly, suture and side-margin more narrowly to behind the middle,
and a small central spot, dark; elytral punctures less close-set. Length 1.7 mm.;
width 1.2 mm. Colorado..... ....**coloradensis** *Horn*
77—Each elytron pale, with all the margins nubilously blackish, more broadly at
base, the pale area feebly oblique and elongate-oval, finely and rather closely
punctate; body elongate-oval, the pronotum piceous, minutely, not very closely
punctulate, the sides not quite continuous. Length 1.65 mm.; width 0.9 mm.
California (southern).....................................**sordidus** *Horn*
Each elytron pale throughout, or just visibly and suffusedly piceous toward the
suture 78
78—Larger, more narrowly oval, the elytra longer than wide, the pronotum finely
and sparsely but evidently punctulate; pubescence rather abundant, suberect,
coarse and conspicuous. Length 1.3–1.6 mm.; width 0.8–1.0 mm. Maryland,
Indiana, Kansas and Texas ....**intrusus** *Horn*
Smaller, shorter and more broadly oval, the elytra not longer than wide, sparsely
punctulate, the pubescence rather sparse, more decumbent and less conspicuous;
pronotum wholly impunctate, sparsely pubescent, the sides continuous with those
of the elytra but more arcuate, feebly convergent. Length 1.1 mm.; width 0.75
mm. Florida (Tampa)..................................**inops,** sp. nov.
79—Prosternal carinæ entire, attaining the apical margin......................80
Prosternal carinæ abbreviated in front; body pale throughout or nearly so....... 87
80—Elytra black, with a transverse post-basal pale band narrowly prolonged along
the suture to the base; body narrowly oblong and parallel, the punctures fine
and sparse; prothorax testaceous. Length 1.4 mm. ; width 0.8 mm. Florida
(Haulover near Jupiter) ....**balteatus** *Lec.*
Elytra black, each with a single small yellow spot slightly in front of the middle, the
apex narrowly pale; body oval; prothorax piceo-testaceous, paler at the sides,
the latter almost continuous with those of the elytra; legs testaceous; size very
small. Length [1.25 mm.]. Florida (Biscayne Bay and Punta Gorda).
                                                       bigemmeus *Horn*

Elytra black or piceous, each with two pale spots...........................81
Elytra black throughout, the apex broadly pale in fourth or fifth, the pale area divided by the rather broadly black suture to the apical angles ; body very small, broadly oval, thé head, prothorax and legs throughout pale testaceous ; prothorax short and transverse, finely punctulate, the sides nearly continuous, strongly convergent and arcuate ; elytra barely as long as wide, very finely, evenly and not densely punctate, the pubescence short but pale and coarse.   Length 1.2 mm. ; width 0.85 mm.   Locality not indicated............. ......**dichrous** *Muls.*
Elytra black, with a broad apical red area which is not divided by the suture ; legs red throughout...83
Elytra black or brown throughout, the apex not, or only very narrowly, paler....84
Elytra pale, with a black spot or design..... 86
81—Form very narrowly oblong and parallel, black, shining, the legs pale ; pronotum pale, infuscate toward the middle ; punctures fine and sparse, the pubescence short, suberect and quite conspicuous ; elytra each with two large pale spots, the anterior at basal third the larger, extending somewhat obliquely and becoming subattenuate toward the humeral callus, the posterior at apical fourth and obliquely suboval.   Length 1.6 mm. ; width 0.9 mm.   Florida (Enterprise) and Louisiana..**4-tæniatus** *Lec.*
Form broadly oval 82
82—Prothorax black, faintly piceous toward the apical angles, the sides nearly continuous, strongly convergent and feebly arcuate ; elytra longer than wide, finely and not very closely punctate, each with a moderate subquadrate spot just before the middle, nearer the suture than the side, and another, smaller and reniform, in the same line at apical fourth ; apex scarcely paler ; pubescence rather coarse and distinct.   Length 1.7 mm. ; width 1.15 mm.   Pennsylvania.

myrmedon *Muls.*
Prothorax pale rufo-testaceous throughout ; head and legs similar in coloration, the hind femora blackish, except at tip ; abdomen pale, blackish toward base ; body stout, oblong-oval ; prothorax short and transverse, finely but distinctly, rather closely punctate, the sides slightly discontinuous, moderately convergent, evenly and strongly arcuate ; elytra subquadrate, as long as wide, very obtuse at apex, black, finely but strongly, evenly and not very closely punctured, each with a very oblique pale line from anterior two-fifths and inner third to and enveloping the entire humeri, subdivided near its middle point, and a transverse broader spot at apical fourth or fifth, narrowly and equally distant from the suture and side margin, the apex very narrowly pale ; pubescence coarse, suberect and distinct. Length 1.7 mm. ; width 1.2 mm.   North Carolina (Asheville).

adulans, sp. nov.
Prothorax yellow, darker in front of the scutellum ; elytra piceous, a narrow apical border and two spots, one small and rounded in front of the middle, nearer the suture than the side, and the other transverse and slightly sinuous, at apical third, touching the side but not the suture [not so drawn in the figure], pale ; legs yellow.   Length [1.25-1.5 mm].   Southern New Jersey.. **liebecki** *Horn*
83—Prothorax black, with the apex narrowly, and the apical angles more broadly, testaceous, the sides not quite continuous, moderately convergent and broadly

arcuate, the punctures fine but strong and moderately close-set ; elytra distinctly longer than wide, finely but strongly, not very closely punctate ; pubescence coarse and distinct. Length 1.3–1.8 mm. ; width 0.9–1.25 mm. New York to Texas and Iowa ; [*femoralis* Lec.]..... ....**terminatus** *Say*
Prothorax flavo-testaceous, with a broad parabolic basal spot of black not extending to the apex, the sides strongly discontinuous, feebly convergent, evenly and strongly arcuate, the punctures fine, strong and very close-set ; elytra broadly oval, not longer than wide, with the apical fourth abruptly pale, the punctures strong but very small, finer and closer than in *terminatus*, the pubescence rather short and fine, darker in color and less conspicuous. Length 1.7 mm.; width 1.25 mm. Texas (Austin)................... ....**partitus**, sp. nov.
84—Form broadly oval, the elytra not longer than wide, black, shining, the pubescence coarse, suberect, cinereous and conspicuous ; head, legs and pronotum pale testaceous, the latter slightly infuscate before the scutellum, the sides continuous, strongly. convergent, evenly and rather feebly arcuate ; elytra minutely, sparsely punctulate, the apical margin narrowly and indefinitely pale, the scutellum black. Length 1.25 mm.; width 0.9 mm. Texas (Columbus).... **houstoni**, sp. nov.
Form narrowly oval, the elytra distinctly longer than wide ; head and legs pale, the elytra narrowly paler at apex, almost imperceptibly so in *brunnescens ;* sides of the prothorax continuous but a little more arcuate than those of the elytra, rather strongly convergent..85
85—Pronotum testaceous, gradually infuscate in the middle toward base, finely but rather strongly and closely punctulate ; elytra black, shining, rather strongly and somewhat closely punctured, the pubescence cinereous, only moderate in length and coarseness. Length 1.2–1.3 mm.; width 0.8–0.85 mm. North Carolina (Asheville), ... **appalacheus**, sp. nov.
Pronotum pale flavo-testaceous throughout, the elytra very pale brown, sometimes slightly darker and picescent in a large triangular nubilous basal region on the suture, rather sparsely and very finely punctate, the pubescence quite long, coarse suberect, bristling and conspicuous ; under surface blackish-piceous, the abdomen paler. Length 1.3–1.6 mm.; width 0.9–1.05 mm. Texas (Brownsville)—Mr. Wickham............. .. ..**brunnescens**, sp. nov.
86—Oval, much longer than wide, shining, pale flavo-testaceous throughout above and beneath, the legs still paler ; head and pronotum subimpunctate, the latter short, the sides continuous but more arcuate, moderately convergent ; elytra distinctly elongate, minutely, sparsely punctate, with a slightly transverse common sutural spot at apical third, which is feebly arcuate anteriorly and semicircular behind ; pubescence only moderate in length. Length 1.3 mm.; width 0.88 mm. Florida.
stigma, sp. nov.
Oval, minute, not much longer than wide, very pale albido-flavate, the legs very pale ; sterna of the hind body, and sometimes the median basal parts of the abdomen, black ; pronotum short and very transverse, scarcely punctulate, the sides not quite continuous, feebly arcuate and moderately convergent, pale, with a short transverse black spot before the scutellum ; elytra scarcely as long as wide, pale, with a sharply defined deep black design, consisting of a large common basal spot semicircularly rounded behind, continued narrowly along the basal margin,

flexed posteriorly at the humeri and continuing narrowly along the side-margin to
the middle, the large basal spot also connected by a short sutural isthmus with a
small rounded common sutural spot just behind the middle ; pubescence long,
coarse and bristling. Length 0.9–1.0 mm.; width 0.65–0.75 mm. Bahamas
(Eleuthera and Egg Islands)—Mr. Wickham...........**bahamicus** sp. nov.
Oblong, much longer than wide, very pale luteo-flavate, the pronotum less pale than
· the elytra but uniform throughout and without a median basal spot, much less
transverse than in *bahamicus ;* sides somewhat discontinuous, feebly convergent,
evenly and feebly arcuate, the punctures minute but visible and rather close-set ;
elytra evidently longer than wide, nearly straight at the sides, very obtuse at apex,
finely but strongly, somewhat closely punctate, the darker design piceous- black
and less abruptly defined than in *bahamicus*, consisting of a large subtriangular
common basal spot, somewhat prolonged in a fine acuminate line at each side of
the suture, but not united to the rounded common sutural spot at apical two-fifths ;
flanks infuscate at the middle and again at the external apical arcuation ; pubescence rather short and inconspicuous. Length 1.15 mm.; width 0.78 mm.
Bahamas (Egg Island)................................**putus**, sp. nov.
87—Larger species, broadly oblong-oval, pale and uniform luteo-flavate throughout, the
abdomen piceous at the middle of the base ; pronotum finely punctulate, the sides
almost continuous but a little more arcuate, strongly convergent ; elytra a little
longer than wide, parallel, very obtusely but circularly rounded behind, finely
but strongly, rather closely punctate, the suture with a parallel nubilous piceous
vitta from the base to rather behind the middle ; pubescence coarse and moderately short. Length 1.55 mm.; width 1.05 mm. Kansas....**dulcis**, sp. nov.
Smaller and more narrowly oval, the elytra not darker on the suture.............88
88—Elytra about as long as wide, not narrowed behind except toward tip........89
Elytra longer than wide, narrowed behind from near basal third ; prothorax well
developed, only moderately transverse, scarcely perceptibly punctulate, the sides
continuous with those of the elytra but rather more arcuate, moderately convergent ; elytra rather narrowly subtruncate at tip, finely but distinctly and rather
closely punctate, the pubescence very short and subdecumbent. Length 1.1–1.2
mm.; width 0.65–0.7 mm. Michigan and Illinois........... **æger**, sp. nov.
89—Prothorax minutely punctulate, the sides continuous with those of the elytra,
rather strongly convergent and very feebly arcuate ; elytra finely and quite closely
punctate, the pubescence very short, abundant and subdecumbent. Length 1.1–
1.3 mm.; width 0.75–0.8 mm. California (Alameda Co.)......**debilis** *Lec.*
Prothorax relatively smaller and more convex, impunctate, the sides evidently discontinuous, feebly convergent, evenly and rather strongly arcuate ; elytra distinctly
and somewhat abruptly wider than the prothorax, obtusely rounded or subtruncate
at tip, with somewhat coarse but very shallow and sparse punctures, the pubescence longer, sparser and more erect than in *debilis*, but still quite short. Length
1.1 mm.; width 0.68 mm. Florida..................... **pusio**, sp. nov.

In the subgenus *Scymnobius* the prosternum is wholly devoid of
carinæ, but there is frequently a fine short groove following the
margin of each acetabulum ; this is a very well-marked group of

species, and may prove to have full generic value. In *Diomus* the prosternal carinæ are as distinctive and charcteristic a feature as in *Pullus* or *Scymnus* proper, and they are by no means obsolete as stated by Dr. Horn ; they are, however, finer and less visible under low powers of amplification. In this group, which is indeed almost entitled to generic rank, the first ventral suture is generally more obliterated toward the middle than in the others. The separation of *Scymnodes* Blackb., from *Scymnus*, upon this character, would not be warranted even if the line of demarkation could be distinctly drawn. In the old world, *Scymnus* proper seems to be about as abundant as *Pullus*, but in America the disparity in numbers is very great, the former being relatively very feebly represented.

Scymnus punctum of LeConte, which is closely allied to the European *punctillum*, belongs to the genus *Stethorus* of Weise, very distinct on account of the deflexed prosternum; it is in no way related to *nanus*, with which it is compared by Dr. Horn.

The following species are omitted from the table because of uncertainty regarding their true position.

S. brullei Muls.—Oval-oblong ; elytra black, each with a rounded red spot in apical third. Length 3.1 mm.; width 1.5 mm. Florida. May be placed before *hemorrhous* but the proportional elongation is much greater.

S. puncticollis Lec.—Broadly oval, black, the head and prothorax finely and densely punctured, the latter with a small yellow spot at the apical angles ; elytra densely punctate, with a narrow testaceous apical margin ; legs pale, the femora piceous. Length 2.25 mm. Upper Mississippi. May be placed just before *agricola* in the table.

S. abbreviatus Lec.—Black throughout, the legs rufo-piceous ; prothorax sparsely punctured, densely toward the sides ; elytra densely and coarsely punctured, the metacoxal plates three-fifths as long as the segment. Length 2.1 mm. Lake Superior (Eagle Harbor). To be placed immediately after *weidti* in the table.

S. flebilis Horn—May be inserted just before *nubes* in the table

S. opaculus Horn—May be placed just after *circumspectus*.

S. bisignatus Horn—To be inserted immediately after *bivulnerus*.

S. amabilis Lec.—To be placed just before *guttulatus*.

S. xanthaspis Muls.--Should appear immediately before *houstoni*.

S. icteratus and *cyanescens* of Mulsant, cannot be placed, and the *atramentarius* and *infuscatus* of Boheman, cannot be certainly identified.

Cephaloscymnus *Crotch.*

The two species thus far discovered are mutually closely allied, but differ in color and sculpture. The *Cephaloscymnus ornatus* of Horn, is in no way related, but belongs to the Scymnillini, where it forms the type of a new and rather isolated genus. The color of the body is uniform and black or piceous.

Black, the elytra sparsely punctured. Maryland and South Carolina.

zimmermanni *Crotch*

Brownish or piceous; elytra more coarsely and quite closely punctured. Southern California and Arizona. .... occidentalis *Horn*

These species are of an oblong-oval form and 1.5–2.0 mm. in length. They may be recognized at once by the very large head and deeply emarginate prothorax, the sides of which are discontinuous with those of the elytra.

RHYZOBIINI.

The insects of this tribe are of a regularly oval, moderately convex form and are clothed throughout with more or less fine semi-erect pubescence, as in Scymnini. They are not, however, closely allied to that tribe, as they possess wider, moderately descending and internally margined epipleuræ, long and slender antennæ, with loosely connected serrate 3-jointed club, entire or subentire and coarsely faceted eyes and entire metacoxal plates, always shorter than the segment, and, in the two genera defined below, the prosternum is flat, moderately or widely separating the coxæ and with two strong entire converging carinæ. The abdomen has six segments, the sixth very small, the maxillary ·palpi normally securiform and the legs perfectly free. The prothorax is very feebly and evenly sinuate at apex, with broadly rounded angles as in Psylloborini. The tarsal claws are well developed, evenly arcuate and slender, with a moderate subquadrate dilatation internally at base, but in the males the anterior and intermediate are thick and bifid, thus forming an exception to the entire family as far as known.' The genera before me may be defined as follows :—

Epistoma transversely truncate and simple at apex ; hypomera nearly simple ; prosternal carinæ arcuate, diverging widely at base, coalescent at apex ; metacoxal plates very short.. .....*Rhyzobius
Epistoma deeply emarginate, the bottom of the sinus transverse and having a membranous margin ; hypomera with a narrow deep groove extending, parallel to the side margin, from the apex nearly to the middle, the prosternal carinæ straight, not coalescent at apex ; metacoxal plates much larger, extending almost to the apex of the segment...Lindorus

The definition of *Rhyzobius*,—the original spelling of which I agree with Wollaston in following,—is taken from the South African *trimeni* Csy.

Lindorus, gen. nov.

The single species is represented before me by two examples, kindly communicated by Dr. Blaisdell, and one taken by myself in Sonoma County, in 1885, which is apparently prior to its introduction by the Agricultural Department.

Broadly oval, pale rufo-testaceous throughout, except the entire clytra, which are black with feeble æneous lustre ; pronotum frequently with a transverse piceous cloud just before the middle, the sides but feebly convergent, slightly arcuate and distinctly discontinuous, the punctures fine and rather sparse ; elytral punctures slightly stronger but not very close-set, the pubescence unevenly directed. Length 2.2-2.7 mm.; width 1.5-2.0 mm.   California (Coronado to Sonoma); [*toowoombæ* Blackb]..............................**lophanthæ** *Blaisd.*

Coccidulini.

A single remarkable genus, apparently confined to the palæarctic and nearctic provinces, demands tribal separation.   The body in *Coccidula* is elongate-oval and moderately convex, pubescent throughout, with the eyes, antennæ, palpi and metacoxal plates as in Rhyzobiini, and the abdomen composed of six segments, the sixth large and distinct.   The mentum is not impressed, as it is in Rhyzobiini, the epistoma truncate, with coriaceous margin, the prosternum tumid in the middle anteriorly, becoming flat and rather widely separating the coxæ at base, bicarinate, the carinæ coalescent before the apex upon the summit of the tumidity, the hypomera simple ; epipleuræ narrow, horizontal, more finely margined within, becoming obsolete at the fourth abdominal segment, the metacoxal plates about half as long as the segment, the legs perfectly free, rather stout, with the claws feebly bifid within at some distance from the apex.   The prothorax is narrowed at base and very feebly sinuate at apex.

Coccidula *Kugel.*

The single species before me resembles the European very closely and may be thus briefly defined :—

Elongate ; body and head black, the prosternum, legs, abdomen, except in the middle at base, and pronotum, testaceous, the latter with a small and transverse dark area at apical fourth ; elytra testaceous, arcuately black at base and along the sides to behind the middle, also with a common transversely oval sutural black spot at two-thirds, the punctures rather coarse, deep, close-set and uneven in size, the larger tending vaguely to lineal arrangement at some parts of the disk ; pubescence very short, almost even. Length 3.0 mm.; width 1.4 mm. Michigan (Detroit) ....**lepida** *Lec.*

Suturalis Ws. (Ann. Belg., March 1895, p. 132), described from Ohio, of which the Californian *occidentalis* Horn, is said by Weise to be a synonym, is not before me at present and is therefore omitted.

APPENDIX.

I.

List of Coccinellidæ taken in equatorial and southern Africa by Messrs. Cook and Currie, and by the author, while a member of the Transit of Venus expedition to the Cape of Good Hope, in 1882.

Lioadalia flavomaculata *DeG.*—Wellington, near Cape Town.

Isora anceps *Muls.*—Wellington.

Stictoleis 22-maculata *Fabr.*—Liberia (Mt. Coffee). The black spots coalesce in some individuals.

Œnopia cinctella *Muls.*—Cape Town.

Verania comma *Thunb.*—Wellington.

Cydonia 4-lineata *Muls.*—Wellington. The specimens are in three varieties. First : the median vitta of the elytra is entire, with a finer external arcuate vitta joining the principal vitta near the base and apex—the normal form, which is rare. Second : the principal vitta is abruptly abbreviated at apical fourth, and, third : the principal vitta extends only to basal third or fourth. Both of the last two varieties are more abundant and have the external vitta wholly obsolete.

Cheilomenes lunatus *Fabr.*—St. Helena, Cape Town and Wellington.

Cheilomenes orbicularis, sp. nov.—Similar in form to *lunatus*, but with the discal spot before the middle of each elytron broadly amalgamated with the humeral elongate spot, the latter narrowly separated at base from the inner basal spot and not fused with it as in *lunatus*. Further, with the transverse blotch at the suture and apical third evidently formed of two spots and not forming a regular arcuate band as in *lunatus*. Both of these species are represented by large series, and the markings are extraordinarily constant in each. Liberia (Mt. Coffee).

Thea variegata *Fabr.*—Wellington.

Epilachna reticulata *Oliv.*—Liberia (Mt. Coffee). The pale ground color between the spots is frequently filled with a black reticulation which never approaches the spots by more than half of their own diameter, the latter becoming ocellated.

Epilachna africana *Crotch.*—Liberia (Mt. Coffee).

Epilachna liberiana, sp. nov.—Somewhat similar to *africana*, but larger and more dilated. Broadly rounded, strongly convex, rufo-testaceous, the elytra, epipleuræ externally and legs throughout, black, the elytra sparsely and rather finely but unequally punctate, each with six large subequal irregular pale blotches, three subsutural and three submarginal, the anterior subsutural not attaining the base and the posterior submarginal not in line with the three subsutural. Length 6.8 mm.; width 6.5 mm. Liberia (Mt. Coffee).

Epilachna occidentalis *Crotch.*—Liberia (Mt. Coffee).

Epilachna peringueyi, sp. nov.—Ovate, the elytra subprominently rounded and widest at basal fifth, black throughout, the epipleuræ pale, margined externally with black, the elytra minutely, not densely punctate, with larger, widely scattered punctures intermingled, black, each with three large subconfluent spots in apical half, two smaller spots in a transverse line at two-fifths, the external of which is broadly confluent with a lunate basal spot extending almost to the scutel'um ; head and pronotum without pale spots at any point. Length 5.8 mm. ; width 4.7 mm.   Cape Town.   Belongs near *infirma.*

Chnootriba erythromela *Widem.*—Cape Town.

Chnootriba assimilis *Muls.*—Liberia (Mt Coffee).

Chnootriba curriei, sp. nov.—Similar to *assimilis,* but shorter and more broadly oval, with the fine punctures of the elytra much sparser and the coarse punctures very much larger, the surface more convex and more shining ; subhumeral spot rounded ; median band—composed of two spots—much less oblique, almost transverse.   Length 5.4 mm. ; width 3.9 mm.   Liberia (Mt. Coffee).   Named in honor of Mr. R. P. Currie.

Lotis neglecta *Muls.*—Broadly rounded, polished, black above ; pronotum finely, closely punctulate toward the sides, the apical angles pale ; elytra each with two large orange spots on the median line, the anterior the larger and extending from one-sixth to two-fifths and from inner fourth to outer third, the posterior from two-thirds to five-sixths and from inner fifth or sixth to outer two-fifths ; limb feebly rufescent ; punctures fine and not close-set ; under surface and legs testaceous, the sterna and median basal parts of the abdomen darker. Length 2.0–2.2 mm. ; width 1.8–2.0 mm.   Cape Town.   The elytral spots are a little larger than indicated by Mulsant.

Lotis distincta, sp. nov.—Similar to *neglecta* in form but alutaceous and with still more minute and obsolete punctures, black throughout above, each elytron with two spots in the same position but smaller, not more than a fifth as wide as the elytron, the posterior elongate-oval ; punctures gradually becoming distinct toward the sides ; surface with obscure and very obsolete impressed longitudinal striiform lines toward the suture ; under surface and legs black throughout, the epipleuræ piceous.   Length 2.3 mm. ; width 2.1 mm.   Cape Town.

Lotis stigmatica, sp. nov.—Slightly smaller and more narrowly rounded behind, polished, black above, with a feeble greenish reflection, the elytral punctures small and sparse but distinct, the spots similarly placed but very small, the anterior rounded, about a seventh as wide as the elytron, the posterior very small, circular, with rather nubilous outline ; under surface and legs black throughout, the epipleuræ piceous.   Length 1.75–2.1 mm. ; width 1.6–1.9 mm.   Wellington.

Lotis nigerrima, sp. nov.—Similar to *stigmatica* in form, size and sculpture, but deep black above, polished and without trace of elytral spots ; under surface black, the legs and abdomen picescent ; epipleuræ pale testaceous, margined with black externally.   Length 2.1 mm. ; width 1.9 mm.   Wellington. Much larger than *nigritula* Cr., and with more distinct punctures.

Xestolotis (gen. nov.) stictica, sp. nov.—Almost circular, very convex, polished, black, the head, pronotum and suffused limb of the elytra dark picco-rufous ; under surface piceous, the legs, palpi and antennæ pale testaceous ; pronotum and elytra strongly and equally punctate, the former closely, the lat-ter sparsely and without trace of impressed lines at any part. Length 1.8 mm. ; width 1.7 mm. Liberia (Mt. Coffee). Taken in abundance by Mr. Cook.

The genus *Xestolotis* is similar to *Lotis* in the structure of the front, but has the clypeal margin more broadly truncate and only very feebly sinuate ; the eyes are not emarginate and the antennæ are rather well developed, with the club flattened, compact and elongate-oval ; the fourth joint of the maxillary palpi is very obliquely securiform, the free apex somewhat prolonged and finely acuminate. The coxæ are all widely separated, the tarsi well developed and subcompressed, and the claws simple, becoming arcuately thickened internally toward base. The abdomen is composed of five segments ; the metacoxal plates at-tain the segmental apex toward the sides and are concave. The fifth ventral is longer than the preceding, as in all genera with true five-segmented abdomen, and, in all my representatives, the tip of the ab-domen is deflexed, this being apparently a normal condition. The epipleuræ are uneven and subfoveolate, the met-episterna remarkably divided at a point opposite the extremity of the straight mesocoxal line, and the third tarsal joint is evidently free. It may be distin-guished from *Sticholotis* (*punctata*) by the characters of the epipleuræ and met-episterna, as well as by the more finely faceted and entire eyes, which, in *Sticholotis*, are nearly as coarsely granulated as in the rhyzobiids and slightly emarginated by the post-antennal parts of the front.

Chilocorus cooki, sp. nov.—Broadly rounded, polished ; head, pronotum, entire under surface and legs pale brownish-testaceous ; elytra black, a large oval basal spot on the suture of the same color as the anterior parts, extending, at the basal margin, two-fifths from the suture, and, on the latter, slightly beyond the middle ; punctures minute and sparse, each surrounded by a fine irregular ring of extremely minute punctulation ; epipleuræ piceous-black, testaceous inwardly. Length 5.4 mm. ; width 4.8 mm. Liberia (Mt. Coffee). Named in honor of Mr. O. F. Cook.

✓ **Exochomus versutus** *Muls.*—Wellington.

✓ **Exochomus flavipes** *Thunb.*—Wellington.

Platynaspis capicola *Crotch.*—Wellington.

Telsimia (gen. nov.) **tetrasticta**, sp. nov.—Broadly elliptical, evenly and moder-ately convex, shining, finely but strongly, sparsely impresso-punctate, clothed rather sparsely throughout with somewhat long suberect and ashy pubescence, black, the legs but slightly picescent ; each elytron with two rounded testaceous spots nearly as in *Lotis*, both near inner third and at two-fifths and three-fourths from the base respectively; flanks regularly declivous to the edge, which is minutely reflexo-beaded. Length 1.5–1.6 mm.; width 1.25 mm. Wellington. Differs from the following in its larger size and maculate elytra.

Telsimia inornata, sp. nov.—Broadly rounded, strongly somewhat compresso-convex, shining, strongly, closely punctate, the pubescence rather short, ashy, suberect and moderately abundant; elytra without ornamentation, the edge slightly more thickly reflexo-beaded than in *tetrasticta ;* metacoxal arcs more apical but still far from the apex of the segment, the tarsi more slender, with the basal joint more elongate. Length 1.1 mm.; width 0.9 mm. Liberia (Mt. Coffee).

The genus *Telsimia* has been sufficiently characterized in the body of the present paper under the head of Telsimiini.

Pharus 6-guttatus *Gyll.*—Wellington.

Pharus inæqualis, sp. nov.—Similar to *6-guttatus* but more oblong and less rounded, with the prothorax relatively narrower, more rounded at the sides and more strongly and closely punctured; elytra with the spot at the middle and inner fourth very much smaller than the other two, and not subequal as in *6-guttatus ;* under surface and legs black throughout. Length 2.4 mm.; width 1.8 mm. Cape Town.

Pharopsis (gen. nov.) **subglaber,** sp. nov.—Broadly oval, very strongly convex, black throughout above and beneath, the legs not paler, minutely but evidently punctate, the elytra sparsely so, polished and glabrous; head and pronotum duller, strongly microreticulate and clothed with very short, rather sparse, decumbent and inconspicuous silvery-gray hairs; basal joint of the tarsi elongate, the claws simple and strongly arcuate. Length 1.45 mm.; width 1.2 mm. Wellington.

This genus has been defined previously in the present paper, under the head of Pharini.

Hyperaspis felixi *Muls.*—Wellington.

Hyperaspis newcombi, sp. nov.—Elongate, suboblong-oval, moderately convex, polished, black throughout above and beneath, the head, except at the basal margin, and the sides of the pronotum in a parallel area nearly twice as long as wide with the inner outline feebly bisinuate, orange-yellow; elytra with a rounded marginal pale spot at apical sixth of the length; anterior legs pale, the two posterior pairs black. Length 2.7 mm.; width 1.8 mm. Wellington. Named in honor of Prof. Simon Newcomb. Differs from *mercki* in the form of the subapical spot of the elytra, which is here much smaller and separated throughout its extent from the margin by the fine black bead, becoming only slightly more distant posteriorly; it is separated from the suture by rather more than its own width.

Cranophorus notatulus *Muls.*—Wellington. The male has the fifth segment broadly sinuato-truncate, with a small suberect liguliform tooth at the middle of the apical edge, the sixth angularly emarginate, with the surrounding surface deeply impressed, and, through the emargination, a small seventh segment can be discerned.

Cranophorus 4-notatus *Muls.*—Cape Town.

Cranophorus trapezium, sp. nov.—Similar to *4-notatus* but more broadly oval, shining, moderately pubescent, finely rather closely punctate, deep black above, the pronotum pale and diaphanous at the apical margin, more broadly laterally, the pale margin extending only to the middle of the length; elytra each with two small rounded pale spots, nearly equal in size, near one-third and two-thirds from the base and both at about two-fifths from the suture; under surface and legs black; male with the fifth ventral feebly sinuate, not denticulate, the sixth sinuato-truncate and broadly impressed. Length 1.7 mm.; width 1.1 mm. Wellington. Abundant.

Cranophorus parvulus, sp. nov.—Similar to the preceding but much smaller, the elytra more finely, sparsely and obsoletely punctate and more truncate at tip, the two spots of each elytron extremely small and nearly on the median line; male with the fifth segment truncate and not modified, the sixth perfectly flat, broadly subtruncate at apex, with a very minute angulate median notch. Length 1.15-1.25 mm.; width 0.75-0.85 mm. Wellington. A single pair.

Stethorus jejunus *Csy.* (Ante, p. 136)—Cape Town.

Scymnus (Scymnus) morelleti *Muls.*—Wellington.

Scymnus (Scymnus) capicola, sp. nov.—Broadly oval, black, the elytral apices narrowly margined with red; abdomen black, the apical margin paler; legs testaceous throughout; head rufo-piceous in the male, black in the female, the pronotum black throughout in both sexes, finely but strongly, not closely punctate, the sides nearly continuous, strongly convergent and moderately arcuate; elytra as long as wide, rounded behind, punctured nearly like the pronotum but less finely; under surface dull, very densely punctate throughout, more finely on the abdomen. Length 1.7-2.0 mm.; width 1.2-1.5 mm. Wellington. The male has the fifth ventral broadly, feebly sinuate at the middle of the apex but not notably impressed.

Scymnus (Scymnus) monroviæ, sp. nov.—Broadly oval, moderately pubescent, finely but strongly, somewhat closely punctate; head black, the pronotum black with the apex nubilously pale toward the sides, the latter strongly convergent, feebly arcuate and rather discontinuous; elytra black, the apical margin narrowly and nubilously pale, each with a rather large, obliquely oval discal red spot just before the middle; under surface blackish, dull, very densely but finely punctate, the abdominal apex slightly paler; legs pale testaceous, the femora somewhat infuscate except toward tip. Length 1.75 mm.; width 1.25 mm. Liberia (Mt. Coffee). A single female.

Scymnus (Nephus) angustus, sp. nov.—Very narrowly oval, about twice as long as wide, moderately convex, minutely and very closely punctate, black, the elytra testaceous, with the suture and side-margin in basal three-fifths blackish, the dark areas broadening toward base and becoming coalescent; under surface and legs piceous or blackish, the knees and tibiæ somewhat paler. Length 1.6 mm.; width 0.8 mm. Wellington.

Rhyzobius trimeni, sp. nov.—Oval, moderately convex, the pubescence ashy, moderately long and abundant; body black, the tarsi and abdominal limb broadly throughout pale; pronotum with the apex at and near the angles pale, the sides reflexed, strongly convergent, evenly, rather strongly arcuate and discontinuous, the base finely margined ; elytra finely but distinctly, sparsely punctate, each with two rather small rounded pale spots, the anterior, slightly the larger, near one-fourth and very slightly nearer the suture than the margin, the posterior not quite at three-fourths and near inner third or two-fifths; abdomen finely, not densely punctulate. Length 2.6–3.0 mm.; width 1.8–2.15 mm. Wellington. Named in honor of Mr. Roland Trimen. The basal angles of the prothorax are slightly more than right, and are not at all rounded but not prominent, the base being oblique and straight from the scutellum to the sides.

II.

The present opportunity is taken to describe a few new members of the Coccinellidæ from regions beyond the United States.

Epilachna parvicollis, sp. nov.—Ovate, very convex, polished, the pubescence short and only moderately dense ; head and pronotum black throughout, the latter finely, not densely punctate, broadly concave and reflexed at the sides, two and one-half times as wide as long, distinctly narrower than either elytron, the sides rather feebly convergent ; scutellum blackish, a little longer than wide ; elytra but little longer than wide, widest at basal third or fourth, where the sides are evenly rounded to the base and gradually less strongly, becoming strongly convergent, to the apex, which is ogival, pale rufo-testaceous in color, the reflexed margins evenly throughout, a small rounded spot on each at the middle and inner two-fifths, and another in the same range near the margin and transverse, black ; sculpture sparse, consisting of very coarse deep punctures, with others, small and feebly impressed, intermingled, the surface subrugose ; under surface, epipleuræ and legs throughout black.   Length 9.6 mm.; width 8.0 mm.   Bolivia.

Some time after this description had been written I received a second Bolivian specimen, agreeing exactly with the type, from Mr. Fruhstorfer, under the name *"rufipennis."* I have been unable to find this name in the literature of the subject, and Mr. Fruhstorfer informs me that he also is unable to recall its origin.

Nephaspis (gen. nov) **gorhami**, sp. nov.—Oval, moderately convex, finely, closely punctate, finely, evenly and abundantly pubescent, the hairs all directed longitudinally on the elytra ; head, pronotum, prosternum, legs and abdominal apex and sides pale testaceous ; elytra piceous-black.   Length 1.2 mm.; width 0.85 mm.   Colombia (Panama).

Nephaspis brunnea, sp. nov.—Similar but more narrowly oval, the minute punctures sparser, the surface more polished, the pubescence similar and subdecumbent but sparser ; body dark piceous-brown throughout, the head, prosternum, legs and abdomen toward tip testaceous ; sterna closely and more coarsely punctured.   Length 1.2 mm.; width 0.8 mm.   Colombia (Panama).

The genus *Nephaspis* is remarkable, among those allied to *Scymnus*—and in fact the entire family,—in the structure of the prosternum ; this widely separates the coxæ, which are obliquely conical and decumbent upon the surface separating them, the latter being thus obliquely biconcave, the elevated part reduced to a mere cusp point anteriorly, the coxæ being subcontiguous at their apices.   The sterna of the hind body are very convex, and the mesosternum is abruptly terminated anteriorly by a deep vertical wall.   The coxal arcs are nearly as in the subgenus *Nephus*, but the tarsal claws are long, feebly arcuate, extremely slender and perfectly simple.   The epipleuræ are extremely narrow, and extend scarcely behind the middle, and the two basal joints of the antennæ are large and compressed, the remainder very small and slender ; the palpi are normally securiform. The eyes are simple and almost entire and are well developed, the clypeus deeply sinuate.   The prothorax is as wide at base as the elytra and, in repose, heads rest upon the body in such a way as to conceal all anterior to the mesosternum.   The abdomen has six segments as in *Scymnus*, the first as long as the next three combined.   The genus will form a distinct tribe in the neighborhood of Scymnini.

　　Zagloba beaumonti, sp. nov.—Broadly oval, shining, finely, rather sparsely punctate and somewhat sparsely clothed with long stiff ashy-yellow hairs, unevenly directed and suberect ; body pale brownish-testaceous in color throughout, the legs more flavate ; sides of the prothorax moderately convergent, very feebly arcuate and distinctly discontinuous with those of the elytra.   Length 1.5 mm.; width 1.1 mm.   Colombia (Panama)—Mr. J. Beaumont, to whom I am indebted also for the two species described above.

This species has the metacoxal arcs incomplete and formed as in the subgenus *Scymnus*, the emargination of the eyes normal and the prosternum wide and flat between the coxæ, not carinate but tumid or beaded laterally along the acetabula ; the tarsal claws are strongly arcuate, and have a large quadrate internal tooth at base.

NEW AFRICAN SESIIDÆ.

By Wm. Beutenmüller.

Sesia gabuna, sp. nov.

♂. Head and antennæ black ; collar black above, white at sides and beneath ; palpi black above and at tip, yellowish-white beneath. Thorax black with a greenish reflection, a yellow line on each patagia and a yellow transverse mark on the posterior end. Beneath with a golden yellow patch on each side. Abdomen blue-black with a golden yellow ring on the posterior edge of the 2, 3, 4, 6 and 7 segments and a yellow line on each side from the base meeting the first ring. Anal tuft blue-black, narrowly edged with yellow on each side. Legs blue-black, anterior coxæ white on each end ; tarsi yellowish on one side. Fore wings transparent bordered with violet black, outer border very broad ; transverse mark rather prominent. Hind wings transparent with a narrow violet black outer border. Fore wings beneath as above but with silvery rays between the veins in outer border. Expanse, 20 mm.

♀. Larger and more robust than the male, with rings on abdomen white and with a white ring at base on hind tibiæ. Anal tuft entirely black. Expanse, 25 mm.

Habitat: Valley of the Ogowé River, French Congo. 1 ♂, 4 ♀ ♀.

Allied to *Sesia basiformis* (*lustrans* Grote).

Sesia africana, sp. nov.

Head, thorax and abdomen above metallic green-black. Abdomen beneath whitish. Collar very slightly orange above ; palpi beneath pure white ; anterior and hind pair of coxæ pure white. Legs green-black, middle and hind pair with white annulations. Fore wings green-black with three small transparent spaces, basal one linear, the one in cell oval, and the one beyond the transverse mark largest and rounded. Hind wings transparent with very narrow outer border. Antennæ long, black. Expanse, 26 and 34 mm.

Habitat: Valley of the Ogowé, River, French Congo. 2 ♀ ♀. Coll. W. J. Holland.

Easily recognized by its uniform color and by the three transparent spaces on the fore wings. It evidently comes near the European *scoliæformis*, but has no anal tuft ; this may have been worn off.

Sesia festiva, sp. nov.

Head black with a minute orange spot at the base of the antennæ, which are black above and brown beneath. Collar golden orange above, white beneath. Palpi white beneath, black above. Thorax brilliant orange red, except in the middle above brownish-black. Abdomen violet black with a white ring on the 4th, 6th and 7th segments. Anal tuft violet black, slightly white in the middle beneath. Legs violet with white tufts; anterior coxæ white. Wings transparent with narrow violet black borders and transverse mark. Hind wings transparent, border very narrow. Wing neath as above. Expanse, 18 mm.

Habitat :   Valley of the Ogowé River, French Congo.   1 ♂. Coll. W. J. Holland.

May be known by the golden orange red thorax and violet abdomen.

Sesia albiventris, sp. nov.

Head black above, front white ; palpi white, tip black.   Antennæ black ; thorax black, white on each side beneath ; patagia tipped with white posteriorly.   Abdomen black with a metallic green reflection and a narrow white ring on the 4th segment ; last segment edged with white, anal tuft black ; underside of abdomen white on the 3d, 4th, and 5th segments.   Legs black, annulated with white, middle femora and anterior coxæ white.   Fore wings violet black with a basal transparent streak and a small spot composed of white scales beyond the middle.   Hind wings transparent, border and fringes violet black.   Wings beneath as above.   Expanse, 11 mm.

Habitat :   Valley of the Ogowé River, French Congo.   1 ♂.   Coll. W. J. Holland.

Sesia olenda, sp. nov.

Wholly bronzy violet-black above and below, except the fore coxæ, ringlets on legs and palpi white ; last joint of palpi black.   Fore wings with a very small rounded transparent mark in the cell and a large transparent area beyond the transverse mark.   Expanse, 15 mm.

Habitat :   Valley of the Ogowé River, French Congo.   1 ♀.   Coll. W. J. Holland.

Sesia nyanga, sp. nov.

Head, thorax and abdomen above and below bronzy black, except the last two segments beneath white.   Anal tuft bronzy black.   Fore. wings largely transparent, borders and transverse mark very narrow, black ; similar beneath with the costa yellowish.   Legs black, middle coxæ white.   Expanse, 17 mm.

Habitat :   Valley of the Ogowé River, French Congo.   1 ♀.   Coll. W. J. Holland.

Sesia nuba, sp. nov.

Head black, front white on each side ; antennæ black, ferruginous beneath ; palpi with loose hairs, black, white at tip.   Thorax and abdomen bronzy black, the latter with a pale, dirty yellowish band on the 2d, 4th, and two last segments, encircling the body ; legs black with white ringlets, hind pairs with loose black hairs ; fore coxæ white.   Fore wings transparent, borders and transverse mark, moderately broad, bronzy black.   Hind wings transparent, border narrow, bronzy-black.   Expanse, 14 mm.

Habitat :   Valley of the Ogowé River, French Congo.   2 ♂ ♂.   Coll. W. J. Holland.

One example differs by having the last four segments of the abdomen ringed ; the transverse mark on fore wings orange outside, a little orange on inside of outer border and the patagia finely lined with yellow. The hind tibiæ and tarsi are clothed with rather long hair.

Sesia malimba, sp. nov.

Head brown-black, face pale orange-yellow ; palpi orange ; antennæ black above, orange beneath ; thorax brown-black with indication of a fine orange stripe along the patagia. Abdomen brown-black with a broad yellow band on the 2d and 4th segments above, brown beneath. Legs orange, femora brown. Fore wings with transparent areas small, the outer one round, border and transverse mark broad brown-black ; rayed with a little orange between the veins on outer part, along inner margin and in the basal transparent area. Underside similar to the above. Hind wings transparent with narrow brown margin. Expanse, 20 mm.

Habitat: Valley of the Ogowé River, French Congo. 1 ♀. Coll. W. J. Holland.

Allied to *S. mellinipennis.*

Sesia brillians, sp. nov.

Head black ; palpi orange ; antennæ black. Underside of abdomen, thorax and legs orange. Thorax and abdomen above blue-black, the former with a transverse orange mark posteriorly, the latter with an orange ring on 2d segment and 4th and 5th segments orange. Anal tuft blue-black. Fore wings orange, basal half and outer border blackish. In the orange field are two very small opalescent spots, and a similar one in the cell but is a little larger than the rest. Hind wings transparent, outer border broad, and gradually narrowing as it reaches the hind angle, bronzy brown marked with orange inside. Fore wings beneath golden orange, outer part brown-black, spots repeated. Hind wings as above. Expanse, 13 mm.

Habitat: Valley of the Ogowé River, French Congo. 1 ♂. Coll. W. J. Holland. May be easily known by the orange fore wings with three opalescent spots.

Sesia tropica, sp. nov.

Head black ; palpi, thorax posteriorly, legs and abdomen above wholly orange ; anal tuft blue-black, beneath golden yellow. Thorax anteriorly black. Antennæ brown-black with a white patch before the tip. Fore wings orange, outer part and fringes brown ; costa narrowly brown ; in the cell is a small triangular space, and an oblong one beyond the transverse mark. Hind wings transparent, border narrow, brown. Fore wings beneath golden yellow at base, gradually becoming brown. Expanse, 13 mm.

Habitat: Valley of the Ogowé River, French Congo. 1 ♀. Coll. W. J. Holland.

THE MEGALOPYGID GENUS *TROSIA*, WITH DESCRIPTION OF A NEW SPECIES.

By HARRISON G. DYAR.

Genus **Trosia** *Hübner*.

1822—*Trosia* HÜBNER, Verz. bek. Schmett. 196.
1855—*Sciathos* WALKER, Cat. Brit. Mus. III, 752.
1856—*Edebessa* WALKER, Cat. Brit. Mus. VII, 1755.
1874—*Isochroma* FELDER, Reise Novara, Lep. IV, pl. 83.

Kirby calls this genus *Sciathos* in the catalogue, omitting Hübner's term. Druce and others also neglect *Trosia*, perhaps from a prejudice against the Verzeichniss names. In the following, species succeeded by a dash are unknown to me, except by description and are referred to this genus on the authority of the authors quoted.

SYNOPSIS OF SPECIES.

Thorax and abdomen discolorous, thorax in part white, abdomen bright red.
Thorax without black marks.
Costa broadly red ; wings in part pure white, at least in a stripe next the red margin...1. **tricolora** *Fab.*
Costa very narrowly red ; wings without any pure white.
2. **obsolescens** *Dyar*
Thorax with a black band in front and on each side. ....3. **purens** *Walk.*
Thorax and abdomen concolorous, abdomen not bright red.
Costal margin red ; a transverse row of black dots. ....4. **diamas** *Cram.*
Fore wings without marks, uniformly ochraceous............5. **ribbei** *Druce.*

1. **T. tricolora** *Fab.*

1787—*Bombyx tricolora* FABRICIUS, Mant. Ins. II, 114.
1790—*Bombyx punctigera* STOLL, Pap. Exot. suppl. pl. 34, fig. 1 (nec Linn.).
1822—*Trosia tricolora* HÜBNER, Verz. bek. Schmett. 196.
1855—*Sciathos punctiger* WALKER, Cat. Brit. Mus. 752.
1856—*Sciathos punctiger* WALKER, Cat. Brit. Mus. VII, 1711.
1874—*Isochroma fallax* FELDER, Reise Novara, pl. 83, figs. 18, 19.
1887—*Sciathos punctigera* DRUCE, Biol Cent.-Am. Lep. I, 212.
1892—*Sciathos punctigera* KIRBY, Cat. Lep. Het. I, 540.

Kirby makes this the *punctigera* of Linnæus (Syst. Nat., 509, no. 67), but the description does not coincide in the least with this species, either in structure, color or habitat.

2. **T. obsolescens,** sp. nov.

Head white in front, red on the vertex behind the antennæ and shading to red below on the palpi. Thorax white above with six clusters of red hairs ; abdomen red, white at tip. Fore wings uniformly pale ocherous, almost white, appearing pinkish from the red scales below ; costa very narrowly red at base, dark ocher at apical portion. A straight row of eight small black spots between the veins beyond the middle of the wing, the sixth spot between veins 4 and 5, the seventh opposite the discal cross-vein and the eighth between vein 6 and the stalk of veins 7 to 10. Hind wings red, fringe ocherous. Below both wings as secondaries above ; body largely white ; coxæ and femora red above, tibiæ and tarsi ringed with black. Expanse, 27 mm.

Nearly allied to *T. tricolora* Fab., which is however an inhabitant of the tropical regions, whereas this comes from the Mexican plateau.

One male, Nogales, Koebele collector, August 15, 1898, U. S. Nat. Mus., type no. 4104. Nogales is a town on the border line between Arizona and Mexico.

3. **T. purens** *Walk.*

1856—*Edebessa purens* WALKER, Cat. Brit. Mus. VII, 1755.

1892—*Sciathos purens* KIRBY, Cat. Lep. Het. I, 540.

Sir G. F. Hampson has kindly examined Walker's types of the species for me and the generic characters correspond with *Trosia*.

4. **T. dimas** *Cram.* —

1775—*Bombyx dimas* CRAMER, Pap. Exot. I, pl. 59 C.

1822—*Trosia dimas* HÜBNER, Verz. bek. Schmett. 196.

1854—*Chrysauge dimas* WALKER, Cat. Brit. Mus. II, 375.

1892—*Idalus* (?) *dimas* KIRBY, Cat. Lep. Het. I, 198.

1894—*Sciathos dimas* DOGNIN, Lep. Loja. 173.

1897—*Sciathos dimas* DRUCE, Biol. Cent.-Am., Lep. Het., II, 440.

5. **T. ribbei** *Druce.* —

1898—*Sciathos ribbei* DRUCE, Biol. Cent.-Am. Lep. Het. II, 441, pl. 88, fig. 1.

NEW SPECIES OF SYNTOMIDÆ.

BY HARRISON G. DYAR.

Pseudapinconoma elegans *Auriv.* var. **curriei**, var. nov.

Under side of thorax entirely crimson, legs white, femora and basal half of hind tibiæ crimson above ; abdomen bluish gray, segmental black bands linear, the basal segments with orange hair and the lateral tufts orange ; a dorsal series of crimson dots. Wings as in *elegans*, but the hyaline patches between veins 2 and 6 large and diffuse, reaching nearly to the termen, with ill defined outer border.

Two males, Mt. Coffee, Liberia (R. P. Currie). U. S. Nat. Mus., type no. 4247.

Cosmosoma sicula, sp. nov.

Black, pectus, frons and abdomen with metallic blue patches, the latter in sub-dorsal and lateral series. Wings hyaline, veins black, an orange red streak below costa and above internal margin, the former reaching three-fourths to apex, the latter almost reaching tornus ; a small orange red patch at base above vein 1 ; outer margin black, very broad at apex but widening gradually and regularly ; a narrow black bar at end of cell and the space between veins 2 and 3 up to cell filled in with black, powdered with red scales as well as the extreme base of the space between veins 3 and 4 ; a small red spot near end of vein 2 below.   Hind wings with black border, broad on the outer margin.   Tegulæ and patagia with orange red scales.

One male, Venezuela.   Expanse, 27 mm.   U. S. Nat. Mus., type no. 4248.

Allied to C. festivum and C. centrale, next to which it comes in Hampson's tables.

Cosmosoma perfenestratum, sp. nov.

Head black, frons and vertex with metallic blue ; antennæ black ; thorax orange red, black below ; legs with patches of blue ; abdomen black with dorsal red stripe not reaching base or extremity and subdorsal series of metallic blue spots. Wings hyaline, the veins and margins black ; fore wing with orange red basal patch and streaks below costa and above internal margin running nearly to termen ; an orange red discal patch cut by the black veins ; an orange red patch filling in the space between veins 2 and 4, but not completely ; terminal band very wide at apex, almost wholly orange red, only the veins and extreme margin black, expanding at tornus and joining the patch between veins 2 to 4.   Hind wing with some red at base, the terminal band black, expanding at apex and tornus, edged within by red scales.

One male received from Staudinger and Haas as "Læmocharis fenestrata."   U. S. Nat. Mus., type no. 4244.

This falls in Hampson's table between C. achemon and C. hypocheilus.

Eriphioides ustulata Feld. var. columbina, var. nov.

Differs from ustulata in having a large discal orange patch on the under side of fore wings, powdery and diffuse and cut by the black veins.   The fore coxæ are white.

One male, received from Staudinger and Haas as " Autochloris columbina."   U. S. Nat. Mus., type no. 4245.

Cyanopepla melinda, sp. nov.

Black, thorax and abdomen strongly shot with metallic blue green, also on the head, palpi and legs and forming a dorsal band and segmental rings on the abdomen ; coxæ, tibæ, tarsi and venter of abdomen powdered with white.   Fore wings with a metallic blue dot at base of costa and a streak in submedian interspace ; a crimson fascia from within end of cell to tornus at vein 1, not reaching costal edge or margin ;

a smaller oblique spot between veins 5 and 7. Hind wings with the basal two-thirds shot with metallic blue ; a rounded submarginal crimson spot between veins 2 and 4, narrowly cut by the black vein 3. Expanse, 41 mm.

Two males, Petropolis, Brazil (F. G. Foetterle). U. S. Nat. Mus., type no. 4246.

PROCEEDINGS OF THE NEW YORK ENTOMOLOG-ICAL SOCIETY.

MEETING OF OCTOBER 18, 1898.

Held at the American Museum of Natural History.

In absence of the President and Vice-President, Mr. Chas. Palm was elected chairman *pro tem.* Twelve members present.

Mr. Beutenmuller proposed Mrs. W. H. Browning for active membership.

Mr. Beutenmuller spoke on his collecting trip to Florida in July last and stated that he was fully satisfied with the results. About two thousand specimens of Coleoptera were taken, amongst which were *Dyschirius schaumii, Holopeltis larvalis, Languria marginipennis, Elater sturmii, Polycesta,* sp., *Actenodes auronotata, Mecas cana, Oedionychus ulkei, Oxacis tæniata, Helops viridimicans, Formicomus scitulus* (?), and many other good species : A large gray Katydid *Cyrtophyllus* allied to *C. concavus* was also taken as well as many species of other insects.

After discussion, adjournment.

MEETING OF NOVEMBER 1, 1898.

Held at the American Museum of Natural History.

President Love in the chair. Ten members present.

Mrs. W. H. Browning was elected a member of the Society. Mr. Rabe proposed Mr. Chas. Wunder, for active membership.

Mr. Davis spoke on *Cicindela consentanea,* which was taken at Manchester, N. J. He thought that it was a valid species and not a variety of *sexguttata.*

Mr. Schaeffer read a paper on *Dineutes.* He called attention to the variability of the apices of the male elytra of *D. hornii,* which are described as rounded, but a large series shows all intergrades from the rounded to projected apices of the female elytra.

Mr. Zabriskie exhibited under the microscope a transverse section of the elytron of *Cyllene robiniæ,* showing faded portion, also a few scales which retained their color. He spoke on coloration of insects and stated that dermal coloration will invariably remain, while hypodemal color will more or less fade after death. He further stated that the brightness of living insects depends greatly upon their emotion.

Mr. Davis stated that he succeeded in preserving the color of gold-fish with a mixture of Epsom Salt and Formaline, while he failed to preserve the color of some insects with this mixture. Dr. Love stated that a 2 % solution of Formaline is sufficient for preserving, but cannot be recommended as the Formaline will evaporate and nothing but water will remain.

Mr. Beutenmuller exhibited a curious abberration of *Pyrameis huntera* and Dr. Love showed a melanic form of *Argynnis aphrodite.*

After a general discussion, adjournment.

REVIEW OF THE AMERICAN CORYLOPHIDÆ, CRYPTOPHAGIDÆ, TRITOMIDÆ AND DERMESTIDÆ, WITH OTHER STUDIES.

By Thos. L. Casey.

The following pages record the results of a number of studies made at various times during the year just coming to an end, and may possibly be of some service to collectors in arranging their cabinets. The descriptions give only the salient characters of each species, and, in a genus such as the corylophid *Gronevus* for example, wherein the species mutually resemble each other very closely, can be appropriately limited to the few apparent differential characters. Further elaboration in such cases would prove to be largely repetition, and serve no really useful purpose in the present preliminary outline sketches, which are only intended to partially and imperfectly point the way.

Fort Monroe, Va., December 14, 1899.

HYDROPHILIDÆ.

Limnebius *Leach*.

The minute species composing this genus have the body elongate-oval and convex, the very small sparse punctures of the upper surface bearing each a fine decumbent hair. The labrum is transverse, with the apex sinuate at the middle. The inferior part of the eye is well developed and prominent, with the individual facets convex, but the superior part is not more convex than the frontal surface, with the facets larger and perfectly flat. The antennæ are partially received in repose in a very narrow groove between the eyes and the buccal opening, and, curving around the lower contour of the eyes, the club is

concealed within the deep depression for the eyes in the anterior part
of the hypomera.   The anterior coxæ are separated by a narrow prominent lamina.   Hind tarsi slender, the first two joints short.

This genus was investigated by the writer some time since under
the name *Limnocharis* Horn, (Bull. Cal. Acad. Sci.).   The male has
the sixth ventral more elongate, sometimes as long as the two preceding combined or even longer, the seventh transversely impressed at
base and the elytral apices transversely rounded.   In the female the
sixth ventral is not longer than the preceding, the seventh smaller and
simple, and the elytral apices are frequently obliquely pointed at tip.
The eighth segment, heretofore noted, is the projecting part of the
dorsal pygidium, and does not belong to the venter.   In most of the
species the male seems to be much less abundant than the female.

The American species of the genus may be defined as follows from
the female throughout :—

Elytra oval in outline, the sides arcuate..2
Elytra conical, truncate at tip, the sides straight...7
2—Elytral apices in the female obliquely subtruncate ; pubescence rather long.......3
Elytral apices in the female rounded ; last joint of the maxillary palpi fusiform......4
3—Piceous to black in color, the sutural angles distinctly rounded ; last joint of the
     maxillary palpi narrowly fusiform, pointed at tip ; pronotum obsoletely but rather
     coarsely micro-reticulate throughout.   California [*politus* Csy.]....**piceus** *Horn*
Black, the sutural angles extremely narrowly rounded and more nearly right, the sub-
     truncate apices slightly less oblique ; last joint of the maxillary palpi cylindric,
     the tip truncate; pronotum not micro-recticulate, except very feebly toward the
     sides ; body relatively narrower and more elongate.   Texas...**angustulus** *Csy.*
4—Black or piceous-black, the seventh ventral obtusely angulate.....................5
Castaneous in color, the seventh ventral longer, trapezoidal, its apex broadly arcuato-
     truncate 6
5—Pubescence of the upper surface long and well developed ; pronotum strongly
     micro-reticulate, the sides very feebly arcuate.   California (coast regions).
                                                                              alutaceus *Csy.*
Pubescence extremely short and inconspicuous ; pronotum very obsoletely and more
     coarsely micro-reticulate, the sides more arcuate ; body smaller and more slender.
     California (coast regions)..**congener** *Csy.*
6—Narrowly oval, moderately shining, rather coarsely micro-reticulate, the pubes-
     cence well developed but very fine ; prothorax strongly transverse, the sides con-
     vergent and feebly arcuate ; elytra scarcely one-half longer than wide, the apex
     unusually broadly rounded ; under surface piceous-black, the legs pale.   Length
     1.2 mm.; width 0.58 mm.   Vermont (Bennington Co.)......**discolor**, sp. nov.
7—Pale piceo-testaceous, the head and pronotum smooth, the elytra micro-reticulate ;
     pubescence long but sparse ; last two joints of the maxillary palpi stouter, the
     scutellum smaller than usual ; under surface and legs normal.   Texas.
                                                                      coniciventris *Csy.*

Alutaceus is the largest species, being fully 1.6 mm. in length, and *coniciventris* the smallest. The latter greatly resembles a species from South Africa, taken some years ago by the writer. *Piceus* is very abundant in the coast regions from Monterey northward, and the female described by me as *politus* does not seem to differ ; it is the only species before me which is represented by both sexes. Individual examples vary but little among themselves in point of size.

STAPHYLINIDÆ.

The genus *Homœusa* of Kraatz, represents an isolated group of the subtribe Aliocharina, containing a number of genera for the most part monotypic as far as known. Those before me may be characterized as follows :—

Antennæ 11-jointed..2
Antennæ 10-jointed... 4
2—Prothorax broadly and evenly rounded at the sides, the apical angles rounded....3
Prothorax broadly angulate at the sides at the middle, the angle rounded, the apical
    angles obtuse but not rounded, the base arcuate, not sensibly sinuate toward the
    angles, which are obtuse but not rounded ; two basal tergites broadly, equally
    and deeply impressed in about basal half ; fine elevated anterior bounding line of
    the metasternum strongly and narrowly arcuate anteriorly at the middle, the
    mesosternal process long and finely acuminate, extending to almost opposite the
    apices of the coxæ ; infraorbital elevated line of the head feeble and obtuse ; an-
    tennæ very strongly incrassate ; basal joint of the hind tarsi distinctly shorter than
    the next two combined ; [type *crassicornis* Csy.]...............**Myrmobiota** *Csy.*
3—Base of the prothorax transverse and broadly bisinuate, the basal angles nearly
    right and not at all rounded ; two basal tergites narrowly, deeply and rather
    abruptly impressed along the basal margin ; antennæ moderately incrassate ; an-
    terior marginal line of the metasternum transverse and only just visibly and very
    broadly arcuate anteriorly at the middle, the mesosternal process as in *Myrmo-
    biota ;* head with the infraorbital ridge very fine, and, between it and the eye,
    having two additional broad feeble and parallel ridges ; basal joint of the hind
    tarsi fully as long as the next two combined ; [type *acuminata* Märk.].
                                                                    Homœusa *Krtz.*
Base of the prothorax arcuate, becoming feebly sinuate near each angle, the latter
    slightly obtuse and distinctly, though narrowly, rounded ; basal tergites not im-
    pressed at base ; antennæ feebly incrassate, the last joint longer than the two pre-
    ceding combined and somewhat compressed apically ; mesosternal process angu-
    late, much shorter than in the two preceding, the metasternal line obscured in
    the type ; infraorbital ridge fine but distinct, the additional ridges of *Homœusa*
    wanting ; basal joint of the hind tarsi distinctly shorter than the next two com-
    bined ; pubescence longer and more conspicuous ; [type *rinitula* Csy. infra.].
                                                                    Soliusa, gen. nov.

4—Prothorax as in *Homœusa*, the base transverse and bisinuate, the angles right, not rounded and somewhat prominent, the sides arcuate and apical angles very broadly rounded ; basal tergites as in *Soliusa*, not impressed at base ; eyes smaller than usual, the antennæ rather strongly incrassate, the tenth joint probably formed by the fusion of two, but not relatively longer than in *Homœusa ;* infraocular ridge fine but abruptly and strongly elevated, the additional ridges wanting ; basal joint of the hind tarsi as long as the last and a little longer than the next two combined ; pubescence moderate in length as in *Homœusa* ; [type *expansa* Lec.]..**Decusa**, gen. nov.

The type of *Soliusa* may be briefly described as follows :—

Moderately stout and depressed, the head nearly three-fifths as wide as the prothorax, with the eyes slightly prominent ; antennæ but little longer than the width of the body ; prothorax three-fourths wider than long, equal in width to the elytra and distinctly shorter, the sides rotundato-convergent anteriorly ; abdomen at base nearly as wide as the elytra, acuminate, the sides straight ; fifth tergite at apex three-fifths as wide as the first ; color throughout pale brown, the abdomen a little darker, the surface rather shining and quite feeble punctulate ; pubescence conspicuous but subdecumbent as usual.   Length, 1.75 mm. ; width 0.63 mm. New York... ....**crinitula**, sp. nov.

Individuals seem to be rare in all of these genera and probably have throughout a more or less complete symbiosis with ants.   *Myrmobiota crassicornis* and *Decusa expansa* have both been sent to me by Mr. Wickham as having been discovered in ant-nests.

Mr. Wasmann ('Tijd. v. Ent. XLI), states that *Myrmobiota* Csy. (Col. Not. V, p. 594) is identical with *Homœusa* Krtz., citing specimens collected by Mr. Wickham and forwarded to him through Mr. Schmitt.   There is manifestly some mistake in identification, however, and my friend's remarks must refer to the species here described under the name *Soliusa crinitula* or to one closely allied thereto ; but if the latter surmise prove to be correct, I am forced to differ in opinion concerning the status of that species, for a study of the basal tergites of the abdomen, form of the mesosternum and thoracic base, and other characters, show that *crinitula*,

Fig. 1.—Prothorax of *Myrmobiota* and *Homœusa*.

also, is generically distinct.   In regard to *Myrmobiota*, there can be no doubt of its wide isolation from *Homœusa*, as an inspection of the small accompanying diagrams of the prothorax will abundantly demonstrate, the upper figure referring to *Myrmobiota* and the lower to *Homœusa acuminata*. That Mr. Wasmann has fallen into an error in identification, is furthermore evident at once from his statement that *crassicornis* (Wasm. nec

Csy.) differs from *acuminata*, among other minor characters, in having finer and denser pronotal punctuation, while, as a fact, the pronotum is much more coarsely punctured in *crassicornis* Csy. than in *acuminata*.

Chitosa gen. nov.

The type of this genus is *Dinarda nigrita* Rosh., which differs from *Dinarda*, as represented by *märkeli* and *dentata*, very profoundly in antennal and tarsal structure, as well as in the entire form of the prothorax and nature of the sculpture. In *Dinarda dentata* the prothorax is broadly and evenly bisinuate at base, and the sides near the basal angles are parallel and nearly rectilinear, the antennæ cylindrical, becoming somewhat acuminate at tip, and the basal joint of the hind tarsi but little longer than the second, the first four joints in fact diminishing only just visibly and quite regularly in length. In *nigrita*, on the other hand, the base of the prothorax is arcuate, becoming emarginate at each side, and the side margin is emarginate near the basal angles; the antennæ are gradually and strongly incrassate, a form wholly foreign to *Dinarda*, and, finally, the hind tarsi are very remarkable in structure and wholly different from any I have seen elsewhere in the Aleocharini. The basal joint is thicker than the remainder, darker in color or more highly chitinized, cylindrical and longer than the next three joints combined, the latter short, gradually diminishing in length and obliquely truncate at their apices, the fifth as long as the preceding three together and more slender. These characters prove that *Chitosa* is a genus quite isolated from any other ; it is however related to *Dinarda*. It occurs in Spain.

SCAPHIDIIDÆ.

This family seems to be very much better represented in America than in Europe, and a number of new forms have been discovered since my revision (Col. Not., V.).

Scaphidium *Oliv*.

The species before me seem to be five in number, *piceum* being quite evidently distinct from the maculate forms; they may be defined as follows:—

Elytra black, each with two pale subexternal spots... 2
Elytra uniform in coloration throughout, with a few discal coarse punctures in short
    series 4

2—Elytra without large punctures in series at any part of the disk, black, the spots small and flavate, the anterior triangular, not extending inwardly to outer third, the posterior very small, transversely oval, at outer fourth, twice as far from the apex as from the side margin. Length 4.7 mm. ; width 2.6 mm. Southern New England to Indiana..**obliteratum** *Lec.*

Elytra with coarse punctures in short series toward base and inner two-thirds, black, the spots large and generally rufescent...3

3—Coarse punctures very few in number, the remainder of the disk with the punctures very sparse and subobsolete ; posterior spot transverse and only very feebly sinuate anteriorly. Length 3.8-4.0 mm. ; width 2.3 mm. Indiana, Iowa and Kansas..**quadriguttatum** *Say*

Coarse punctures numerous and close-set in the series, the general punctuation usually more evident ; posterior transverse spot strongly arcuate, its anterior margin deeply sinuate ; body slightly smaller and distinctly narrower. Length, 3.8 mm. ; width 2.1 mm. Colorado? (Cab. Levette)....................**ornatum,** sp. nov.

4—Body deep black throughout, rather narrowly oval ; impressed area of the metasternum in the male sparsely punctate, the fulvous hairs longer. Length 4.0-4.2 mm. ; width 2.4-2.5 mm. Rhode Island to Indiana and Iowa.

piceum *Melsh.*

Body castaneous, more broadly oval, the metasternal area of the male larger, more closely punctured and clothed with shorter hairs ; sculpture similar to that of *piceum* and *quadriguttatum.* Length 4.5 mm. ; width 2.7 mm. Indiana.

amplum, sp. nov.

In *quadriguttatum* the first two, of the five joints constituting the antennal club, are equal in size and smaller than the last three ; in *ornatum*, however, the seventh joint is distinctly larger than the eighth.

Cyparium *Erichs.*

The two species now known to me may be distinguished as follows :—

Broader and somewhat oblong-oval, castaneous in color, the legs paler ; antennæ pale throughout. Length 3.2-3.3 mm. ; width 1.9-2.0 mm. North Carolina. [*substriatum* Reit]...**flavipes** *Lec.*

Narrower and evenly oval, the body black throughout, the head rufescent and the legs rufo-piceous ; antennæ pale, the 5-jointed club blackish ; eyes rather less widely separated on the front; punctures of the six abbreviated elytral series much smaller. Length 3.5 mm. ; width 2.0 mm. Texas (Brownsville).

ater, sp. nov.

The characters given by Reitter to distinguish *substriatum* (Verhand. Nat. Ver. Brünn, XVIII) are completely those of *flavipes*, and the name must therefore be relegated to synonymy.

Bæocera *Erichs.*

The known species of this genus have materially increased in number of late, and those in my cabinet may be arranged as follows :—

Scutellum not visible behind the basal lobe of the pronotum when the latter is normally adjusted to the base of the hind body...2
Scutellum visible though always very minute and transverse................................9
2—Basal marginal stria of the elytra entire.. 3
Basal stria either much abbreviated externally or interrupted about the middle of each elytron...7
3—Larger species, 1½ mm. in length or more...4
Very small species, scarcely exceeding 1 mm. in length................................6
4—Third antennal joint distinctly shorter than the fourth ; punctures along the dilated posterior margin of the intermediate acetabula small or moderate in size...5
Third antennal joint fully as long as the fourth ; marginal punctures very coarse ; color black throughout, the elytra feebly rufescent at the apical margin and the abdomen paler toward tip ; legs dark rufous ; upper surface not distinctly punctate at any part. Length 1.6 mm.; width 0.9 mm. Iowa (Keokuk).
                                                                            speculifer *Csy.*
5—Body broadly oblong-oval, the third antennal joint as long as the second, black, the elytra picescent posteriorly, very obsoletely punctulate ; legs dark rufous. Length 2.0–2.6 mm.; width 1.3–1.75 mm. Pennsylvania, District of Columbia, Illinois, and Iowa ....**concolor** *Fabr.*
Body narrowly oval, black throughout, the elytra rather abruptly rufous in apical sixth or seventh ; legs dark rufous ; elytra sparsely and very obsoletely punctulate ; third antennal joint much shorter than the second. Length 1.9–2.2 mm.; width 1.2–1.25 mm. New York (Long Island), North Carolina (Asheville), and Indiana ....**congener** *Csy.*
6—More narrowly oblong-oval, black, the elytra more distinctly rufescent at apex ; antennæ shorter, not as long as the width of the body. Length 1.15–1.25 mm.; width 0.7–0.75 mm. Rhode Island (Boston Neck), and Michigan
                                                                            apicalis *Lec.*
Broadly and evenly elliptical, the median line of the body very much more arcuate in profile, deep black, the elytral apices scarcely paler ; antennæ relatively distinctly longer, as long as the width of the body. Length 1.1–1.15 mm.; width 0.7–0.72 mm. Texas (Columbus)..................................**robustula** *Csy.*
7—Marginal line of the prothorax evenly and moderately arcuate when viewed laterally ; mesepimera narrower ; species very small in size................................8
Marginal line, when viewed laterally, nearly straight, becoming rather abruptly and strongly bent downward through basal third ; mesepimera broader and not extending quite so far toward the coxæ ; metepisternal suture coarse, straight as usual ; size large, the body black, the elytra feebly rufescent at tip, not distinctly punctulate ; antennæ long and slender. Length 2.3–2.5 mm.; width 1.3–1.35 mm. Rhode Island(Boston Neck), Virginia and Arkansas........**deflexa** *Csy.*
8—Body moderately convex longitudinally, the metasternum more elongate, the episternal suture not very coarse ; color pale flavescent throughout, the pronotum gen-

erally shaded a little darker; basal stria of the elytra fine, disappearing completely somewhat before attaining the middle of the width. Length 1.15 mm.; width 0.68 mm. Pennsylvania (near Philadelphia)..............**pallida,** sp. nov.

Body strongly convex longitudinally, smaller in size, deep black, the elytra rufescent toward tip ; under surface blackish, the abdomen and legs fulvous ; metepisternal suture much shorter and very coarsely excavated ; basal stria of the elytra stronger and only interrupted for a short space just beyond the middle of the width. Length 0.88 mm.; width 0.62 mm. Massachusetts (Tyngsboro).

abdominalis, sp. nov.

9—Basal stria of the elytra entire ; body larger, blackish throughout, the elytral apices very narrowly rufescent ; antennæ moderately long and slender, bristling with rather long stiff setæ ; basal lobe of the pronotum rather feeble and broadly rounded ; scutellum distinct. Length 1.7 mm.; width 0.95 mm. Texas (Columbus)..**texana** *Csy.*

Basal stria of the elytra much abbreviated externally ; size minute....................10

10—Basal angles of the prothorax normally acute..11

Basal angles obtusely truncate at tip ; metepisternal suture arcuate 14

11—Scutellum very short and indistinct ; abdomen concolorous ; metepisternal suture straight.......... 12

Scutellum distinct and longer than usual ; body deep black throughout, the entire abdomen abruptly pale rufous ; size very minute............. 13

12—Metepisternal suture coarse ; scutellum extremely small, short and very transverse ; body very smooth and polished, rufous throughout, the pronotum piceous ; antennæ moderate. Length 1.3 mm. ; width 0.72 mm. Michigan.

discolor, sp. nov.

Metepisternal suture finer ; scutellum less abbreviated, but little more than twice as wide as long ; body rufo-piceous to blackish in color, the abdomen pale at tip. Length 1.15-1.25 mm. ; width c.65-0.7 mm. Pennsylvania (near Philadelphia) and Michigan............... ....·..**picea** *Csy.*

13—Rather narrowly oval, highly polished and impunctate ; metepisternal suture feebly arcuate, fine and rather distinctly punctured ; mesepimera rather small and narrow, scarcely extending more than half way to the coxæ. Length 0.95 mm. ; width 0.55 mm. Rhode Island (Boston Neck).........**rubriventris,** sp. nov.

14—Rather stout, polished, black throughout above and beneath, the legs feebly rufescent. Length 0.9-1.0 mm. ; width 0.58-0.68 mm. Rhode Island (Boston Neck), Massachusetts, Michigan and Texas (Austin)...................**nana** *Csy.*

Nana is a very widely distributed species of minute size, and is quite aberrant in the form of the basal angles of the prothorax and in the strongly arcuate metepisternal suture, but it does not differ generically.

Scaphiomicrus, gen. nov.

The species described by LeConte under the name *Scaphisoma pusilla,* must form the type of a distinct genus because of the shorter and thicker antennæ, situated at a greater distance from the eyes,

which are notably smaller, the shorter tarsi, and especially, because of
the radically different form of the post-coxal plates of the abdomen.
These plates in *Scaphisoma* are very short and only developed internally, the bounding arc extending outward externally, very gradually
approaching the base of the segment, while in *Scaphiomicrus* the plates
are more nearly semi-elliptic, having the outer part of the bounding
curve directed upon the base without change of direction toward the
sides of the body, somewhat, in fact, as in the subgenus *Pullus* of the
Coccinellidæ. The species are all very much more minute than in
Scaphisoma, and those which are represented before me may be distinguished by the following characters : —

Abdominal plates almost evenly parabolic in form, the apex more broadly rounded
    and the outer side more arcuate and approaching the base scarcely less obliquely
    than the inner side ; sutural line of the elytra not flexed outward basally.........2
Abdominal plates more narrowly rounded at apex, the external branch of the bound-
    ing curve much less arcuate than the internal, and directed almost perpendicu-
    larly upon the base ; sutural line of the elytra flexed outward at base, parallel to
    the basal margin ; elytra blackish, gradually and broadly pale toward tip.........5
2—Elytra bicolored, black in about basal half, the remainder rufous......................3
Elytra pale throughout...4
3—Abdominal plates extending much beyond the middle of the segment, the
    punctures and the reticulations of the segment almost effaced ; form rather short
    and stout. Length 0.8–0.95 mm.; width 0.55–0.65 mm. North.Carolina (Ashe-
    ville) ....**pusillus** *Lec.*
Abdominal plates not quite extending to the middle of the segment, the surface of
    which is distinctly reticulate and finely, sparsely punctulate ; metasternum strongly,
    though sparsely, punctate; body slightly larger, the elytra destinctly longer when
    compared with the prothorax. Length 1.1 mm.; width 0.7 mm. Rhode Is-
    tand (Boston Neck)...**dimidiatus,** sp. nov.
4—Entire body and legs pale fulvo-testaceous throughout, the form more narrowly
    oval ; abdominal plates broadly rounded, not extending quite to the middle of
    the segment ; metasternal punctures minute and very feeble. Length 0.9 mm.;
    width 0.6 mm.; Michigan,—Mr. Schwarz.....................**flavescens,** sp. nov.
5—Sutural line of the elytra extending outward along the base to inner third.........6
Sutural line extending nearly to the middle of the width ; body more minute and less
    oval...7
6—Abdominal plates relatively larger, extending to the middle of the segment ; body
    blackish, the legs, antennæ and apical half of the abdomen pale; elytral punctures
    sparse but rather distinct, effaced as usual toward base. Length 0.85 mm.;
    width 0.57 mm. Lake Superior ....**lacustris,** sp. nov.
Abdominal plates very small, extending but little beyond basal third of the length,
    narrowly rounded at apex ; body in coloration and sculpture nearly similar to
    lacustris, the outline a little more broadly oval. Length 0.9 mm.; width
    0.65 mm. Iowa (Keokuk)..**nugator,** sp. nov.

7—Minute in size, blackish, the elytra gradually rufescent behind the middle, sparsely, finely and very obsoletely punctate, the punctures almost effaced; legs yellow; abdominal plates well developed, extending almost to the middle. Length o. 7 mm.; width 0.47 mm.   Oregon....................**exiguus,** sp. nov.

CORYLOPHIDÆ.

The Corylophidæ constitute a small family, evidently allied to the Silphidæ, as shown by antennal structure, and, like them, display great variety in external habitus; they are, however, remarkably homogeneous among themselves in sternal and abdominal structure. In *Orthoperus* a relationship with Scaphidiidæ can be observed, and there are some characters, such as the 4-jointed tarsi with the third joint small, the post-coxal plates of the Corylophini and the projecting rounded pronotum of the Parmulini—homologous with Cranophorini,—which proclaim an indubitable relationship with the Coc-

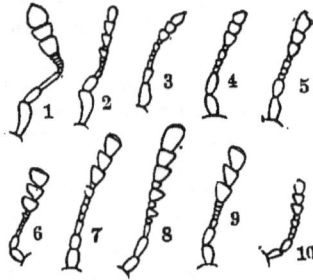

FIG. 2.—Antennæ of CORYLOPHIDÆ—1 *Bathona* (*Corylophodes* is similar, except that the third joint is shorter than the second); 2 *Gronevus* (also nearly of *Rypobius*); 3 *Sericoderus*; 4 *Orthoperus*; 5 *Eutrilia*; 6 *Molamba lunata*; 7 *Molamba obesa*; 8 *Sacium montanum*; 9 *Arthrolips nimius*; 10 *Œnigmaticum californicum.*

cinellidæ. The chief difference in tarsal structure between these two families resides, indeed, simply in the freedom of the third joint, this being generally anchylosed to the fourth in Coccinellidæ. The anterior coxæ are narrowly separated, displaying variations which serve to define tribal groups, and the cavities are broadly closed behind; the intermediate are more widely separated and the posterior mutually very remote. The scutellum is always distinct, though small, the abdomen hexamerous, the first segment being much the longest and the palpi short, stout and acuminate. The American species may be assigned to four tribes as follows : --

Prothorax widest at base............. 2
Prothorax narrowed at base ; body narrower and lathridiiform..........................3
2—Anterior coxæ long and narrow, inclosed within deep oblique fossæ and attached
    more externally, the intermediate and posterior generally with distinct post-coxal
    plates ; body rounded or oval and convex, generally glabrous ; antennæ 11- or 9-
    jointed...CORYLOPHINI
Anterior coxæ short, oblong ; body pubescent, the pronotum covering the head, con-
    vex, the edges not explanate and the hind angles greatly produced posteriorly
    and acute ; antennæ 10-jointed ; abdomen more extensile, with the basal segment
    shorter than usual ; post-coxal plates wanting...........................SERICODERINI
Anterior coxæ larger, less deeply imbedded and globular ; body more depressed, ob-
    long or oval, pubescent, the pronotum completely concealing the head, explanate
    at the margin, the hind angles not produced posteriorly ; antennæ 11- to 10- or
    possibly 9-jointed................. ....PARMULINI
3—Head completely exposed from above ; anterior coxæ small, oval ; antennæ
    9-jointed ; integuments sparsely and feebly pubescent................ÆNIGMATICINI

All of these tribes occur on both sides of the continent, but in the
first tribe the genera with 9-jointed antennæ are the only ones which
have thus far appeared in the Pacific district.

CORYLOPHINI.

The species of this tribe may be readily known by their rounded or
oval convex form and shining glabrous integuments. The genera
may be separated as follows :—

Antennæ 11-jointed, inserted between and near the eyes, widely separated at base,
    the eyes larger and coarsely faceted ; epipleuræ rather wide and inflexed.........2
Antennæ 9-jointed, inserted more anteriorly and more distant from the eyes, which
    are smaller and less coarsely faceted ; basal joint shorter ; epipleuræ extremely
    narrow or subobsolete, not at all inflexed ; labrum broadly rounded ; prothorax
    emarginate at apex, the head in great part exposed ; post-coxal plates very short ;
    tarsi slender...5
2—Head very deeply inserted within the prothorax, the anterior margin of which is
    evenly rounded and strongly descending ; post-coxal plates large, with rounded
    outline, the subbasal discal line of the metepisterna very oblique ; labrum rounded
    or subquadrate..3
Head partially protruded and less concealed by the overhanging margin of the pro-
    thorax ; post-coxal plates very short, the subbasal line of the metepisterna nearly
    transverse ; labrum small, triangular, with the apex acuminate ; tarsi dilated......4
3—Post-coxal plates of the abdomen more strongly rounded, the external part of the
    bounding line directed upon the base well within the sides ; third antennal joint
    elongate, longer than the second ....**Bathona**
Post-coxal plates of the abdomen less arcuate posteriorly, the bounding line extending
    to the sides of the body ; third antennal joint elongate but shorter than the sec-
    ond....... ....**Corylophodes**

4—Head entirely concealed from above, the prothorax almost evenly rounded anteriorly and more descending, the margins not distinctly thickened and the hind
angles acute and somewhat more posterior than the median parts of the base;
maxillary palpi moderately stout, regularly acuminate.....................**Gronevus**
Head partially visible from above, the prothorax sensibly sinuate at apex, the margins
with a distinct thickened bead and the hind angles right; maxillary palpi very
stout.............. ....**Rypobius**
5—Anterior tibiæ flattened, the external edge acute, the axial line feebly arcuate
throughout; body larger..**Eutrilia**
Anterior tibiæ slender, inwardly bent distally; body very minute.........**Orthoperus**

I have restored the original spelling of *Rypobius*, although it may
not be etymologically correct.   The European *Moronillus* of Du Val
is identical, having similar structure and habits.

Bathona, gen. nov.

In this genus the body is broadly oval, convex, polished and glabrous, with the edges of the pronotum subexplanate and diaphanous,
and the hind angles not posteriorly produced.   The tarsi are long and
are compressed toward base.   The species may be defined as follows :—

Body moderately convex, the sides and apex of the pronotum widely subexplanate;
antennal club large..2
Body very strongly, globularly convex, the limb of the pronotum very narrowly subexplanate; antennal club rather less developed...3
2—Elytral punctures small and sparse but very distinct, impressed, each bearing, as
usual, an excessively minute fine decumbent hair, black, the pronotum piceous-
black, the edges broadly transparent and hyaline; under surface paler, the legs
and antennæ rufescent.   Length 1.35 mm.; width 0.9 mm.   North Carolina
(Asheville)...**carolinæ**, sp. nov.
Elytral punctures smaller and almost effaced; body smaller, black, the pronotal limb
broadly transparent and hyaline; under surface and legs paler.   Length 1.1 mm.;
width 0.75 mm.   Virginia (Norfolk)............................**virginica**, sp. nov.
3—Impunctate, piceous in color, the edges of the pronotum narrowly transparent
and hyaline, the disk gradually darker toward the middle and base.   Length 1.0
mm.; width 0.78 mm.   North Carolina (Asheville) ....**convexa**, sp. nov.
Smaller and with feeble but visible traces of punctuation, piceous or testaceous in color,
the under surface and legs more flavate.   Length 0.8 mm; width 0.7 mm.
Pennsylvania (near Philadelphia)...............................**sphæricula,** sp. nov.

Individuals are rare, but *virginica* is represented before me by a
number of examples, which exhibit no noteworthy variability.

Corylophodes *Matth.*

As in the preceding, the antennæ in this genus have five small com-
pacted and gradually wider joints between the third and the first joint

of the club, and there is no vestige of an enlargement of the second joint before the club as there is in *Rypobius, Gronevus* and *Ortho-perus.* The structure of the shaft differs in fact very radically, and, in this way, these genera are widely isolated. *Corylophodes* resembles *Bathona* in general structure, but, besides the characters indicated in the table, it differs in the generally more narrowly oval form of the body, shorter and less developed prothorax, finely margined along the basal lobe, and more slender and less coarctate five antennal joints immediately succeeding the third. The tarsi are nearly similar, but the anterior are feebly dilated in the male. The three species before me may be thus distinguished among themselves :—

Elytral punctures sparse but rather coarse, deeply impressed and very distinct, blackish-piceous, the limb of the pronotum broadly transparent and hyaline ; legs and
antennæ paler. Length 0.9 mm. ; width 0.72 mm.   Rhode Island, Pennsylvania and North Carolina (Asheville)............................**marginicollis** *Lec.*
Elytral punctures extremely minute and subobsolete...2
2—Form narrowly oval, rufo-piceous in color, the pronotum with broad hyaline
margin as in the preceding and succeeding species. Length 0.85 mm. ; width
0.65 mm.   Florida ....**impunctatus**, sp. nov.
Form more broadly oval, black, the legs, trophi and antennæ pale ; prothorax transverse, the basal angles obtusely blunt as usual. Length 0.8–0.9 mm. ; width
0.7–0.75 mm.   Texas (Brownsville)....................**subtropicus**, sp. nov.

As pointed out by Mr. Matthews, the distinguishing feature of *Cory-lophodes* is the slender third antennal joint shorter than the second, but the author makes no allusion at all to the remarkable post-coxal plates. The genus as extended by its author in the "Monograph" is very composite, and I am unable to place the *C. schwarzi*, described therein from California.

Gronevus, gen. nov.

This and the succeeding genus differ very greatly from the two pre-ceding, in the very short and almost obsolete post-coxal plates, the meso-coxal being even much less developed than in *Orthoperus*, but the subtransverse line at the base of the metepisterna is present as in that genus ; the comparatively wide and steeply inflexed epipleuræ distinguish them at once however from *Orthoperus* and *Eutrilia.* They also differ quite radically in antennal structure, and from all others of the tribe, in the shorter and slightly dilated tarsi. In *Gronevus* the limb of the pronotum is hyaline and moderately widely subexpla-nate, the base not margined, and the hind angles are acute and distinctly though not abruptly produced posteriorly. The European *Peltinus* and *Corylophus* differ in having very narrow horizontal epipleuræ and more slender tibiæ.

The species are somewhat abundant but closely allied; those in my cabinet may be recognized by the following characters :—

Elytra finely and sparsely but more or less distinctly punctate......................................2
Elytra impunctate..5
2—Elytral punctures very minute throughout...3
Elytral punctures strong, especially on the descending flanks ; body smaller, more rounded, very strongly subglobularly convex, blackish, the pronotum paler, with narrow hyaline limb. Length 0.8 mm. ; width 0.65 mm. Iowa.

sticticus, sp. nov.

3—Antennal club blackish ; body more strongly and globularly convex, black, the pronotum slightly piceous, with narrow colorless hyaline margins ; scutellum twice as wide as long, very broadly rounded. Length 0.9 mm. ; width 0.7 mm. Canada (Ottawa)..**fuscicornis,** sp. nov.
Antennal club very pale, not differing in color from the shaft, body rather less convex...4
4—Blackish, the pronotum rufo-piceous, sometimes entirely pale from immaturity ; elytra but little more than twice as long as the prothorax ; scutellum twice as wide as long. Length 0.8-0.9 mm. ; width 0.63-0.75 mm. Massachusetts, Rhode Island, New York and New Jersey..**truncatus** *Lec.*
Paler, piceo-testaceous, the prothorax still paler, form more elongate-oval, the elytra much more than twice as long as the prothorax ; scutellum less transverse and somewhat ogival ; elytral punctures still finer, almost completely effaced posteriorly ; size a little larger as a rule. Length 0.95 mm. ; width 0.8 mm. Iowa and Nebraska..**hesperus,** sp. nov.
5—Blackish, the pronotum and elytral suture rufescent ; elytral margin at and near the humeri more widely subexplanate than in the preceding species. Length 0.78-0.85 mm. ; width 0.65-0.75 mm. Virginia (Norfolk and Fort Monroe)

lævis, sp. nov.

Individuals of the various species are much more abundant than in the two preceding genera, as is also the case in *Rypobius*.

Rypobius *Lec.*

The body in this genus is evenly oval and rather strongly convex, the pronotum evenly declivous toward the limb, which is not reflexo-explanate and not transparent or hyaline at the edges ; the hind angles being right and the apex sensibly sinuate indicates a closer affinity with *Orthoperus*. The integuments are minutely reticulate, and each of the very minute sparse punctules bears a small and very fine decumbent hair. The scutellum is less than twice as wide as long and is parabolic in form. The tarsal claws are rather long, slender and arcuate, with a feeble internal dilatation at base.

The genus *Glæosoma*, of Wollaston, occurring in the Island of Madeira, which has been considered to be identical with *Rypobius*, is altogether distinct, not only in its 10-jointed antennæ and type of elytral sculpture, but in its habits and gait, the single species of *Glæosoma* taking refuge under stones and running with great velocity when disturbed—habits wholly foreign to *Rypobius*. It may however be placed near *Rypobius* in a tabular arrangement of the genera of the family. A Spanish specimen sent me by Mr. Keitter under the name *Rypobius velox*, differs from the true *Rypobius* also in the elytral epipleuræ, which are inclined upward and not at all inflexed, and also in the hind angles of the prothorax, which are acute and sensibly produced posteriorly. I am unable to count the joints of the antennæ with certainty in this example.

The two species of *Rypobius* before me may be distinguished very readily as follows :—

Base of the prothorax almost rectilinear, piceous, the prothorax and under surface paler ; micro-reticulation of the upper surface very deep, the lustre somewhat alutaceous. Length 1.0–1.2 mm.; width 0.7–0.85 mm. Rhode Island, New Jersey and Virginia (near the ocean beaches)...........................**marinus** *Lec.*

Base of the prothorax distinctly bisinuate, blackish, the pronotum rufescent ; legs, antennæ and trophi flavate, polished, the micro-reticulation of the upper surface almost completely effaced ; size very much smaller. Length 0.68 mm.; width 0 5 mm. Texas (Columbus)..**minutus**, sp. nov.

In both these species the first abdominal segment is as long as the next three combined. *Minutus* must bear some resemblance to the Central American *guatemalensis* Matth., but differs in sculpture.

Eutrilia, gen. nov.

The single representative of this genus resembles a very large, broadly oval *Orthoperus*, and is evidently very closely related, being identical in the form of the prothorax and in the structure of the head, coxæ and under surface. It however differs in the form of the anterior tibiæ, as indicated in the table, and in the virtual absence of any trace of epipleuræ, these being indicated only by a slight thickening of the elytral margins due to the very minute marginal bead. The meso-coxal plate is rather well developed, but the metacoxal plate is extremely short as in *Orthoperus*. The intermediate tibiæ are slightly thickened externally just beyond the middle with arcuate outline, the posterior straight and the tarsi slender, the claws small, arcuate and very slender. The first ventral segment is as long as the next four

66 JOURNAL NEW YORK ENTOMOLOGICAL SOCIETY. [Vol. VIII.

combined. The integuments are micro-reticulate and very finely punctate, each puncture bearing a small decumbent hair, these however being a little longer and more conspicuous than in *Rypobius*. The scutellum is well developed, with broadly parabolic outline, and the elytral suture is not at all margined. The wings are well developed, the fringing hairs very short. As in some species of *Orthoperus*, especially *scutellaris*, there is a feeble impressed longitudinal line on each elytron near outer fifth or sixth, extending from the base for a short distance :—

Oval, convex, moderately shining, brownish-testaceous in color throughout, the prothorax moderately developed, the base broadly parabolic, the sides strongly convergent, moderately arcuate and continuous in curvature with those of the elytra; punctures not distinctly visible under a hand lens of moderate power. Length 0.95 mm.; width 0.8 mm. California........................ ...**bruonea,** sp. nov.

Orthoperus *Steph.*

The species of this genus are among the most minute of the Coleoptera, and may be readily recognized by their oval, moderately convex form, exposed head and 9-jointed antennæ, the fifth joint being generally notably longer and sometimes thicker than the sixth. The epipleuræ are represented by a narrow side margin of the descending flank, delimited by a fine line. The integuments of the body are more or less shining, micro-reticulate and virtually glabrous. The pronotum is very finely and feebly margined at base, the flanks not greatly descending, becoming very narrowly and feebly reflexo-explanate toward the basal angles, which are nearly right, not at all produced posteriorly and narrowly rounded. The scutellum is distinct though small and generally parabolic in shape. The species are rather numerous but closely allied among themselves as a rule ; those before me may be recognized as follows :—

Elytral suture not margined except posteriorly................... 2
Elytral suture finely margined to the scutellum.................................,.............8
2—Elytra without trace of punctuation of any kind, the minute reticulation distinct ; body much larger than usual, evenly elliptical, glabrous, pale brown in color throughout, the metasternum impunctate and with very sparse and microscopic hairs. Length 0.82 mm.; width 0.62 mm. Colorado........**princeps,** sp. nov.
Elytral punctures wanting but replaced by very small, sparse and V-shaped scratches ; size rather large...3
Elytral punctures normal, being well-defined points but always very minute and sparse ; size small or very minute.. 4
3—Form somewhat oblong and narrowly oval, black throughout, the legs and antennæ rufescent ; micro-reticulation distinct ; metasternal punctures sparse but distinct.

Length 0.7 mm.; width 0.45 mm. Lake Superior, Northern Illinois and Cali-
fornia (Siskiyou Co.)..**scutellaris** *Lec.*
  Var. A—Piceous and more broadly oval with more arcuate sides, the micro-
  reticulation less distinct and the scratch-like punctulation more visible. New
  York and Ohio...**piceus, v. nov.**
  Var. B—Similar to *piceus* but with the punctules sparse and the elytra more
  rapidly narrowed toward tip. Washington State (Spokane)..**lucidus, v. nov.**
4—The punctures strong deep and very distinct, more especially so toward the suture
  and base of each elytron and toward the base of the pronotum, piceous in color,
  the legs luteo-flavate ; form rather narrowly oval. Length 0.65 mm.; width 0.4
  mm. California (Sta. Cruz Co.)................................... **cribratus** *Matth.*
The punctures extremely fine throughout and only visible under strong amplification ;
  size small..5
5—Form oblong-oval..6
Form evenly oval with more arcuate sides...7
6—Piceous-black, the micro-reticulations finer and stronger, giving a feebly subalutaceous lustre. Length 0.6 mm.; width 0.4 mm. New Jersey, Pennsylvania,
  Delaware, North Carolina and Florida.................................**glaber** *Lec.*
Paler, piceous, smaller and more polished, the reticulation coarser and less visible ;
  suture more strongly margined posteriorly. Length 0.5 mm.; width 0.35 mm.
  Florida (Enterprise)...**suturalis** *Lec.*
7—Reticulations feeble, the surface more highly polished, piceous in color, the eyes
  separated on the front by nearly three times their own width. Length 0.6 mm.;
  width 0.4 mm. Texas (Austin)....................................**texanus, sp. nov.**
Reticulations strong ; body smaller and paler in color, the eyes distinctly larger, separated on the front by but little more than twice their own width ; metasternum
  more coarsely reticulate. Length 0.5 mm.; width 0.35 mm. Illinois.
                                                              micros, sp. nov.
8—Scutellum more transverse, ogival or rounded ; body more oblong-oval............9
Scutellum scarcely wider than long, triangular, the sides straight.......................10
9—Reticulations feeble and finer, the surface polished ; elytral punctures excessively
  fine and scarcely visible ; color testaceo-piceous ; head well developed. Length
  0.65 mm.; width 0.48 mm. Arizona (Tucson) and southern California.
                                                              arizonicus, sp. nov.
Reticulations strong, the lustre somewhat alutaceous ; punctures extremely fine and
  sparse but more visible ; head smaller, the coloration darker, piceous-black.
  Length 0.6 mm.; width 0.4 mm. Texas (Columbus)......**alutaceus, sp. nov.**
10—Body oval, pale testaceous in color, polished, the reticulation very feeble, the
  punctures almost completely obsolete but simple ; sutural margin strongly defined,
  extending unbroken by the scutellum and along the base to beyond the middle of
  the width ; head well developed, meso-coxal plates shorter, with the bounding
  line more rectilinear and transverse. Length 0.55 mm.; width 0.35 mm.
  Bahamas (Harbor Island)..**bahamicus, sp. nov.**

Elongatus of LeConte, belongs to the Ænigmaticini. The *crotchi*
of Matthews I, have not seen.

SERICODERINI.

This tribe is well differentiated from the preceding in the oval pubescent body, with more extensible abdomen and absence of distinct post-coxal plates, and from the following in the non-explanate limb of the pronotum; from both it may be distinguished by the 10-jointed antennæ and shorter basal segment of the abdomen, this, in the extended condition, scarcely equaling in length the next two together. There is but a single genus.

Sericoderus *Steph.*

The species of this genus are so closely allied among themselves that it is scarcely possible to detect structural differences of any kind, and the names given below might be considered to represent subspecies of a single type form. The head is completely concealed from above and moderately deeply inserted, the pronotum broadly rounded at apex and with the hind angles acute and considerably produced posteriorly. The antennæ are slender, with the basal joint narrowly oval and inserted in shallow frontal foveæ at a slight distance from antero-internal margin of the eyes, the latter usually well developed and coarsely faceted. The frontal margin is feebly sinuato-truncate and the labrum short and broadly rounded. The tibiæ and tarsi are slender, and the elytral epipleuræ narrow, becoming strongly inflexed toward base. The following forms seem to be worthy of distinctive names :—

Species of the Atlantic and Gulf regions.. 2
Species of the Pacific slope...5
2—Elytra more strongly narrowed from base to apex.......................... 3
Elytra feebly narrowed, the form more quadrate.....................................4
3—Larger, pale luteo-flavate in color, the usual nubilate subapical spot of the pronotum piceous. Length 0.9 mm.; width 0.65 mm. New York to Lake Superior ....**flavidus** *Lec.*
Smaller, the elytra generally piceous, the pronotum flavate with the subapical spot darker. Length 0.75-0.85 mm.; width 0.6-0.65 mm. Massachusetts, Pennsylvania, and North Carolina...**obscurus** *Lec.*
4—Color pale flavate throughout, the elytra never darker, smaller in size than *flavidus* and more southern in distribution. Length 0.75 mm.; width 0.6 mm. Texas (Brownsville and Austin), Florida and Illinois...............**subtilis** *Lec.*
5—Larger, very broadly oblong, coarsely pubescent, dark rufo-testaceous, the usual subapical spot of the pronotum darker; elytra but feebly narrowed from base to apex; metasternum coarsely imbricato-reticulate but not distinctly punctured. Length 0.85 mm.; width 0.68 mm. California (Monterey).
quadratus, sp. nov.

Very small, the elytra more rapidly narrowed from the base, piceous-brown in color, the pubescence rather less coarse ; metasternum distinctly punctured, especially toward the sides. Length 0.7 mm.; width 0.55 mm. California (Sonoma Co.).

debilis, sp. nov.

When discovered, individuals are rather abundant. *Sericoderus* is said in the "Biologia" to have the antennæ 11-jointed, but these organs are quite evidently 10-jointed in our species, and the details given by Mr. Matthews for the antennæ of *S. latus* show that it should properly form a genus distinct from *Sericoderus*.

PARMULINI.

The numerous species of this tribe can be recognized at once by the oblong or oval and less convex pubescent body, more or less widely subexplanate at the lateral and apical limb of the pronotum, and the rectangular thoracic angles. The genus *Sacium* of LeConte, is said by Heyden, Reitter and Weise to be the same as *Parmulus* of Gundlach, but, as the LeContean *Sacium* is composite, I am in doubt as to which if any is the true *Parmulus* and have therefore not adopted the name for any one of our genera ; it is, however, retained for the tribal designation. The three genera before me may be identified by the following characters : —

Antennæ 11-jointed ; prosternum well developed in front of the coxæ, the posterior margin of the buccal opening deflexed at the middle, forming a broad inferiorly vertical liguliform process..2
Antennæ 10-jointed ; prosternum extremely short in front of the coxæ, the buccal margin not at all deflexed..3
2—Body elongate and subparallel ; antennæ more elongate, the club relatively longer, looser and more serriform, the fifth and seventh joints both enlarged ; basal joint of the hind tarsi shorter, scarcely as long as the next two combined, the basal joints thicker...**Sacium**
Body oval, with more arcuate sides ; antennæ shorter, with a more slender shaft and stouter and more compact club, the seventh joint enlarged, the fifth normal ; hind tarsi with the basal joint elongate and more slender, always much longer than the next two combined..**Molamba**
3—Body oval or oblong-oval, rather more convex but nearly as in *Molamba ;* antennæ moderately elongate, the club loose and well developed, the elongate third joint followed by four small subequal joints ; posterior tarsi slender, with the basal joint elongate, the anterior more or less dilated at base ; species generally minute ....**Arthrolips**

The epipleuræ are horizontal, moderately wide, narrowing gradually and disappearing behind the middle, the first ventral segment very long, equaling the next three or four combined.

These genera are all widely distributed over the continent, but *Sacium* has not yet been found near the Pacific coast line, although occurring in Utah ; it is more northern in habitat than the other two. *Molamba* may perhaps prove to be the same as *Parmulus*, but at present I have no means of determining this.

Previous authors appear to have entirely overlooked the very conspicuous vertical ligula at the hind margin of the buccal opening in *Sacium* and *Molamba*, a character wholly wanting in *Arthrolips*.

Sacium Lec.

In this genus, and the next, the limb of the pronotum is broadly reflexo-explanate antero-laterally, and is scarcely at all so at the middle of the apical margin, while in *Arthrolips* the edge seems to be more narrowly and evenly reflexed throughout the external circumference. *Sacium*, as understood by Mr. Matthews in the "Monograph," is composite, and *lugubre* should have been assumed as the type. The genus includes some of the largest species of the family known to me, and the four represented in my cabinet may be thus defined in brief :—

Prothorax as long as wide or very nearly ; elytral punctures and pubescence rather
    sparse 2
Prothorax shorter and more transverse, much wider than long in both sexes ; elytral
    punctures and pubescence dense.......... 4
2—Elongate, twice as long as wide or very nearly, the elytral punctures smaller and
    rather sparser, piceous to blackish in color, with the usual two pale patches at
    the apex of the pronotum. Length 1.75–2.1 mm.; width 0.9–1.0 mm. Colo-
    rado ....**montanum**, sp. nov.
Less elongate, always distinctly less than twice as long as wide, the elytral punctures
    stronger and less sparse...3
3—Piceous in color to blackish, the elytral suture sometimes slightly paler. Length
    1.6–1.8 mm.; width 0.85–0.9 mm. Lake Superior..................**lugubre** Lec.
Deep black, the suture not paler ; body a little larger and relatively broader. Length
    1.7–2.0 mm.; width 1.0–1.1 mm. Utah (southwestern).......**nigrum**, sp. nov.
4—Body rather smaller, piceous-brown in color, the under surface and legs still paler.
    Length 1.65 mm.; width 0.85 mm. Lake Superior.... ....**obscurum** Lec.

The elytra are uniform in color throughout and there is an entire absence of the paler maculation so characteristic of the two following genera.

The genus represented by *Clypeaster maderæ* Kr. (*pusillus* Woll., nec Gyll.) is radically distinct from *Sacium* in the structure of the antennæ as figured by Wollaston, and I would propose the generic name **Clypeastodes** for that species .

Molamba, gen nov.

The species are much more numerous than those of *Sacium* and are generally of smaller size, though *obesum* is as large as any representative of that genus discovered thus far.   Those before me may be conveniently arranged as follows:—

Elytra uniform in coloration or devoid of distinctly defined pale maculation.............2
Elytra dark in color, with rufous or flavous maculation.......................................3
2—Pubescence long, distinct and rather abundant; body large, piceous, the pronotum
    pale anteriorly, but darker at the middle as usual; punctures fine, rather close-set;
    elytra distinctly longer than wide, sometimes very feebly rufescent behind the
    middle in a small area but not obviously so. Length 1.9 mm.; width 1.2 mm.
    California..**obesa**, sp. nov.
Pubescence virtually wanting, each of the very minute sparse punctures having an ex-
    ceedingly minute hair, only visible under high power, the surface polished; body
    much smaller, black, the pronotum piceous-brown, with two apical albescent
    areas; elytra perfectly uniform black, scarcely longer than wide, the abdomen
    much extended behind them in the single type. Length (to extremity of elytra),
    1.0 mm.; width 0.7 mm. Texas (Columbus)................**specularis**, sp. nov.
3—Elytra each with a small pale spot on the median line well behind the middie......4
Elytra with a transverse pale band, sometimes failing to attain the sides or interrupted
    at the suture................... 6
4—The pale spot nubilate; elytra black or paler from immaturity; punctures fine and
    moderately close, the vestiture distinct. Length 1.3-1.6 mm.; width 0.8-0.9
    mm. Colorado and Utah, one specimen not specifically distinguishable labeled
    " New York."... ....**biguttata** *Lec.*
The pale spot clearly defined; size smaller..5
5—Antennæ moderately long, nearly as in *amabilis;* punctures fine, rather sparse, the
    pubescence distinct; metasternum finely but distinctly punctured and with short
    and stiffer hairs. Length 1.1-1.3 mm.; width 0.7-0.8 mm. Massachusetts
    and Maryland to Iowa and Missouri.....................................**lunata** *Lec.*
Antennæ very short, the club moderate but as long as the entire preceding part omit-
    ting the basal joint; body almost exactly as in *lunata* but much smaller, the meta-
    sternum more minutely punctured but with longer and finer hairs. Length
    0.88 mm.; width 0.6 mm. Florida (Lake Monroe).............**parvula**, sp. nov.
6 —The pale band at the middle of the length consisting of a transverse discal spot on
    each, the elytra each with two large subconfluent basal spots also; pubescence
    dense and conspicuous, the color piceous; pronotum darker along the median
    line. Length 1.5 mm.; width 0.8 mm. Texas ....**lepida** *Lec.*
The pale band just visibly behind the middle and formed as in *lepida*, the body
    throughout as in that species but wholly devoid of pale spots at the elytral base;
    punctures minute but rather close-set. Length 1.4-1.6 mm.; width 0.8–
    0.85 mm. Iowa..**ornata**, sp. nov.
The pale band much behind the middle, more conspicuous and attaining the sides of
    the elytra 7
7—The pale band nubilously interrupted at the suture; body nearly as in the two pre-

ceding species, the elytra indistinctly paler at or near the base subexternally ; pubescence distinct and rather dense, the punctures minute. Length 1.3 mm.; width 0.8 mm.   California ....**amabilis** *Lec.*
The pale band very conspicuous, not at all interrupted at the suture ; pubescence coarser and not so dense........ 8
8—Piceous-black, the pronotum pale, clouded with blackish at the middle anteriorly; elytra each with a pale spot at the humeral callus, the punctures distinct, together much longer than wide.   Length 1.4–1.7 mm.; width 0.85–1.0 mm.   District of Columbia.. ....**fasciata** *Say*
Black, the pronotum as in *fasciata*, the elytra but little longer than wide, the transverse pale band still wider and more conspicuous, wholly devoid of a subbasal pale spot, the punctures more minute and rather sparser ; body shorter and stouter. Length 1.5 mm.; width 1.0 mm.   Texas (Columbus) ....**decora, sp. nov.**

The *Sacium balteatum*, of Matthews, described from North Carolina, I have not seen ; it has on the elytra a straight transverse fascia behind the middle not extending to the sides, and also the apices, yellow. *Lepida* was placed in *Arthrolips* by Mr. Matthews but incorrectly.

Arthrolips *Erichs.*

This genus resembles the last in the outward habitus of the species, but these are in general much more minute and more narrowly oval or oblong-oval and perhaps a little more convex.   In the structure of the anterior parts of the prosternum and of the antennæ it is radically different.   The species are nearly as numerous as those of *Molamba*, and are equally widely disseminated over the more southern parts of the United States ; as far as known to me they may be distinguished by the following characters :—

Elytra dark in color, with paler maculation behind the middle............................2
Elytra unicolorous.. 6
2—Elytra each with an isolated spot which does not attain the suture ; larger species 3
Elytra with a broad band crossing the suture ; size very minute........................5
3—The spot behind the middle large, oblique and more or less rounded ; body piceous, the pronotum paler, becoming broadly diaphanous at apex as usual, the median line remaining dusky ; elytra finely, very closely punctate, distinctly pubescent, without basal or subbasal pale marking ; legs pale.   Length 1.2–1.35 mm.; width 0.78–0.82 mm.   Southern California........................**nimius, sp. nov.**
    Var. A—Similar but stouter and more rounded at the sides, the spot less rounded and obliquely elliptical ; elytra more rapidly rotundato-convergent behind and narrower at tip.   Length 1.4 mm.; width 0.9 mm.   California (Owens Valley) ....**robustulus, v. nov.**
The spot transverse and crescentiform, being broadly sinuate anteriorly................4
4—Elytra black, minutely and not so closely punctate, the limb not paler and without a humeral pale area ; pronotum not paler, except at the apical limb and very

narrowly at the sides to the base ; integuments polished ; under surface blackish, the legs pale.   Length 1.2 mm.; width 0.75 mm.   Virginia (Fort Monroe).

cinctus, sp. nov.

Elytra pale piceous-brown, the entire external limb flavescent, broadening over the humeral regions ; body narrowly oblong-oval ; under surface pale, the legs flavate. Length 1.0 mm.; width 0.65 mm.   Florida (Tampa)............**mollinus** *Schz.*

5.—Pubescence of the elytra moderately abundant and quite distinct, the hairs coarse ; elytra blackish, the pale band broad, sinuate anteriorly at the suture ; pronotum and legs pale.   Length 0.62–0.7 mm.; width 0.42-0.48 mm.   California.

scitulus *Lec.*

Pubescence almost wholly wanting, each of the very minute sparse punctures bearing an extremely minute hair only visible under high power ; surface polished, the elytra piceous, the band transverse ; pronotum pale, clouded with piceous at the middle of the disk.   Length 0.8 mm.; width 0.45 mm.   Florida (Tampa).

splendens *Schz.*

6—Elytra highly polished, without trace of reticulation...7

Elytra minutely reticulate, the punctures minute and less visible ; pubescence short but rather abundant and quite distinct ; body pale luteo-flavate in color throughout, the elytra sometimes slightly darker.   Length 0.7–0.8 mm.; width 0.4–0.53 mm. California and Iowa.   [*Sacium californicum* Mattb.]...............**decolor** *Lec.*

7—Punctures small but strong and distinct throughout above, quite close-set, the pubescence moderately long and abundant, coarse and very distinct ; color black-ish, the pronotum paler.   Length 0.95 mm.; width 0.65 mm.   Massachusetts and Pennsylvania...**misellus** *Lec.*

Punctures minute and sparse, scarcely visible except under high power, each bearing an exceedingly minute hair as in *splendens*, blackish, the pronotum paler, the apical whitish spots small and rather widely separated.   Length 0.8 mm.; width 0.55 mm.   Texas (Columbus)...............................**sparsus,** sp. nov.

There may be some closely allied species included in the material before me which is referred to *decolor*, but in any event they would be so doubtfully distinct that there could be no advantage gained in separating them ; there can be little or no doubt of its identity with the *Sacium californicum*, of Matthews.

The antennæ in *Arthrolips* occasionally appear to have only nine joints, the slightly elongate third joint followed by three instead of four minute joints, but this appearance may be due simply to the difficulty in observing these organs in their natural position.

ÆNIGMATICINI.

The general habitus of the few species comprising this tribe is wholly different from that of the preceding tribes, the body being narrow and somewhat as in *Corticaria*.   There has been but one genus characterized thus far :

Ænigmaticum *Matth.*

The body is parallel, moderately convex, the head entirely exposed and but slightly inclined, the eyes moderate in size, convex, with rather coarse convex facets as usual, the antennæ inserted at some distance from their antero-internal margin in angulate emarginations of the front. The prothorax is narrowed at base and truncate at base and apex, the apical angles very obtusely rounded, the basal more distinct but obtuse, the elytral suture strongly and widely margined, the line extending along the well-developed and transversely triangular scutellum nearly to outer third of each elytron. The anterior coxæ are small and distinctly, though not broadly, separated, and the prosternum is largely developed in front of them ; the middle coxæ are rather narrowly, the posterior very widely, separated, the legs slender, with the tibiæ somewhat clavate and the tarsi rather slender, with the basal joints small. The basal segment of the abdomen is about as long as the next three combined. The two species known to me are minutely reticulate and subglabrous, each puncture bearing a very small but distinct cinereous hair ; they may be defined as follows :—

Prothorax broadly subangulate and widest at the middle, the sides straight or very feebly sinuate thence to the basal angles ; body dark castaneous, finely and sparsely but very distinctly punctate. Length 0.75–0.9 mm.; width 0.42–0.5 mm. California (San Francisco).........................**californicum** *Csy*.

Prothorax almost evenly rounded at the sides, becoming much more convergent toward apex, the latter scarcely more than half as wide as the base, the disk widest behind the middle and narrowed but slightly at base ; punctures sparse and very minute, those of the pronotum more visible and having the form of minute transverse arcs, enclosing each a very minute hair ; color dark brown ; size smaller and rather more slender in form than *californicum*. Length 0.6 mm.; width 0.25 mm. Florida.. **elongatum** *Lec.*

Elongatum was described as an *Orthoperus* by LeConte, and the type is not before me at present, but a drawing from this type made by me some years ago seems to show that the prothorax differs so greatly in outline from that of *californicum* as to indicate some divergence of a generic nature.

In the recent posthumous " Monograph of the Corylophidæ and Sphæriidæ," page 35, Mr. Matthews has fallen into a singular misapprehension, as my letter to him will undoubtedly show. My language was not by any means intended to imply that his *Ænigmaticum ptilioides* was identical with *Orthoperus elongatus*, but simply stated my conviction that the *elongatus* of LeConte is an *Ænigmaticum*. There are many minor errors throughout this important monograph, which would doubtless have been avoided had the author lived to conduct it through the press. *Sphærius politus*, for example, on page 214, is attributed to the author as a new species, whereas it was in reality described by Dr. Horn many years ago.

CRYPTOPHAGID.E.

Under this name have been grouped two closely allied types of so-called Clavicornia, comprising numerous genera and species. The body is small to minute in size, oblong or oval, more or less convex and generally clothed with coarse subdecumbent pubescence, with additional longer and more erect hairs arranged serially on the elytra in many genera, similar to those of the Tritomidæ. The tarsi are pentamerous, becoming heteromerous in the males of certain genera as in certain Cucujidæ, and the anterior coxæ are oval, moderate in size, smaller and more deep-set than in Tritomidæ, becoming decidedly transverse in the Ephistemini, and having an external trochantin. It is this form of the coxæ which principally distinguishes the family from the Cucujidæ, where the anterior coxæ are still smaller, equally or still more deeply inserted and subglobular. The family is also unmistakably allied in many characters, especially evident in the Atomariinæ, to the Scydmænidæ. Among these resemblances may be mentioned the basal impressions of the pronotum, so characteristic of the Cryptophaginæ, the side margins of the latter in *Cænoscelis*, and the recurved ventral sutures of that and some other genera, the elongate form of the trochanters, alternating long and short joints of the antennal shaft and slender pentamerous tarsi. The only serricorn character which is especially evident is the asymmetric antennal club of *Ephistemus*.

Probably the most essentially peculiar structural feature of the Cryptophagidæ, although a distinguishing character of the Silvaninæ as well, is the modification of the lateral edges of the prothorax by serratures or nodular thickenings, and the various forms assumed afford excellent subsidiary criteria for the definition of genera. Another peculiarity is the narrow and feeble dehiscence of the elytra at or very near the apex, there being but few genera, such as *Diplocælus* and *Loberus*, in which this character virtually disappears. The eyes are rounded and convex, usually rather well developed and coarsely faceted, but somewhat variably so. The antennæ are always 11-

jointed, with a loose club which is generally 3-jointed, but sometimes purely 2-jointed, and, in one case—*Anchorius*—4-jointed, a character remindful of Tritomidæ. The anterior coxal cavities are generally widely open behind, but are completely and rather broadly closed in *Diplocœlus*, completely but less broadly in *Cryptophilus*, narrowly but almost completely in *Haplolophus*, and about half closed in *Setaria*, proving that no useful generalization in the definition of the family can be drawn from the form of the cavities. The Biphyllini, as stated by Reitter, are evidently a perfectly natural part of the present family, this being proved by general organization, tarsal structure and especially by the radiating straight lines of the first ventral segment, also occurring in *Cryptophilus*, and, in an arcuate form, in *Tomarus*.* The tribe is quite out of place in the Tritomidæ, to which it was assigned by LeConte and Horn.

The Cryptophagidæ comprise two distinct subfamilies as shown by the following characters :—†

Antennæ inserted under the acutely margined sides of the front and remotely separated at base, the palpi mutually dissimilar, the maxillary elongate and slender, with the fourth joint elongate and more or less acuminate toward the tip, the labial short, with the last joint enlarged, oval to securiform ; pronotum generally bifoveolate at base...CRYPTOPHAGINÆ

Antennæ inserted on the front and more or less approximate at base, the palpi mutually similar, short, stout and acuminate, the last joint of both small, narrow and subuliform ; trochanters always narrow and elongate, bearing the femora distally ; pronotum never bifoveolate at base, though generally impressed ; anterior coxal cavities always widely open behind, the tarsi invariably simple and filiform ; eyes always basal, the first abdominal segment never having radiating lines... ....ATOMARIINÆ

In tarsal structure these two subfamilies are linked together by way of the Cryptophagini and Cænoscelini. The insertion of the antennæ in *Antherophagus* seems to suggest also a slight drift toward the Atomariinæ, but this is very feeble and more apparent than real. In the mode of antennal insertion, and especially in palpal structure, the two subfamilies are radically distinct ; perhaps species may be discovered showing intermediate characters, but it is more probable that these bonds have long ago become extinct.

---

* These lines also occur in the subfamily Silvaninæ of the Cucujidæ.

†Names to which an asterisk is affixed apply to tribes or genera which do not occur within the limits of the American fauna as far as discovered.

CRYPTOPHAGINÆ.

The body in this subfamily is generally larger, more oblong, less convex and more pubescent than in the Atomariinæ, possessing at the same time much more variety in tarsal structure and in the form of the anterior coxal cavities. These variations are important, being always accompanied by a peculiarity of general structure and habitus, and necessitate the erection of a considerable number of distinct tribes as follows :—

Trochanters moderate in size, very obliquely attached at the side of the femoral base, the latter attaining the coxæ ; anterior coxal cavities completely and broadly closed behind ; first ventral segment very much shorter than the next two combined, and having two fine straight cariniform lines diverging from the inner margin of the coxal cavities ; antennal grooves before the eyes narrow and deep, the buccal processes narrow ; front short and without trace of clypeal sutures ; eyes basal and coarsely faceted ; body subdepressed, the pronotum generally with fine longitudinal raised lines at least visible toward the sides ; abdominal sutures fine, broadly arcuate ; tarsi pentamerous in both sexes, the fourth joint small, the third strongly and the second more feebly, lobed beneath..............BIPHYLLINI.

Trochanters very short but less obliquely joined to the femoral base throughout the width of the latter ; anterior coxal cavities narrowly and partially closed behind by an inward projection of the side pieces of the prosternum ; tarsi stout, pentamerous in both sexes, densely clothed beneath with coarse hairs, the fourth joint small ; eyes basal or subbasal and coarsely faceted ; front prolonged, more or less prominently convex above the antennæ and with a short oblique clypeal suture at each side ; middle coxæ narrowly separated ; pronotum never lineate, the basal foveæ minute or subobsolete ; elytra with confused punctuation but becoming regularly seriato-punctate in *Leucohimatium* ; basal segment of the abdomen short or moderate, never lineate, the sutures nearly straight ; antennal grooves before the eyes narrow and deep in *Setaria*, obsolete in *Haplolophus* and *Leucohimatium ;* antennal club 2-jointed in *Setaria*..*SETARIINI.

Trochanters elongate, bearing the femora obliquely attached distally ; middle coxæ smaller and less narrowly separated ; front short before the antennæ, without trace of clypeal sutures ; antennal grooves obsolete ; antennal club always loosely 3-jointed...................... 2

2—Anterior coxal cavities completely, though not very broadly, closed behind ; first ventral segment but little longer than the second and with two straight diverging cariniform lines as in Biphyllini, the tarsi pentamerous in both sexes, with the fourth joint small, the joints toward base having simple brushes of hair beneath ; first joint of the posterior as long as the next two combined in *Cryptophilus* ; eyes basal and coarsely faceted ; pronotal foveæ very minute..........*CRYPTOPHILINI.

Anterior coxal cavities broadly and completely open behind ; basal segment of the abdomen variable in length..3

3—Tarsi pentamerous in both sexes, with the fourth joint small, the third joint strongly, and the second less strongly or obsoletely, lobed beneath, the lobes

narrow and pubescent ; eyes always basal ; first antennal joint relatively small ;
last joint of the labial palpi usually flattened, obtusely truncate or securiform ;
mesosternum flat or feebly concave between the coxæ ; elytra with serial punctures
in *Loberus*.......... ....TELMATOPHILINI.

Tarsi always filiform, simple and never lobed beneath, pentamerous in the females and
heteromerous in the males, the penultimate joint similar in form to the preceding ;
last joint of the labial palpi oval, convex, narrowly truncate at tip ; abdominal
sutures straight throughout the width ; prosternal process acute, freely passing
over the mesosternum, which is generally concave ; eyes variable ; elytra never
margined at base and never having distinctly serial punctuation..CRYPTOPHAGINI.

The tribe Setariini is erected for three isolated European genera
Setaria, *Haplolophus* and *Leucohimatium*, and there is no American representative known thus far. The European genus *Cryptophilus* also
necessitates the creation of a distinct tribe. All the other tribes are
common to the two hemispheres.

BIPHYLLINI.

This is a small tribe, comprising a relatively large number of generic
types. The body is oblong-oval or elongate-oval and generally notably depressed, pubescent and with the abdominal sutures very fine.
The tarsi are pentamerous in both sexes, with the fourth joint small
and simple, the fifth generally much elongated and the subbasal
thicker and lobed beneath. The pronotum generally has some elevated longitudinal lines at least toward the sides ; the scutellum is
short and transverse and the antennæ rather short, with well developed
club, very widely separated at base and inserted under the sides of the
front, the basal joint moderate or small in size. The last joint of the
maxillary palpi is slender, that of the labial large and securiform. The
truncate posterior edge of the prosternum passes freely over the surface
of the mesosternum, and the first ventral segment has two straight
carinæ diverging from the middle of the base and extending to the
apical margin or very nearly. The posterior sutures are flexed backward at the sides to a greater or less degree. The genera before me
may be defined as follows :—

Last joint of the labial palpi broadly oboval, thick and convex, with the apex broadly
truncate and excavated ; antennal club broad, oval, rather compact and 4-jointed,
the eighth joint, however, very small, transverse and obtrapezoidal ; pronotum
longitudinally lineate throughout its width...............................**Anchorius**
Last joint of the labial palpi broadly securiform and flattened, with the apical edge
fine ; pronotum only lineate toward the sides....................................2

2—Antennal club 3-jointed, narrower and more loosely connected, the ninth joint sensibly smaller than the tenth, the eighth small and similar to the seventh, the last
subtransversely oval and generally somewhat narrower than the tenth. [*Marginus*
Lec.].. ....**Diplocœlus**
Antennal club 2-jointed, the eighth and ninth joints small and perfectly similar to the
seventh, the tenth abruptly large, rectilinearly obconic in form and somewhat
wider and longer than the eleventh, which is transversely suboval and obtusely
pointed...***Biphyllus**

In this tribe the joints of the antennal shaft are equal among themselves, showing little if any of the alternating inequality so prevalent
elsewhere in the family.*

Anchorius, gen. nov.

In this genus the body is oblong-oval, rather depressed, the upper
surface feebly and evenly convex. The legs are very much stouter
than in *Diplocœlus*, the femora broadly oval and the antennæ still
shorter. The minute dense punctulation of the under surface is devoid
of larger punctures, which is not the case in either *Diplocœlus* or *Biphyllus*. There is but one species before me at present, which may be
described as follows :—

Uniform dark brown throughout the upper surface, densely dull and devoid of lustre,
extremely minutely and densely punctulate and minutely, densely pubescent ; antennæ as long as the width of the head, the eyes large, convex and very coarsely
faceted ; prothorax twice as wide as long, with the sides moderately convergent
from base to apex, evenly and distinctly arcuate, the apex broadly sinuate, with
the angles bluntly rounded ; basal angles obtuse but not obviously rounded ; surface with ten fine and entire subelevated longitudinal lines, those toward the sides
rather more widely spaced and somewhat more strongly elevated, the intervals
feebly concave and with scattered coarser punctures ; elytra one-half longer than
wide, three times as long as the prothorax but not at all wider, the sides feebly
arcuate, not continuous in curvature with those of the pronotum, evenly, rather
strongly rounded behind ; striæ composed of unimpressed series of fine punctures,
the intervals feebly elevated along the middle, the crest having a single series of
short coarse and somewhat paler hairs, similar to those along the crests of the

---

* I follow DuVal in writing and adopting *Biphyllus* Stephens, instead of the emen-
dation *Diphyllus* Redt. Lacordaire writes *Diphyllus*, with the statement that
Biphyllus is inconsistent with the laws of etymology. This would be perfectly
correct if generic words were subject to the laws of etymology—but they are not.
They are simply pronounceable, and, first of all, constant, combinations of letters
having latiniform endings. They cannot, when once established, be changed under
any circumstances. They are not supposed to have a meaning—that is as an essential
quality. Specific names, on the contrary, always have a meaning, and are therefore
subject to the rules of etymology.

pronotal lines; under surface minutely, densely and evenly punctulate throughout, the surface somewhat shining. Length 3.3 mm.; width 1.4 mm. Arizona.

lineatus, sp. nov.

In the type, the fourth ventral segment has a small and very shallow, transversely oval erosion at the middle and near the hind margin,
the fifth much longer than the fourth, unmodified on the disk, and
and very evenly rounded behind. This species is allied to the Cuban
Diplocœlus costulatus but differs in its larger size and in having ten,
and not eight, longitudinal pronotal lines; it also seems to differ from
the *mus* of Reitter, in the latter character.

Diplocœlus *Guér.*

The species are few in number and are widely isolated structurally
among themselves, in fact constituting several subgenera; they may
be outlined as follows :—

Prothorax parallel or feebly narrowing from base to apex, broadly, evenly convex, the
basal angles not prominent but not at all rounded, the surface with three longitudinal elevated lines at each side, of which the inner is feeble and incomplete;
eyes moderately coarsely faceted; elytra with very feebly impressed series of
close-set punctures, the intervals each with a series of suberect hairs, which are
short and inconspicuous,—becoming subobsolete in the European *humerosus*—
brown or blackish in color, elongate-oval, feebly convex, feebly shining, minutely,
closely punctured and densely pubescent, with coarse punctures interspersed on
the pronotum, sterna and near the sides of the abdomen. Length 3.1 mm.; width
1.18 mm. Indiana. [Diplocœlus, in sp.]...........................**brunneus** *Lec.*
Prothorax subparallel and arcuate at the sides, finely serrulate as usual, the basal angles
minutely prominent and acute, the surface convex and devoid of sublateral elevated lines except feeble traces, an arcuate excavated line at the apex and lateral
fourth quite distinct; eyes very coarsely faceted; elytra pointed behind, having
feebly impressed series of very coarse punctures, the intervals polished and with
single series of very long erect setæ; general vestiture short and sparser, the surface strongly shining; sterna coarsely punctate, the abdomen finely, closely and
evenly. Length 2.0–2.3 mm.; width 0.8–0.92 mm. Florida, Mississippi and
Texas (Houston). [Marginus *Lec.*].................................. **rudis** *Lec.*
Prothorax narrowed in front, the sides nearly straight, the hind angles prominent externally, covering the base of the elytra; surface coarsely punctured, with three
elevated lines at each side, becoming subobliterated in front; elytra with series of
coarse punctures, the single interstitial pubescent lines composed of short and
coarser hairs. Length [3.25 mm.] Michigan. [Subgenus nov.?]
angusticollis *Horn*

The last of these species is unknown to me, but the prominent
basal angles of the prothorax seem to be foretold in *rudis*. *Brunneus*
is a close derivative of the European *fagi*, but is more slender in form.

TELMATOPHILINI.

In this tribe the body is elongate-oval and convex, with slender antennæ, moderate in length and having a narrow loose 3-jointed club,
with the ninth joint notably smaller than the tenth in *Telmatophilus*
and *Loberus*, and subequal to the latter in *Tomarus*. The basal segment of the abdomen is only moderately elongate, and the elytral
suture is margined. The pronotum has two small deep and widely
separated isolated foveæ at the basal margin. The abdominal segments
are perfectly mobile as in Cryptophagini, and the fourth tarsal joint
is very small. The three genera differ considerably among themselves
in general habitus and may be defined as follows :—

Elytra not margined at base, feebly margined along the suture, the surface pubescent
    and closely and irregularly punctured ; prosternal process acute at tip ; eyes well
    developed and coarsely faceted ; tarsi thick and strongly lobed and pubescent
    beneath, the basal joint of the posterior not much longer than the second, the
    claws thick and strongly arcuate but not obviously dentate ; abdominal sutures
    flexed abruptly backward at the extreme sides, the first segment as long as the
    next two combined or longer ; pronotum having a very fine excavated line along
    the marginal basal bead throughout the width.....................**Telmatophilus**
Elytra with a thickened basal marginal bead, along which there are several small
    deep foveæ in *Tomarus*, the suture finely and more or less strongly margined
    throughout ; prosternal process truncate at tip ; abdominal sutures straight
    throughout, the basal segment shorter ; body sparsely and feebly pubescent to
    glabrous..2
2—Pronotum broadly but feebly impressed at base between the foveæ, the elytra
    evenly striato-punctate ; eyes large, convex and very coarsely faceted ; tarsi
    thicker, strongly lobed, the basal joint of the posterior but little longer than the
    second, the claws dentate within at base ; first abdominal segment without diverg-
    ing lines ; segments one to four decreasing gradually and but slightly in length.
                                                                                                    Loberus
Pronotum not impressed at base, the elytral punctures sparse and irregular in dis-
    tribution ; eyes rather small and not very coarsely faceted ; tarsi very slender,
    feebly lobed and only on the third joint, the first joint of the posterior nearly as
    long as the next three combined, the claws very slender, arcuate and perfectly
    simple ; first abdominal segment with two very widely diverging arcuate lines,
    homologous with the straight and less diverging lines of Biphyllini and Cryp-
    tophilini..**Tomarus**

No representative has as yet been discovered in the Pacific coast
fauna, but the tribe is much better developed in America than in
Europe. *Cryptophilus*, which is placed near *Telmatophilus* by Reitter,
is entirely out of place, the completely closed anterior coxal cavities
betraying a greater affinity with *Diplocœlus*.

Telmatophilus *Heer*.

This genus is widely extended in range through all the palæarctic and nearctic provinces, but has not yet occurred on the Pacific coast of America.   We have but one species, as follows :—

Piceous-black, rather shining, densely and deeply but finely punctured throughout above and beneath, the pubescence short, ashy, the elytra in addition with imperfect single series of slightly longer hairs ; antennæ and legs rufous, the former scarcely as long as the head and prothorax ; eyes convex and prominent ; prothorax barely two-fifths wider than long, the sides parallel, evenly arcuate, the edges finely serrulate and single ; apex broadly arcuate and as wide as the base, which is broadly bisinuate ; basal angles acute, the apical obtusely rounded ; disk feebly convex, finely and very densely punctate ; elytra but little wider than the prothorax in the male, and less than three times as long, relatively larger in the female, obtusely rounded at tip, the humeri not exposed at base ; punctures not coarser and much less dense than those of the pronotum ; legs stout.   Length 2.4–2.7 mm. ; width 0.9–1.0 mm.   Canada, Massachusetts, New York, Iowa and Colorado (Greeley)........................**americanus** *Lec.*

The male is a little shorter and stouter than the female and has a deep oval pit at the apex of the fifth ventral segment, and the hind tibiæ strongly dentate externally near the base ; the mesosternum is very feebly concave between the coxæ.   The European *caricis*, which resembles *americanus* very closely, has a very feeble impression at the middle of the fifth ventral of the male, and the hind tibiæ of that sex are much more feebly and obtusely swollen externally near the base.

Loberus *Lec.*

This genus appears to be exclusively American and will prove to be tolerably rich in species.   The resemblance to certain crepidoderid Chrysomelidæ has been alluded to by LeConte and Horn, and is sufficiently striking, the body is however narrower than in the great majority of Crepidoderæ.   The broad and shallow transverse depression extends between the pronotal foveæ but is semi-independent of them.   The species before me may be defined as follows :—

Basal depression of the pronotum broadly impressed and transverse, almost adjacent to
    the basal margin ; elytral margins very narrow and equal........................2
Basal depression more deeply and acutely impressed and somewhat anteriorly arcu-
    ate, being more distant from the basal margin on the median line ; elytral margins
    more broadly reflexo-explanate at basal third........................4
2—Punctures of the elytral series rather coarse, each bearing a moderately long and
    very distinct recurved silvery hair, the intervals glabrous and impunctate.
    Body elongate-oval, convex, polished, dark rufo-testaceous to blackish in color,
    the head and pronotum sparsely pubescent ; antennæ testaceous, with the club

darker, extending slightly beyond the base of the prothorax, the latter very
slightly narrower than the elytra in the male, more distinctly so in the female,
nearly twice as wide as long, the sides parallel and feebly arcuate, abruptly
sinuato-convergent near the base, the basal angles right and not blunt, the apical
obtusely rounded, the base broadly bisinuate ; punctures sparse and rather
coarse ; elytra fully three and one-half times as long as the prothorax, parallel
and broadly arcuate at the sides, the apex rather narrowly rounded ; humeri
somewhat exposed at base ; disk more or less impressed at the suture on the posterior declivity before the apex.   Length 1.8-1.9 mm.; width 0.75-0.8 mm.
Middle States and Rhode Island (Boston Neck)..................**impressus** *Lec.*
Punctures of the elytral series each bearing an infinitesimal hair only visible under
great amplification, the surface throughout polished and apparently perfectly
glabrous ; antennæ and prothorax similar to those of *impressus*, the elytra simil-
larly between three and four times as long as the prothorax, narrowly rounded at
apex and with somewhat basally exposed humeri............................ 3
3—Body dark rufous or rufo-piceous in color, the punctures of the head and prothorax
fine and very sparse, the superciliary ridges fine and scarcely at all flexed inward
anteriorly ; elytral series scarcely at all impressed, the punctures more or less
small in size.   Length 1.9-2.1 mm.; width 0.78-0.8 mm.   New Jersey.
                                                          subglaber, sp. nov.
Body black or blackish in color, the punctures of the head and prothorax coarse but
sparse, the superciliary ridges coarse and strongly bent inward anteriorly ; elytral
series sensibly impressed, the punctures coarse and deep.   Length 1.8 mm.;
width 0.75 mm.   Florida...**imbellis,** sp. nov.
Body rufo-testaceous in color, smaller and more slender in form ; superciliary ridges
very fine, feeble and not flexed inward at their anterior end ; punctures of the
head and pronotum fine but deep and very sparse ; elytral series not or scarcely
at all impressed, the punctures fine but distinct.   Length 1.6-1.75 mm.;
width 0.65-0.7 mm.   Bahamas (Egg Island) and Cuba (Bahia Honda).
                                                          insularis, sp. nov.
4—Body elongate-elliptical, rather less convex, the elytra more strongly narrowed behind from about the middle, polished, blackish-piceous in color, the elytral
humeri and apical fourth testaceous ; antennæ, head and prothorax nearly as in
impressus, the latter sparsely clothed with longer pubescence, finely and rather
less sparsely punctate and much less declivous toward the sides ; elytra quite distinctly wider at or just before the middle than at base ; slightly wider than the
prothorax and nearly four times as long, subacute at apex, the humeri but slightly
exposed at base, the series unimpressed, composed of rather small but distinct
punctures, the intervals also with uneven series of smaller, still more widely
spaced punctures, all the punctures bearing distinct subdecumbent hairs, the
entire surface being sparsely pubescent.   Length 2.0 mm.; width 0.9 mm.
Mexico (Frontera in Tabasco).   Prof. C. H. T. Townsend.
                                                          puberulus, sp. nov.

The lateral edges of the prothorax in all the species are distinctly
thickened and bear a few very minute widely spaced serrules, one behind the apex being especially constant ; the thickened margin is

flexed inward for a short distance at the apical angles, and, along the
base, forms a margin which becomes very feeble or obsolete along the
broad median lobe.   The only species in which sexual characters are
noticeable is *impressus*, and here the male has a very minute shallow
fovea, accompanied by a tuft of loose longer hairs, at each side of the
median line and near the middle of the length of abdominal segments
two, three and four.

Tomarus *Lec.*

The body in this genus is smaller and relatively shorter than in
Loberus, and has a markedly different general habitus.   The lateral
edges of the prothorax are very finely double, the outer edge more or
less distinctly and unevenly undulated, the border flexed inward for a
short distance at apex, and, at base, as far as the foveæ, where the
margin becomes very fine along the basal lobe.   There is a fine superciliary ridge as in *Loberus*, but the antennæ differ in having the basal
joint of the club about as large as the second.   I have not noticed
any distinctive sexual characters in the male.   The three following
are the only species known to me at present :—

Body subglabrous, the head and pronotum clothed sparsely with very short subdecumbent hairs, the elytra glabrous, each with three discal and one marginal series of
   very widely spaced erect setæ; sides of the prothorax very obsoletely undu-
   lated 2
Body clothed throughout with coarse, sparse, subdecumbent hairs in addition to the
   series of elytral setæ; sides of the prothorax more strongly and quite distinctly
   undulated.................... 3
2—Body rather narrowly oval, convex, polished, the anterior part feebly alutaceous,
   flavo-testaceous to blackish throughout, the elytra broadly, suffusedly paler toward
   the humeri and in a transverse band interrupted at the suture, near apical third ;
   antennæ but little longer than the head and prothorax, the club well developed ;
   head and pronotum finely but strongly, rather closely punctured, the punctures
   finer toward the sides of the latter, which is three-fifths to two-thirds wider than
   long and much narrower than the elytra, with the sides parallel and arcuate and
   the apex very nearly as wide as the base ; elytra two and two-thirds to three
   times as long as the prothorax, subinflated and widest at two-fifths, gradually
   narrowed to the acute apex, the humeri feebly denticulate externally and obliquely exposed at base ; erect setæ, moderately long and distinct, the punctures
   fine and sparse, with series of rounded areolæ shining through the translucent
   chitin from the under surface.   Length 1.25–1.6 mm.; width 0.65–0.72 mm.
   New York and Rhode Island to Iowa and Mississippi............**pulchellus** *Lec.*
Body and antennæ nearly similar to the preceding, the former rather shorter and less
   acute behind, pale flavo-testaceous in color, the head and pronotum more alutaceous, the elytra polished and almost similarly maculate, with the erect setæ very

short, those of the discal series extremely short and almost obsolete ; head and
pronotum finely and feebly punctate, the latter shorter and more transverse, almost twice as wide as long, the sides similarly parallel and arcuate, the feeble
punctures becoming almost completely obsolete toward the sides; elytra nearly similar to those of *pulchellus* but shorter and more obtuse, three times as long as the
prothorax and distinctly wider. the punctures sparse and very fine.   Length
1.2 mm.; width 0.59 mm.   Arizona (Tucson)..................**obsoletus**, sp. nov.

3—Body smaller, shorter and more broadly oval than in the preceding, convex,
polished, dark rufous, the elytra black, pale in the basal regions and broadly at
apex, except transversely at apical fourth ; antennæ longer than the head and
prothorax, the latter short and strongly transverse, finely and sparsely but
strongly punctate, the sides parallel and arcuate ; elytra short, but little longer
than wide, ogival at apex, coarsely, rather closely and conspicuously punctured.
Length 1.0–1.1 mm.; width 0.5–0.55 mm.   Florida (Tampa)...**hirtellus** *Schz.*

A small specimen from North Carolina may possibly represent a
distinct species or subspecies of *pulchellus;* it is smaller, more obtuse
behind and somewhat differently colored.   The strong basal margin
of the elytra enclosing a series of foveæ along its posterior edge, is a
marked feature of this genus and it is this which causes the minute
denticulation of the elytral humeri mentioned above.

CRYPTOPHAGINI.

This tribe differs from all those which precede primarily and very
radically in the structure of the tarsi, which, instead of being shorter
and stout, frequently lobed beneath, with the fourth joint very small
and pentamerous in both sexes, are here more or less slender and filiform, never lobed beneath, with the fourth joint similar to the preceding and pentamerous in the females and heteromerous in the males,
as in the Cænoscelini of the next subfamily.   From the Telmatophilini
they differ besides, as a rule, in a coarser and denser sculpture and
vestiture, stouter antennæ, with less loosely connected club and less
coarsely faceted eyes than in *Telmatophilus* and *Loberus.*   The first
segment of the abdomen is usually more elongate, being subequal to
the next two combined, and never has diverging lines ; the sutures are
straight throughout the width, differing in this respect from *Cænoscelis.*
The genera are rather numerous, those before me being readily recog-
nizable by the following characters :—

Eyes ante-basal, small, rather finely faceted and not prominent ; frontal margin deeply
emarginate and impressed at the middle, especially in the male, the front not at
all prolonged beyond the antennæ, the basal joint of the latter large and globular, the second similar to the third and following, the club rather feebly de-

fined in the male but parallel and loosely 3-jointed as usual ; pronotum not impressed but finely, strongly margined at base, the foveæ minute and almost completely obsolete, the sides even, with a rather thick margin, which becomes gradually very thick at the apical angles but continuously so, the apical callus not posteriorly delimited or truncated—as it is in *Cryptophagus* ; elytral suture margined except toward base ; mesosternum rather more concave between the coxæ than usual, the tibiæ compressed and somewhat triangular, the tarsi and claws slender as usual.   [Subtribe ANTHEROPHAGI].............**Antherophagus**
Eyes basal, convex, prominent and more or less coarsely faceted ; tibiæ slender......2
2—Front declivous and concave between the antennæ, the edge not beaded over the insertion of the latter...3
Front declivous but evenly, longitudinally convex anteriorly, the edge sharply angulate but not beaded over the antennæ, the frontal margin with a broadly, posteriorly angulate smooth space, probably homologous with the emargination of *Antherophagus* ; antennæ moderate, the club loosely 3-jointed, with the last joint obliquely and obtusely narrowed from near the base, the first joint small and globular ; prothorax with a broad flat marginal bead at base, before which the surface throughout is feebly impressed, the foveæ wholly obsolete ; sides with a thickened nodal point at the apical angles but otherwise perfectly even, the nodal points projecting anteriorly, the apex broadly emarginate between them as in *Emphylus* ; punctures fine and irregular, the pubescence short, coarse and closely decumbent ; subsutural lines of the elytra not extending to the base ; mesosternum not at all impressed between the coxæ.   [Subtribe SPANIOPHÆNI.]
                                                                    *****Spaniophænus**
Front flat and not more declivous anteriorly ; antennal club normal and 3-jointed, its first joint not differing in form from the second though frequently smaller in size; body strongly punctured and rather coarsely pubescent.   [Subtribe CRYPTOPHAGI]...4
3—Antennæ stout, almost similar to those of *Antherophagus* but with the second joint wider than the third, the 3-jointed club narrow and feebly delimited, and with its basal joint smaller than the second though similar in form ; prothorax not impressed at base, the sides even, with a fine acute edge, the apical angles broadly, obliquely truncate and prominent but only slightly thickened, the apex broadly sinuate between, the basal foveæ and transverse impression obsolete ; elytral suture margined toward tip ; body subglabrous and strongly alutaceous, very finely, feebly and moderately closely punctured.   [Subtribe EMPHYLI].............**Emphylus**
Antennæ slender, the basal joint elongate-oval, not very thick, the second still narrower, elongate, broader than three to eight, which are very slender and elongate, the club narrow, loosely 3-jointed, gradually formed, the ninth joint being slender elongate and obconical, altogether dissimilar in form to the tenth and unique in the family ; prothorax very feebly impressed transversely at base between the large but feeble foveiform depressions, the sides broadly and feebly triundulate, the edge thickened but not very prominent at the undulations, which are at the apex and near apical and basal third, the apical angles not modified, the apex broadly arcuate from side to side ; elytral suture margined, very obsoletely so toward base ; body coarsely sculptured and pubescent, nearly as in Cryptophagi.
[Subtribe PARAMECOSOMÆ].................................*****Paramecosoma**

4—Prothorax triundulate at the sides—at the apex and near apical and basal third,— the undulations similar among themselves, the apical angles not more thickened ; elytral suture feebly margined toward tip.

Body short, broadly oval and convex, with long herissate vestiture, the prothorax finely bifoveate at base, the foveæ connected by a fine deep transverse impression, the disk also having a fine cariniform line at each side at some distance from the side margin and parallel thereto ; lateral undulations moderate in development, rounded and finely serrate ; eyes very small and extremely prominent..**Crosimus**

Body elongate and less convex, clothed with denser, shorter and more decumbent pubescence as in *Cryptophagus*, the prothorax without a sublateral line, having two small basal foveæ connected by a very feeble and broadly impressed line, the lateral undulations simple but very prominent and dentiform ; eyes as in *Cryptophagus*..**Salebius**

Prothorax with thickened and obliquely truncate apical angles, the edge even, excepting a minute acute tooth at about the middle and sometimes minute serrulations thence to the basal angles, the basal foveæ very small and feeble, connected by a fine feeble impression along the basal margin, the apex truncate or feebly bisinuate ; elytral suture only margined posteriorly. [Subgen. Mniononus Woll.]
Cryptophagus

Prothorax with thickened and obliquely truncate apical angles, the edges thence evenly, feebly arcuate, slightly converging and evenly, finely serrulate to the base, the basal foveæ distinct and mutually connected by a larger deep basal impression ; elytral and other characters nearly as in *Cryptophagus*.....*Micrambe**

Prothorax not thickened at the apical angles or undulated at the lateral edges, the latter perfectly even from apex to base and serrulate ; elytral suture margined very nearly to the scutellum

Body oval, convex, coarsely sculptured and pubescent, the prothorax with two small but deep basal foveæ connected by a very deep and conspicuous groove ; serrulation of the lateral edges more or less coarse and distinct.
Henoticus

Body oblong, parallel and strongly depressed, finely, more closely sculptured and pubescent, the prothorax with two very small but distinct basal foveæ, the connecting impression or groove wholly obsolete ; lateral edges very minutely serrulate...**Pteryngium**

The definition of *Emphylus* is taken from the Europern *glaber*, and, as I have not seen the American representative—*americanus* Lec., of the catalogue,—the genus will not be further dwelt upon ; its affinity with *Antherophagus* is much more pronounced than with *Cryptophagus*, and the sinuation of the thoracic apex—due to the prominence of the apical angles—which has been hitherto advanced as a differential character, is one of the least important.

Antherophagus *Latr.*

This is one of the most isolated genera of the family and contains by far the largest species, *Haplolophus* being the only other which approaches it in this respect. The emargination of the clypeus, very deep in the male but feeble in the female, is apparently a unique character in the family, and the antennæ are peculiarly thick and compact in the male, though bearing some resemblance to those of *Emphylus ;* the female antennæ are much shorter, more slender and with relatively larger club. The eyes are almost without parallel in the family in their position upon the side of the head and in their relatively slight convexity, the convexity and prominence of these organs being one of the most characteristic features of the family. The body is oblong, rather convex, very finely, densely punctate and clothed, often densely, with very short subappressed pubescence. The elytra in some of the paler forms clearly show the regular series of areolæ on their under surface, shining through the diaphanous chitin and perhaps of significance in indicating that the family may be derived from seriately punctate archetypes ; at present these series of areolæ are not connected in any way with the punctuation of the surface, which is altogether irregular, but there are frequently very feebly impressed superficial lines which appear on the exposed surface above them. The species are few in number and those in my cabinet may be thus characterized :—

Body large, more broadly oblong, densely clothed with pubescence which nearly conceals the surface, the eyes smaller, the mandibles more prominent ; tibiæ rapidly enlarged from base to apex ; basal angles of the prothorax more or less obtuse...2
Body smaller, the sides of the prothorax parallel and straight, the basal angles right and not at all blunt; body smaller, the eyes moderately large, rather more convex and less finely faceted, the mandibles smaller and less prominent, the pubescence quite sparse, not at all concealing the surface ; tibiæ but feebly enlarged from base to apex...3
2—Body broadly oblong, testaceous throughout, the antennæ of the male except at base and apex, and the tibiæ toward base, blackish ; antennæ of the male thick, almost as long as the head and prothorax, the second joint much shorter than the third though equal in width ; prothorax distinctly less than twice as wide as long, parallel and almost straight at the sides, but slightly rounding and convergent at apex and base, the punctures fine and dense ; elytra not wider than the prothorax, a third longer than wide, obtuse at apex, very densely and finely punctate. Length 4.1–4.5 mm.; width 1.7–1.9 mm. New York to Minnesota.
ochraceus *Melsh.*
Body less broadly oblong and slightly smaller, equally densely but still more minutely punctate and densely clothed with short cinereous pubescence, pale flavo-testaceous, the tibiæ and antennæ colored as in *ochraceus*, the latter thick in the male and much shorter than the head and prothorax, the second joint equal in length

and width to the third ; prothorax shorter, scarcely visibly less than twice as wide
as long, the sides parallel and evenly, distinctly arcuate ; elytra two-fifths longer
than wide, not wider than the prothorax and less obtusely rounded at apex.
Length 3.2-4.25 mm.; width 1.2-1.7 mm.   Utah (southwestern)—Mr. Weidt.

pallidivestis, sp. nov.

♂—Body narrowly oblong-oval, pale rufo-testaceous throughout, the antennæ and legs
concolorous, polished, the elytra slightly alutaceous ; antennæ moderate in the
female ; prothorax less than twice as wide as long, parallel and straight at the
sides, finely but deeply, not very densely punctate ; elytra subangularly dilated at
two-fifths and wider than the prothorax, the base equal to that of the latter, the
apex obtusely rounded ; punctures very fine, feeble and rather sparse.   Length
3.3 mm.; width 1.35 mm.   Wisconsin (Bayfield) —Mr. Wickham

convexulus *Lec.*

The stout mandibles are bifid at tip, and the antennæ are inserted
within very small foveæ on the vertical sides at a great distance from
the eyes ; they differ very obviously in the sexes, as indicated above.
Suturalis of Mäklin, I have not seen.

Crosimus, gen. nov.

In the general structure of the body this genus is allied to *Salebius*,
and especially in possessing three lateral projections at each side of
the prothorax, and in the same positions, but here the nodes are not
thickened and take the form of broadly rounded and rather feeble undulations of the edge, the salients being spiculato-serrulate.   It differs
greatly from *Salebius* or *Cryptophagus* in the short stout, very convex
and oval form of the body, long hirsute sparse vestiture, in having the
elytral punctures arranged in uneven unimpressed double series, in
having a fine raised line near each side of the pronotum extending
from base to apex, and in the more longitudinally convex prosternum,
the process being elevated far above the coxæ from an under view, the
process more strongly margined at the sides ; the antennæ, oral organs
and legs are throughout as in *Cryptophagus*.   The basal foveæ of the
pronotum are connected by a very deep channel along the basal margin, which is never interrupted at the middle by a carina, and the
callous discal spots of *Cryptophagus* appear to be obsolete.   The tarsi
are very slender and as long as in *Salebius*.   The eyes are unusually
small, absolutely basal and extremely convex, not very coarsely
faceted.   The two species before me may be described as follows : —

Body more narrowly oval, polished, black, the legs and antennæ testaceous, the elytra
bright rufous, black at the apex, at the middle of the flanks and transversely behind the base near the suture ; pubescence moderately long and sparse ; prothorax

about two-thirds wider than long, the sides in general form nearly straight and
strongly convergent from base to apex, continuing the sides of the elytra; punctures fine but deep and not very close-set, the surface shining; submarginal line
rather feeble; elytra oval, before the middle much wider than the prothorax,
scarcely three times as long as the latter, the punctures fine and sparse, the double
series ill-defined. Length 1.6 mm.; width 0.78 mm. New York.

obesulus, sp. nov.

Body throughout in form and coloration as in *obesulus*, but a little stouter, the prothorax nearly four-fifths wider than long, with the sides feebly convergent, nearly
straight in general form but not continuing the sides of the elytra, the surface less
finely, very deeply and very closely punctate, the submarginal line parallel to the
edge fine but strong; elytra nearly as in *obesulus* but more broadly oval and with
more prominent humeral callus, the punctures larger and less sparse, the pubescence longer, more abundant and with very long erect subserial hairs in addition. Length 1.7 mm.; width 0.85 mm. Iowa (Iowa City)—Mr. Wickham.

hirtus, sp. nov.

These species are mutually very closely allied but appear to be distinct. The genus is probably confined to the Atlantic regions of the
continent.

Salebius, gen. nov.

This genus, with *Crosimus,* is distinguished from *Cryptophagus* by
having three subequal obtusely dentiform nodal points along each side
of the prothorax —at the apex and near apical and basal fourth of the
length, instead of a single nodal point, with a submedian spicule as
in that genus. The node at the apical angles in *Salebius* is merely
thickened, convex and more or less pubescent, but the two posterior
often have a deep puncture at the middle of the summit analogous to
the central puncture of the flattened apical node so prevalent in *Cryptophagus.* The tarsi are long and slender and nearly all the other
anatomical structures are similar to those of *Cryptophagus,* except that
only the anterior two of the pronotal callous spots are visible, and the
impression along the basal margin is feebler, with the median carina
always distinct. The five species in my cabinet may be recognized
as follows :—

Punctures very fine but deep as usual, those of the pronotum very dense; body dark
piceous, blackish beneath, the antennæ and legs castaneous; pubescence short,
even, decumbent and rather abundant, more distinct on the pronotum along the
sides and median line; prothorax parallel and slightly rounded at the sides, not
more than one-half wider than long; elytra two-thirds longer than wide, only
slightly wider than the prothorax and fully three times as long, the punctures fine
and rather close-set; hind tarsi nearly as long as the tibiæ (♀). Length 2.4
mm.; width 0.9 mm. Queen Charlotte Islands (Massett)—Mr. Keen.

6-dentatus, sp. nov.

Punctures strong and moderately coarse more or less close-set on the pronotum.........2

2—Tarsi moderately elongate, the posterior distinctly shorter than the tibiæ in both
  sexes..3

Tarsi more elongate, the posterior as long as the tibiæ. 5

3—Eyes rather large and well developed, more than half as long as the head; body
  much stouter, dark rufo-testaceous throughout, the vestiture much longer though
  sparse; prothorax relatively small, three-fifths wider than long, parallel and
  nearly straight at the sides, the teeth very large, subacute and conspicuous; elytra
  large, parallel, evenly rounded behind, three-fifths longer than wide, fully a
  fourth wider than the prothorax and much more than three times as long, the
  punctures coarse and not very close-set (♀).   Length 2.6 mm.; width 1.1 mm.
  California...**minax**, sp. nov.

Eyes smaller but not more prominent, scarcely half as long as the head; body darker
  in coloration, the pubescence much shorter 4

4—Body oblong-oval, moderately slender, shining, blackish-piceous in color, the legs
  paler; pubescence moderately short, coarse, somewhat abundant and distinct;
  prothorax rather strongly transverse, about two-thirds wider than long, strongly,
  densely punctate, parallel and broadly arcuate at the sides, the teeth well developed but less so than in *minax;* elytra elongate, two-thirds longer than wide,
  only slightly wider than the prothorax and more than three times as long, quite
  coarsely, but not very densely, punctate (♂).   The female is larger but virtually
  similar in every way, the prothorax not relatively much smaller.   Length 1.9–
  2.5 mm.; width 0.75–0.9 mm.   California (Siskiyou and Sta. Cruz Cos.).
                                                                              lictor, sp. nov.

Body nearly similar in form and coloration but less elongate, the prothorax large,
  much less transverse, barely one-half wider than long, the vestiture much shorter
  and inconspicuous, the sides parallel and evenly arcuate, the teeth pronounced;
  elytra shorter, three-fifths longer than wide, slightly wider than the prothorax and
  two and three-fourths times as long, the punctures decidedly less coarse and
  rather more close-set, the pubescence much shorter, even, decumbent and not
  very close (♂).   Length 2.0 mm.; width 0.8 mm.   California (Lake Tahoe).
                                                                          montanus, sp. nov.

5—Body narrowly oval, rather depressed, shining, pale rufo-ferruginous throughout,
  almost similar in the sexes, the female larger; eyes rather small, not quite half as
  long as the head, the antennæ moderate as usual; prothorax parallel and broadly
  arcuate at the sides, but little more than one-half wider than long, the teeth
  strongly developed and serriform; elytra three-fifths to two-thirds longer than
  wide, relatively a little broader in the female, rather coarsely, but not very
  densely, punctate, rather arcuate at the sides and narrowly rounded behind,
  slightly (♂) or distinctly (♀) wider than the prothorax and three times as
  long,—or slightly more in the female, the pubescence short, even, rather sparse
  and suberect.   Length 1.75–2.3 mm.; width 0.7–0.85 mm.   California (southern)—Mr. Fall...**tarsalis**, sp. nov.

The species are sufficiently numerous and individually abundant on
the Pacific coast, to which region the genus appears to be confined.
I place here provisionally the Sitkan *Cryptophagus 8-dentatus* of

Mäklin, who states that the prothorax is quadridentate at each side ; this would not apply to the *6-dentatus*, described above, unless the author included the basal angles and these are in no respect dentiform in the latter species.

Cryptophagus *Hbst.*

This is a large genus, including some of the larger and more conspicuous species of the family ; they are easily separable among themselves but rather difficult to classify in a satisfactory manner.   The body is oblong-oval, convex, strongly punctured and always coarsely, distinctly, though not densely, pubescent, the elytra having in addition some longer hairs, which are frequently very conspicuous and always subserial in arrangement, although the punctuation may, and usually does, exhibit no trace of series.   The antennæ are moderate in length, thick, with the club abrupt, parallel and loosely 3-jointed.   The prothorax is wider than long, subparallel anteriorly and narrowed toward base from about the middle, where there is a more or less distinct acute and reflexed marginal tooth, and the apical angles are thickened and obliquely truncate, the oval truncature sublateral, polished, generally flat or rarely concave and foveate at the middle ; the lateral edges between the submedian denticle and the well-defined and sometimes subprominent basal angles is generally obsoletely serrulate ; the disk is deeply, though finely, bifoveate at the base, the foveæ connected by a fine groove following the basal margin and often subinterrupted at the middle by a fine longitudinal carina.   There are also quite generally visible two small impunctate and feebly callus-like spots at each side near lateral third.   The maxillary palpi are well developed, the last joint elongate and gradually, somewhat obliquely and obtusely acuminate, the last joint of the labial moderately stout, oval and truncate at tip, the mentum large, transverse, the basal parts concave and punctured and separated from the deflexed apical parts by a strong, transversely arcuate carina, which is prolonged anteriorly on the median line to the extreme apex.   The anterior coxæ are obliquely oval, rather large and deep-set, and the intercoxal process is prolonged posteriorly, with its free tip ogivally acuminate and dorsally margined.   The mesosternum is broadly and feebly concave.   The tarsi are slender, and the abdominal segments two to four decrease gradually in length, the first longer, generally exceeding the next two combined, the fifth about as long as the second and rounded in both sexes, the sutures transverse, perfectly free and virtually straight throughout. The elytra have sometimes—as in *plenus*—a smooth callous discal spot near the apex of each, which may be homologous with the smooth polished mirror-like sexual spots of the melyrid genus *Eurelymis*. Sexual differences in the form of the body are occasionally very pronounced, the male being shorter and stouter than the female, with relatively broader prothorax and shorter elytra.

The species before me may be tentatively characterized in the following manner :—

Lateral spicule of the prothorax situated at or near the middle of the length ; front not constricted between the antennæ ; species general in distribution..............2
Lateral spicule situated far behind the middle, the sides just posterior to them frequently arcuately prominent ; front narrowed by the very large antennal foveæ. Pacific coast 32
2—Sides of the prothorax broadly and conspicuously angulate at about the middle, the spicule at the apex thereby rendered more prominent and separated from the truncature of the anterior angles by a pronounced sinus..............................3
Sides of the prothorax in the form of a continuous and generally evenly arcuate curve, from the truncature of the apical angles to the base, the submedian spicule abruptly projecting from the limb and frequently extremely small...................15
3—Eyes large, generally one-half as long as the head or more ; elytra finely and rather closely punctured....................................... 4
Eyes smaller but more strongly convex, always much less than one-half as long as the head.. 6
4—Nodes of the thoracic angles very prominent and posteriorly unciform, the prothorax much wider anteriorly than at the middle, rather finely but deeply, densely punctate, the discal callous spots obsolete ; elytra elongate, between three and four times as long as the prothorax.   Length 1.9–2.5 mm. ; width, 0.75–0.9 mm.   Europe, Siberia and Northern America.................**acutangulus** *Gyll.*
Nodes moderate in development, acute but not unciform posteriorly, the prothorax equally wide anteriorly and at the middle..5
5—Pubescence long, coarse and very conspicuous, the serial hairs of the elytra distinct ; elytra distinctly more than three times as long as the prothorax.   Length 2.2–2 5 mm. ; width 0.9–1.0 mm.   Europe and Northern America.
cellaris *Scop.*
Pubescence short and more decumbent, less coarse and very much less conspicuous, the serial hairs subobsolete ; pronotum finely but deeply, only moderately densely punctate, the callous spots feeble ; elytra more oval and less elongate, about three times as long as the prothorax.   Length 1.9–2.1 mm. ; width 0.8–0.85 mm. California...**debilis** *Lec.*
6—Elytral punctures fine, the pubescence very short, inconspicuous and decumbent, the subserial hairs subobsolete or very short, the pronotal callous spots obsolete or scarcely traceable ; nodes of the thoracic angles sharply truncated, the prothorax as wide at the middle as at the apex...7

Elytral punctures more or less coarse and much less close-set, frequently quite sparse, the surface strongly shining throughout ; pronotal callous spots generally conspicuous 8

7—Elytral punctures moderately close-set, the surface strongly shining ; prothorax evenly convex, rather strongly and closely punctate, the nodes of the apical angles moderately prominent, much shorter than the sinus separating them from the median denticles ; antennal club moderately broad ; elytra two-thirds longer than wide Length 2.1 mm. ; width 0.8 mm. Alaska......**bidentatus** *Mäkl.*

Elytral punctures extremely dense, the entire surface rather dull in lustre ; prothorax less convex and more uneven, two-thirds wider than long, the truncated nodes large and more prominent, though not unciform behind, and but little shorter than the sinus between them and the denticles ; elytra more than three times as long as the prothorax and a little wider, three-fourths longer than wide ; antennal club well developed and rather broad. Length 2.3 mm. ; width 0.88 mm. Colorado...**confertus,** sp. nov.

8—Truncate node of the thoracic angles very large, though only moderately prominent, distinctly longer than the sinus separating it from the median spicules, the truncature elliptical, flat and sharply defined, the prothorax equally wide at apex —that is between the posterior angles of the truncatures—and at the middle, the median tooth short and broad, unciform posteriorly, the punctures rather coarse, deep as usual and only moderately close-set, the callous spots rather distinct, especially in the male ; elytra much larger in the female than in the male, not wider than the prothorax in the latter ; pubescence rather long but sparse, the subserial hairs long, suberect and conspicuous. Length 2.3-2.7 mm.; width 0.9-1.15 mm. New Jersey..**nodifer,** sp. nov.

Truncate node of the thoracic angles rather small, always much shorter than the sinus separating it from the denticles, the truncature narrow, convex and very acute posteriorly...9

9—Elytra rather oblong-oval, more elongate and less strongly rounded at the sides, never more than slightly wider than the prothorax...10

Elytra oval, relatively more convex, more narrowly rounded behind and always very much wider than the prothorax..14

10—Node of the thoracic angles very small and not acute or angulate posteriorly from a vertical viewpoint; body pale ferruginous throughout ; prothorax rather finely but deeply, only moderately closely punctate, a little narrower at apex than at the middle, the callous spots large and conspicuous, though not much elevated ; elytra only moderately coarsely and rather sparsely punctate ; pubescence long, coarse and conspicuous throughout, pale ochreo-cinereous in color (♀). Length 2.3 mm.; width 1.0 mm. Indiana ?................................**parvinoda,** sp. nov.

Node of the thoracic angles better developed, with the posterior extremity very acutely prominent and unciform from a vertical point of view, the prothorax subequally wide at the middle and apex............. 11

11—Pubescence moderately long and suberect, sparse......................................12

Pubescence short and more closely decumbent, even, the longer hairs subobsolete....13

12—Body blackish-piceous in color, the pronotum rather paler and the elytra dark testaceous ; pronotum evenly convex, rather coarsely and closely punctured, the callous spots very distinct, rather small and scarcely elevated ; elytra about two-thirds longer than wide, broadly rounded behind, coarsely and unusually sparsely punctate (♀).   Length 2.45 mm.; width 1.00. Pennsylvania (Westmoreland Co.)..**infuscatus,** sp. nov.

Body nearly as in *infuscatus* but smaller and rather less elongate, pale rufo-ferruginous in color throughout, the elytra rather more strongly narrowed and less broadly rounded behind, similarly sculptured but with the antennal club shorter and relatively broader and more compact (♀).   Length 2.25 mm.; width 0.9 mm. District of Columbia..**plectrum,** sp. nov.

13—Body parallel, rufo-ferruginous throughout, the prothorax large, three-fifths wider than long, fully as wide as the elytra, strongly, moderately densely punctured, the callous spots small but very conspicuous and distinctly elevated ; elytra three fifths longer than wide, rather abruptly and very obtusely rounded behind, the punctures coarse and rather sparse, but much closer and rather more perforate than in the two preceding species (♂).   Length 2.4 mm.; width 0.95 mm.   New York—Mr. H. H. Smith..**cicatricosus,** sp. nov.

14—Prothorax as wide at the apex as at the middle ; body pale flavo-testaceous in color throughout, the pubescence long, erect and hispid, very conspicuous though unusually sparse ; prothorax small, transverse, strongly, but not very coarsely or closely, punctured, the callous spots all very distinct ; elytra oval, just before the middle nearly two-fifths wider than the prothorax, the punctures very coarse, deep and sparse, but, as usual, small or obsolete toward apex, each with an elongate callous median space near the tip (♀).   Length 1.9 mm.; width 0.8 mm.   North Carolina..**politus,** sp. nov.

Prothorax much narrower at the apex than at the middle ; body broadly oval, strongly convex, highly polished, dark piceo-rufous in color throughout, the pubescence moderately long, sparse, coarse and ashy ; pronotum evenly convex, not very densely punctate, the callous spots small and subobsolete ; elytra inflated, scarcely one-half longer than wide, quite pointed at apex, the punctures very coarse, sparse and conspicuous toward the base and sides (♀).   Length 1.85 mm.; width 0.8 mm.   Lake Superior......................**difficilis,** sp. nov. (*Lec.* MS)

15—Truncature of the anterior thoracic angles forming a broadly oval, sharply defined, flat or feebly concave disk, having a large subcentral foveiform puncture, and from a vertical viewpoint, oblique and perfectly rectilinear.........................16

Truncature irregular, narrow, sometimes nearly flat but generally more or less convex..25

16—Elytral pubescence semi-erect, the longer subserial hairs distinct and more or less bristling..17

Elytral pubescence short, decumbent and even, the longer subserial hairs almost or completely obsolete 22

17—Species of the Atlantic coast ; eyes small and strongly convex.....................18

Species of the Pacific coast, the eyes still smaller, extremely convex and subparabolic in outline from above...21

18—Body very short and stout, not more than twice as long as wide, oblong, convex, blackish-piceous in color, the head, pronotum, antennæ and legs dark testaceous; prothorax large, very nearly as wide as the elytra, three-fifths wider than long, strongly and densely punctate, uneven, the callous elevations distinct ; angular truncatures sensibly shorter than the distance thence to the spicules ; elytra very

short and obtusely rounded, less than one-half longer than wide, about two and
one-half times as long as the prothorax, coarsely and rather closely punctate ;
pubescence rather long, suberect, coarse and bristling but not dense ; antennal
club moderate in development ; each elytron has a large embosed rounded impunctate spot near the apex at inner two-fifths (♂).   Length 2.0 mm.; width
1.0 mm.   Florida.. **plenus,** sp. nov
Body more elongate, more than twice as long as wide, more or less pale ferruginous
in color throughout...19
19—Subserial setæ of the elytra very long and conspicuous; body large, coarsely
punctured, the elytra not very closely and sometimes subserially ; sexual differences
very marked, the male stout, with the antennæ very thick, the elytra one-
half longer than wide, but little wider than the prothorax and barely three
times as long, the female much narrower and more elongate, with thinner antennæ and smaller prothorax, having rather more prominent but otherwise similar
angular nodes, the elytra three-fourths longer than wide, much more than three
times as long as the prothorax and distinctly wider, the punctures somewhat more
sparse and more inclined to serial arrangement ; callous spots of the pronotum small
but obvious ; angular truncatures large but barely as long as the distance thence
to the spicules.   Length 2.6–2.7 mm.; width 1.1–1.2 mm.   North Carolina.
(Asheville)...**amputatus,** sp. nov.
Subserial setæ only moderately distinct ; body much smaller ; angular nodes of the
prothorax well developed but not prominent, the callous spots small but distinct;
antennal club broad and well developed..20
20—Body parallel, dark rufo-ferruginous in color, the pubescence rather abundant
and conspicuous though only moderately long ; prothorax scarcely narrower
than the elytra in the male, distinctly narrower in the female, strongly and closely
punctured, the angular nodes large and conspicuous, as long as the distance
thence to the spicules, or even longer in the male ; elytra, in the latter sex, one-
half longer than wide, less than three times as long as the prothorax, coarsely
and closely punctured, in the female decidedly more elongate, more than three
times as long as the prothorax and less closely though equally coarsely, punctured ;
a feebly eroded adventitious second, line parallel to the posterior subsutural stria,
is sometimes evident.   Length 2.0–2.2 mm. ; width 0.85 mm.   South Carolina
to Illinois ; [*crinitus* Zimm.]...**croceus** Zimm.
Body more oval and rather more convex, pale flavo-ferruginous throughout ; prothorax
smaller and more rounded at the sides than in *croceus*, the angular lobes much
smaller, distinctly shorter than the distance thence to the spicules, the punctures
strong and rather close-set, the posterior callous spots more obvious than the anterior ; elytra three-fifths longer than wide, rather strongly rounded at tip, about
three times as long as the prothorax, the punctures only moderately coarse, deep,
decidedly close-set and inclined to subserial arrangement, the pubescence shorter
and less conspicuous than in *croceus* (♂).   Length 1.9 mm. ; width 0.75 mm.
Pennsylvania (Westmoreland Co.)..............................**laticlavus,** sp. nov.
21—Antennal club broader, with its basal joint scarcely smaller than the second as
usual ; body rather small, oblong, dark testaceous in color, the pubescence
moderately long and sparse ; prothorax very nearly as wide as the elytra, strongly
transverse, three-fourths wider than long, strongly and closely punctured, the

callous spots feeble, the angular nodes well developed and as long as the interval thence to the spicules, which are very minute ; elytra two-thirds longer than wide, more than three times as long as the prothorax, rather finely but deeply, not very closely punctate (♀).   Length 2.0 mm. ; width 0.88 mm.   California (near San Francisco) ....**inscitus**, sp. nov.
Antennal club narrow, with its basal joint distinctly smaller than the second ; body small, oblong, compact and convex, shining, dark rufo-testaceous in color, the vestiture rather long and abundant, suberect and distinct ; prothorax large, about as wide as the elytra, strongly, very closely and deeply punctato cribrate, nearly even, the posterior of the callous spots alone distinct ; sides parallel, very feebly narrowed at base, the angular nodes well developed and as long as the adjacent sinus, the spicules strong and distinct ; elytra about three-fifths longer than wide and two and three-fourths times as long as the prothorax, not very coarsely, but deeply and quite closely, punctate (♂).   Length 1.8 mm. ; width 0.78 mm. California (Mokelumne Hill, Calaveras Co.)—Dr. Blaisdell.
                                                                  cribricollis, sp. nov.
22—Body normally convex, pale ferruginous in color throughout, the nodes of the thoracic angles well developed but not prominent and not unguiculate behind..23
Body sensibly depressed, blackish in color, the legs piceous, the head and pronotum rufo-piceous ; thoracic nodes smaller but much more prominent, strongly acute and unciform behind 24
23—Body oblong-oval, rather stout, the antennal club moderately wide, with the middle joint just visibly the widest, the pubescence very short and even, somewhat sparse ; eyes rather large and not very convex ; prothorax well developed, the angular nodes moderate in size, not more than a fifth of the total length and very much shorter than the distance thence to the spicules, which are small but distinct, the sides between them and the nodes broadly sinuate ; punctures close-set, the anterior callous spots small but abruptly elevated and very distinct, the posterior almost obsolete ; elytra large, distinctly wider than the prothorax and more than three times as long, the punctures moderately coarse and somewhat close-set (♀).   The male is shorter, with more finely and densely punctate prothorax and elytra.   Length 2.0–2.4 mm. ; width 0.88–1.0 mm.   California (Sta. Cruz Co.).. ....**brevipilis**, sp. nov.
Body nearly similar in form and color but with the eyes rather smaller and more convex, the thoracic lobes larger, about a fourth of the total length and but little shorter than the sides thence to the spicules, which portion is straight, the callous spots less distinct; elytra unusually finely and quite closely punctured; pubescence rather longer than in *brevipilis* but nearly even and decumbent (♀).   Length 2.25 mm.; width 0.9 mm.   California (exact locality not recorded).
                                                                  lepidus, sp. nov.
24—Broadly oblong-oval, feebly shining, the antennal club moderate and the eyes quite small and strongly convex ; prothorax relatively rather small but not very transverse, about three-fifths wider than long, unusually finely and very densely punctate, the callous spots very small and inconspicuous, the sides parallel, arcuately narrowing toward base, the spicules broad and truncate, unciform behind, the angular lobes rather small but very prominent, obliquely, rectilinearly truncate from above ; elytra large, black, evenly rounded behind, nearly a fourth wider

than the prothorax and three and one-half times as long, rather finely but strongly, only moderately closely punctate, the pubescence very short, decumbent, even and sparse, with scarcely a trace of longer subserial hairs (♀). Length 2.6 mm.; width 1.0 mm. Lake Superior.—A male specimen from Siskiyou Co., California, is attached for the present and may belong to this species, as the differences presented are all in directions shown by other series to be sexual in origin...**depressulus**, sp. nov.

25—Nodes of the thoracic angles small but very prominent, forming a narrow uneven convex surface sharply pointed behind ; body rather narrow, elongate, moderately convex, black, the head and pronotum slightly piceous, the antennæ and legs rufo-piceous ; pubescence rather sparse, coarse, suberect, the subserial bristles long and distinct ; eyes small, the antennal club moderate ; prothorax not very transverse, one-half to three-fifths wider than long, not very coarsely but deeply and densely punctate, the callous spots subobsolete ; meuian denticle short but rather broad, acute posteriorly, the sides thence moderately convergent and broadly arcuate to the base, the width across the denticles slightly greater than at the apical nodes ; elytra elongate-oval, slightly wider than the prothorax and more than three times as long, the punctures moderately coarse, deep and ,not very close-set (♀). Length 2.3 mm.; width 0.83 mm. Utah (southwestern) and Colorado...**porrectus**, sp. nov.

Nodes of the thoracic angles not notably prominent...................................26

26—Antennæ long, rather stout and unusually developed, about half as long as the body, the club moderate ; eyes moderate ; body quite short and stout, oblong-oval, pale rufo-ferruginous in color throughout, the pubescence conspicuously long, coarse, suberect and bristling, rather abundant and subeven ; prothorax large, as wide as the elytra, three-fourths wider than long, evenly convex, coarsely, but not very closely, punctured, the callous spots subobsolete ; nodes‾ large, almost as long as the distance thence to the acute denticles, narrow and shining sublaterally ; elytra one-half longer than wide, between two and three times as long as the prothorax, coarsely but not very closely and in part subserially, punctate, very minutely so toward apex (♂). Length 1.8 mm.; width 0.8 mm. Michigan...**antennatus**, sp. nov.

Antennæ much shorter, always much less than half as long as the body..............27

27—Eyes very small and strongly convex, scarcely a third as long as the head ; body large, rather stout, elongate-oval, pale flavo-testaceous in color throughout, the pubescence only moderately long but suberect, very abundant and conspicuous, the subserial hairs but little longer and not very distinct ; antennal club moderate ; prothorax well developed, two-thirds to three-fourths wider than long, evenly convex, finely, deeply and closely punctured, the callous spots subobsolete, the sides parallel and evenly rounded from base to apex, the spicules very minute, the nodes well developed but narrow, polished, with a small central puncture ; elytra elongate-oval, rather narrowly rounded behind, two-thirds longer than wide, obviously wider than the prothorax, the punctures fine and unusually close-set ; male and female almost completely similar in form throughout, the former very slightly less stout. Length 2.6–2.9 mm.; width 1.05–1.2 mm. Pennsylvania (Westmoreland Co.)—Mr. Schmitt...**valens**, sp. nov.

Eyes moderate in size and relatively less prominent, nearly half as long as the head..28
28—Nodes of the apical angles well developed and longitudinally convex, but little
   shorter than the distance thence to the spicules ; antennal club unusually long
   and narrow, the last joint longer than wide ; body rather large, narrowly elongate-
   oval, rather dark rufo-testaceous throughout, the pubescence somewhat short and
   moderately abundant but with the subserial setæ quite long and obvious ; prothorax moderately transverse, densely and deeply punctured, the callous spots
   small but rather distinct, the lateral spicules small ; elytra unusually elongate,
   rather acutely ogival at apex, distinctly less elongate in the male, quite coarsely
   and deeply but not very closely punctured.   Length 2.6–2.8 mm. ; width 1.05
   mm.   Utah (southwestern)—Mr. Weidt.........................**histricus**, sp. nov.
Nodes of the apical angles small, very much shorter than the distance separating
   them from the spicules...29
29—Pubescence moderate in length and subdecumbent, the punctuation rather fine ;
   body oblong, shining, dark testaceous throughout ; prothorax well developed,
   moderately transverse, but little narrower than the elytra, the apical nodes extremely small, feeble, very oblique, narrow and convex sublaterally, with a
   minute posterior spicule ; submedian spicule very minute, slightly behind the
   middle and separated from the nodes by between two and three times the length
   of the latter, punctures small but deep, moderately close-set, the callous spots
   small and rather feeble but distinct ; elytra oval, rather obtusely rounded at tip,
   three-fifths longer than wide, the punctures rather fine but deep and not very close-
   set (♀).   Length 2.2 mm. ; width 0.9 mm.   Indiana ; [Carolina—Zimm.]
                                                                 fungicola *Zimm.*
Pubescence long and bristling, the body coarsely punctured.   Sonoran regions......30
30—Lateral spicules of the prothorax distinct and moderately large, the sides behind
   them thickened and laterally subprominent half the distance to the base, then
   sinuate to the angles ; body rather stout, oval, convex, polished, piceous in color,
   the elytra blackish, nubilously paler at apex ; prothorax moderately transverse,
   very coarsely, closely punctate, uneven, the callous spots large, conspicuous and
   unusually approximate longitudinally ; apical nodes very oblique, spiculate behind ; elytra oval, distinctly arcuate at the sides, moderately obtuse at apex,
   scarcely more than one-half longer than wide, and, at the middle, slightly wider
   than the prothorax, the punctures very coarse, deep and somewhat sparse, as
   usual very small toward tip (♀).   Length 2.1 mm. ; width 0.9 mm.   Arizona
   (Tucson)..**discedens**, sp. nov.
Lateral spicules extremely minute and sometimes apparently obsolete, the sides be.
   hind them arcuately converging, sometimes sinuate very near the angles, in
   which case the latter are acutely prominent.....................................31
31—Narrowly oblong-oval, testaceous, the elytra frequently infuscate ; prothorax
   rather short and transverse and slightly narrower than the elytra in both sexes,
   the punctures moderately coarse, deep and somewhat close-set, the callous spots
   large and normally placed, the lateral edges rather widely reflexed ; sides
   strongly convergent behind the middle ; elytra more than three times as long as
   the prothorax in the female, much shorter in the male, rather narrowly obtuse
   behind, very coarsely, but only moderately closely, punctate.   Length 1.9–2.0
   mm. ; width 0.78 mm.   Utah (southwestern)—Mr. Weidt.
                                                          fumidulus, sp. nov.

Rather broadly oblong-oval, flavo-testaceous throughout, otherwise similar to the preceding, except that the punctures are less coarse and more close-set, the sides of the pronotum more narrowly reflexed and the lateral spicules still more minute and frequently almost invisible ; elytra broader and more broadly rounded behind (♀). Length 2.0 mm.; width 0.88 mm. California (southern); [*pilosus, hirtulus* l.cc.]...**lecontei** *Harold*
32—Tarsi moderate in length, the posterior distinctly shorter than the tibiæ in both sexes 33
Tarsi more elongate, the posterior fully as long as the tibiæ in the male and but little shorter in the female ; body oblong-oval, convex, shining, dark rufo-testaceous in color throughout; pubescence short, even, decumbent, yellowish, and not very dense ; antennæ slender, the club moderate, the second and third joints both elongate and longer than the first, which is subglobular; eyes moderate ; prothorax well developed, one-half (♂) to three-fifths(♀) wider than long, nearly as wide as the elytra in both sexes, the nodes elongate-oval, flat and centrally punctate, the spicules small, the sides behind them prominently rounded and convergent ; punctures moderately coarse, deep and dense, the callous spots visible ; elytra nearly similar in the sexes, about three-fifths longer than wide, the punctures moderately coarse and not very close-set. Length 1.9–2.3 mm. ; width 0.75–0.85 mm. Queen Charlotte Islands (Massett)—Mr. Keen......**hebes**, sp. nov.
33—Body similar to that of *hebes* in form, sculpture and vestiture but smaller, with the prothorax more transverse and the antennæ less elongate and relatively stouter, the third joint obviously shorter and more slender than the second ; elytra three-fifths (♂) to two-thirds (♀) longer than wide, three times as long as the prothorax in the latter sex but much shorter relatively in the male, but little wider than the prothorax in either sex. Length 1.7–2.2 mm. ; width 0.72–0.82 mm. California (Coast regions from Humboldt to San Diego)..**lyraticollis**, sp. nov.
Body nearly similar to the two preceding in general form, but differing in the finer and closer punctures of the elytra, and, from *lyraticollis*, in the more slender and somewhat more elongate antennæ, the club rather narrow, loose and parallel as usual ; prothorax only moderately transverse, less so than in *lyraticollis* but otherwise nearly similar, the callous spots large and very distinct ; elytra large, much wider than the prothorax and distinctly less than three times as long, one-half longer than wide (♀). Length 2.25 mm. ; width 0.95 mm. California (Mendocino Co.)..**otiosus**, sp. nov.

I have been unable to identify the *4-dentatus* of Mannerheim, from the Island of Sitka, or the Alaskan *tuberculosus, punctatissimus* and *4-hamatus* of Mäklin. The last named must be very closely allied to *depressulus*, of the table, but differs somewhat in coloration, and especially in its much smaller size. I fail to identify the European *lapponicus* among our species, and the *nodulangulus* of Zimmerman, is also unrepresented in my cabinet. The *8-dentatus* of Mäklin, is a *Salebius* without much doubt, and the *californicus* of Mannerheim, belongs to the genus *Henoticus*. *Humeralis* of Kirby, was placed in *Triphyllus* by Leconte, but in reality forms the type of a new Melandryid genus, which will be described further on in the present paper, and the *concolor* of the same author, I have been unable to trace.

Henoticus *Thoms.*

The general structure of the body, prosternum, legs and tarsi, trophi and antennæ are here almost precisely as in *Cryptophagus*, but the converging sides of the front above the antennæ are finely reflexo-marginate, and the structure of the sides of the prothorax wholly different, there being no trace of thickened nodal point, apical or otherwise ; the edge is regularly spiculato-serrulate throughout, except for a short distance near the basal angles ; it also differs in having the fine subsutural line entire or subentire.   The deep groove near the basal margin of the pronotum connecting the conspicuous basal foveæ is similar to that of *Crosimus* and without trace of medial interrupting carina. The elytral punctures are arranged wholly without order, the pubescence short and the pronotum without trace of callous spots.   The species known thus far are two in number, and are both very abundant in individuals ; they may be outlined as follows :—

Black or blackish in color throughout when mature, the legs and antennæ paler, polished, oblong, convex and moderately stout in form, the pubescence short, very sparse, even and reclined ; eyes well developed though scarcely half as long as the head ; prothorax moderately transverse, the sides very nearly parallel, broadly and evenly arcuate, the serratures even and moderately developed, some eight to ten in number; punctures not coarse but deep, moderately close-set, the surface rather convex ; elytra oblong, distinctly wider than the prothorax and three times as long or a little less, obtusely rounded behind, the punctures coarse and decidedly sparse. Length 1.7–2.1 mm.; width 0.65–0.85 mm.   Entire northern America, Siberia and northern Europe. [*Paramecosoma denticulata* Lec.]
                                                                                serratus *Gyll.*
Pale testaceous in color throughout, shining, the pubescence not quite so short, rather abundant and suberect, bristling with slightly longer hairs toward the sides ; eyes small and extremely convex ; prothorax strongly transverse, distinctly widest slightly before the base, the sides thence rather strongly convergent and straight and provided with some eight very strong, acute and equal serratures to the apex ; surface feebly convex, more finely, rather closely punctate, evidently impressed near the lateral margins toward apex ; elytra but little wider than the base of the prothorax and three times as long or more, very finely, though not very coarsely, punctate.   Length 1.75–2.1 mm.; width 0.78–0.9 mm.   California (San Francisco to Monterey)................ ....**californicus** *Mann.*

The latter of these was assigned to *Cryptophagus* by its author. The *Paramecosoma inconspicua* Lec., i. litt., is unknown to me, but is probably founded upon a very small example of *serratus*.

Pteryngium *Reitt.*

Among the close allies of *Cryptophagus*, the two species of this genus may be instantly recognized by the rather narrow, strongly depressed and planulate body, with parallel sides, finely, densely punctured surface, short pubescence and entire subsutural lines. In this last feature, as well as the evenly arcuate and minutely, evenly serrulate sides of the prothorax, they resemble *Henoticus*, but differ in the depressed body and in the very minute basal foveæ of the pronotum, connected by a very fine and feeble basal groove, which is finely interrupted at the middle. In the structure of the legs, prosternum, trophi and antennæ they perfectly resemble *Cryptophagus*, but differ from that genus, as well as *Henoticus*, in the somewhat shorter and thicker tarsi, and especially in the much more elongate basal segment of the abdomen, this being as long as the next three combined; the sutures are free and perfectly straight throughout, as usual in the tribe. The frontal margin above the antennæ is very obsoletely and indistinctly margined. The species may be thus characterized :—

Body parallel, depressed, rather feebly shining, pale rufo-ferruginous throughout, the pubescence short, even, subdecumbent and rather abundant; eyes moderate in size and prominence, not very coarsely faceted, the antennæ scarcely as long as the head and prothorax, notably stout, the club parallel and broad, its first two joints equal and very strongly transverse, joints one to three rapidly decreasing in size; prothorax about one-half wider than long, the sides parallel, evenly, distinctly arcuate from base to apex and minutely, evenly serrulate, the apex broadly arcuate, the punctures fine but deep and very close-set; elytra but little more than one-half longer than wide, about equal in width to the prothorax and two and one-half times as long, parallel, obtusely rounded behind, finely but deeply, very closely punctate. Length 1.8 mm.; width 0.72 mm. Lake Superior and Europe...**crenatum** *Gyll.*

Body similar in general form and coloration but smaller, narrower and more shining, the antennæ distinctly less stout, with the club less robust; prothorax similar in form but a little more transverse, finely, strongly punctured but only moderately closely, the surface more shining; elytra similar in general form but more elongate, scarcely wider than the prothorax but almost three times as long, the punctures fine, strong and rather close-set but much less dense than in *crenatum*, and, as in that species, having the surface broadly, transversely impressed at some distance behind the base, but here the impression bears traces of longitudinal striiform lines, which are wanting in *crenatum*; the pubescence, also, is still shorter, sparser and less evident throughout. Length 1.65 mm.; width 0.6 mm. Queen Charlotte Islands (Massett)—Mr. Keen.................**malacum**, sp. nov.

These two species are each represented before me by a single example in which the hind tarsi are 4-jointed. It is presumable, of

course, that the female has these tarsi 5 jointed. In each case the three
basal joints are short, stout and equal and together scarcely longer
than the last.

ATOMARIINÆ.

The genera of this subfamily may be readily recognized by the palpal
structure and position of the antennæ, these organs being inserted
upon the front and more or less approximate at base, the foveæ being
either small and exposed or deep cavities, separated above by a short
angular extension of the upper surface, and particularly developed in
Cænoscelis and *Sternodea*. The tarsi are always slender and filiform,
as in the Cryptophagini, of the preceding subfamily, and, as in that
case, there is frequently a feeble thickening of the anterior in the
males. The body is much smaller as a rule than in the Cryptophaginæ, and may be either narrow and parallel, as in *Agathengis*, or oval
and more convex, as in the great majority of genera. The subfamily
may be resolved into the four following rather widely differentiated
tribes :—

Prosternal process free, the tip passing over the flat or feebly concave surface of the
    mesosternum ; antennæ free, the grooves before the eyes wholly obsolete, the
    club loosely 3-jointed ; basal segment of the abdomen not modified behind the
    coxæ except in *Tisactia*...2
Prosternal process broader and flatter, generally prominent, forming a continuous sur-
    face with the mesosternum, its tip broadly arcuate and received closely within a
    corresponding depression at the apex of the mesosternum ; body more compact,
    the prothorax more closely fitted to the elytra ; buccal processes obsolete ; elytra
    never margined at base ; tarsi pentamerous 3
2—Prosternal process prolonged and acute at tip, the mesosternum concave ; prothorax
    with a double lateral margin ; tarsi pentamerous in the female and heteromerous
    in the male ; first abdominal segment well developed, longer than the next two
    combined, the sutures bent strongly backward for a short distance at the sides ;
    antennal cavities large, narrowly separated ; buccal processes long and promi-
    nent, the eyes very coarsely faceted ; body elongate, only moderately convex,
    generally coarsely sculptured and sparsely pubescent.. ....CÆNOSCELINI
Prosternal process shorter and truncate, the mesosternal surface generally flat ; pro-
    thorax with a fine single lateral edge ; tarsi pentamerous in both sexes ; abdominal
    sutures straight throughout the width ; antennal cavities small and superficial, the
    buccal processes extremely short and inconspicuous, the eyes much less coarsely
    faceted ; body variable in form, sculpture and vestiture....... ....ATOMARIINI
3—Antennæ free, the club 2-jointed in *Sternodea*, the cavities very large and deep,
    contiguous ; first ventral segment as long as the next three combined, without
    post-coxal plates, the sutures broadly, feebly reflexed toward the sides ; proster-
    num extremely prominent along the middle, with acute lateral margins extend
    ing to the anterior margin ; tibiæ feebly claviform ; scutellum well developed and
    transverse ; anterior coxæ almost rounded ....*STERNODEINI

Antennæ variable, the club loosely 3-jointed, the foveæ small, more widely separated on the front and superficial ; eyes somewhat less coarsely faceted ; basal segments of the abdomen relatively rather shorter and generally with a short and broadly arcuate post-coxal plate, the sutures straight throughout ; prosternum broader and less prominent, the acute lateral margins not extending to the anterior margin ; tibiæ slender, the scutellum very small ; anterior coxæ transverse, the intermediate very widely separated ; body broadly oval, convex and generally glabrous ....EPHISTEMINI

The Sternodeini are peculiar to the palæarctic provinces, but the other tribes are well represented in America, the Ephistemini, however, by no means so extensively as in Europe.

CÆNOSCELINI.

This tribe is composed at present of the single genus *Cænoscelis*, which is very well developed in the northern parts of America, and, to a less degree, apparently, in the palæarctic region ; its species are the largest and most conspicuous of the subfamily, and compare very closely in this respect with *Cryptophagus*, but the body is narrower and more elongate as a rule.

Cænoscelis *Thoms.*

This is one of the best defined and more isolated genera of the family, distinguished by the elongate, strongly punctured and pubescent body, with double lateral margin and broadly impressed basal parts of the pronotum, convex, coarsely faceted and sparsely setulose eyes and well developed stout antennæ, with the basal joint unusually large and obconical, the second and third diminishing in size and four to eight still narrower and alternately shorter and longer, as usual in the Atomariinæ ; the basal joint of the club is small, the last two well developed. The tarsi are very slender and the posterior are 5-jointed in the female and 4-jointed in the male, there being otherwise but little sexual disparity ; the male is usually rather narrower, with relatively larger, and occasionally somewhat less transverse, prothorax. The prosternal process is narrower, the tip prolonged, free, concave toward tip and acuminate, the mesosternum being appreciably concave. The abdominal sutures differ greatly from the usual type and are strongly reflexed for a short distance at the sides, especially posteriorly. The American species appear to be far more numerous than the European as described thus far, and those before me may be outlined as follows :—

Body ferruginous in color throughout...2
Body piceous-brown to black in color; pronotum broadly impressed at base, parallel
    and evenly distinctly arcuate at the sides.......... 9
2—Prothorax less transverse, never so much as one-half wider than long; body narrow
    and much elongated...3
Prothorax one-half or more wider than long, the body stouter and more oval in form..8
3—Prothorax strongly arcuate at the sides, the pronotal punctures fine and close-set,
    the subbasal impression medial only...4
Prothorax feebly arcuate at the sides, the punctures coarse though generally close-set,
    the subbasal impression arcuate, deep and extending almost from side to side...5
4—Antennæ stout; the club robust and densely clothed with fine gray down-like
    pubescence, the joints increasing in size from the base and forming a gradual
    transition to the shaft; prothorax one-third wider than long, convex, the basal
    impression median and feeble, the sides evenly rounded, more convergent an-
    teriorly, so that the apex is notably narrower than the base, the double margin
    narrow and feeble, not much more distinct toward base; elytra oval, two-thirds
    longer than wide, nearly two-fifths wider than the prothorax; body elongate-oval
    in form, the pubescence distinct, fine and sparse on the elytra, with the irregular
    series of longer hairs characterizing the genus. Length 1.8 mm.; width 0.7 mm.
    Alaska (Kenai)...**ferruginea** *Sahlb.*
Antennæ much less stout, the club similar in structure but narrower; body narrower,
    more parallel and more depressed, the pubescence finer and rather denser, the
    prothorax one-third wider than long, less rounded at the sides, the apex not nar-
    rower than the base, the lateral margin and basal impression similar, the latter a
    little stronger; elytra a third or fourth wider than the prothorax, three-fourths
    longer than wide, closely and finely punctate. Length 1.7–1.75 mm.; width
    0.55–0.65 mm. Colorado—Mr. Schmitt.......................**ochreosa,** sp. nov.
5—Elytra finely and rather sparsely punctured................ 6
Elytra strongly and more closely punctured; body smaller, elongate-oval...............7
6—Body narrow and parallel, the elytra very feebly arcuate at the sides, fully four-
    fifths longer than wide and only slightly wider than the prothorax, the latter
    · quadrate, but very slightly wider than long, the sides parallel, and evenly, feebly
    arcuate throughout, the apex scarcely narrower than the base, with prominent
    angles, the double edge slightly inflexed and notably wider toward base; an-
    tennæ moderate in length, the three basal joints well developed, the first as wide
    as the club, which is unusually narrow, sixth and eighth joints very small and
    subglobular, notably narrower than the fifth, seventh and ninth; the latter scarcely
    larger than the seventh, the club virtually 2-jointed (♂). Length 2.0 mm.;
    width 0.75 mm. Colorado............... ....**paralella,** sp. nov.
Body similar in size, sculpture and color, but less parallel, the elytra not quite so
    elongate and more rounded at the sides, fully two-fifths wider than the prothorax,
    which is otherwise similar to that of *parallela*, but more distinctly wider than
    long, with the parallel sides a little more arcuate; antennæ similar but not so
    thick toward base, the first joint not so thick as the virtually 2-jointed club (♂).
    Length 2.1 mm.; width 0.8 mm. Locality not recorded.
                                        angusticollis, sp. nov.

7—Antennæ stout, the three decreasing basal joints moderate in development, four
to eight globular and moniliform, ninth distinctly larger, obviously transverse,
the club rather stout and notably wider than the first joint ; prothorax one-third
wider than long, the sides very slightly converging from base to apex, evenly and
feebly arcuate, the apex quite distinctly narrower than the base ; elytra fully three-
fourths longer than wide, only slightly wider than the prothorax, the pubescence
fine and short (♀).   Length 1.8 mm.; width 0.63 mm.   California (Siskiyou
Co.)..**shastanica,** sp. nov.
Antennæ more slender and rather longer, more than two-fifths as long as the body,
relatively a little more thickened toward base, the first joint but little narrower
than the last two, the ninth joint not wider than long, the club very small ; prothorax more transverse, two-fifths wider than long, the sides parallel and very
feebly arcuate, the apex scarcely narrower than the base, the double side margin
more inflexed and wider toward base than in the preceding ; elytra relatively
much wider and more oval, two-thirds longer than wide, nearly two-fifths wider
than the prothorax, the pubescence rather coarser and sparser (♀).   Length
1.7–1.8 mm.; width 0.65–07 mm.   New York and Pennsylvania.

                                                            macilenta, sp. nov.
8—Basal joint of the antennæ unusually developed, one-half as long as the width of
the head and subequal in width to the last two joints, eight to ten, increasing
gradually in width, the ninth larger than usual when compared with the tenth,
the latter only moderately transverse ; prothorax one-half wider than long, the
sides subparallel, evenly and rather strongly arcuate, the disk broadly impressed
at base, as usual, and coarsely, deeply and closely punctured ; elytra oval, narrowed and strongly rounded behind, but somewhat sparsely, subseriately
punctate, much wider than the prothorax, the pubescence sparse and rather
coarse, but subeven and not long.   Length 2.2–2.25 mm.; width 0.8–0.83 mm.
New York..**basalis,** sp. nov.
Basal joint of the antennæ normally developed, much less than one-half as long as
the width of the head, the club moderate ; body smaller in size, the pronotum
quite coarsely, deeply and closely punctate, as usual, three-fifths to two-thirds
wider than long, parallel and strongly, evenly rounded at the sides ; elytra about
two-thirds longer than wide, oval, slightly narrowed behind, quite distinctly
wider than the prothorax, rather coarsely, deeply but unusually sparsely, irregularly
punctured, the pubescence rather long and coarse but sparse.   Length 1.5–1.8
mm.; width 0.65–0.78 mm.   South Carolina and Kentucky....**testacea,** *Zimm.*
9—Body larger, the antennæ more elongate, with joints four, six and eight longer
than wide...10
Body small or moderate and relatively stouter, the antennæ shorter, with the fourth,
sixth and eighth joints not longer than wide; pronotum coarsely, deeply and
more or less closely punctured, the elytra also strongly and more or less sparsely
so...12
10—Elytra inflated at the middle, fully two-fifths wider than the prothorax ; body
elongate-oval, rather convex, rufo-piceous, the elytra blackish ; pubescence
coarse, moderately long, sparse as usual ; antennæ moderately slender, distinctly
less than one-half as long as the body, the club moderate, scarcely wider than the
first joint, the ninth joint intermediate in width between the eighth and tenth,

the latter two-fifths wider than long ; prothorax two-fifths wider than long, rather strongly and closely punctured ; elytra three-fifths longer than wide, evenly oval, a little more than three times as long as the prothorax, rather finely but deeply, moderately closely and irregularly punctate (♀). Length 2.0 mm.; width 0.85 mm. North Carolina..**ovipennis**, sp. nov.

Elytra not inflated and but little wider than the prothorax in either sex.................11

11—Antennæ slender, about half as long as the body, the basal joint moderately developed and not as long as the next two combined, the third unusually elongate, the club rather narrow though wider than the basal joint, the ninth not at all wider than long but intermediate in width between the eighth and tenth, the latter only slightly transverse ; prothorax scarcely two-fifths wider than long, closely and moderately coarsely punctured ; elytra about three times as long as the prothorax, rather finely but strongly, moderately closely, subscriately punctured, the vestiture as usual (♂). Length 2.25 mm.; width 0.85 mm. Kentucky...**macra**, sp. nov.

Antennæ stout, not quite half as long as the body, the basal joint unusually developed, stout, fully as long as the next two combined, the third joint not unusually elongate and shorter than the second, the club rather stouter, the ninth joint slightly wider than long and the tenth more transverse ; prothorax shorter, more than one-half wider than long, strongly, but rather less densely, punctured ; elytra more than three times as long as the prothorax, rather finely but strongly, quite sparsely and irregularly punctured, the pubescence rather shorter and less coarse, a fifth wider than the prothorax in the male and a fourth in the female (♂ ♀). Length 1.8–2.1 mm.; width 0.75–0.82 mm. Pennsylvania (Westmoreland Co.) —Mr. Schmitt..**elongata**, sp. nov.

12—Antennæ stout, the first joint rather well developed and subequal in length to the next two combined, the club moderate ; prothorax rather short and transverse, strongly rounded at the sides, three-fifths wider than long ; elytra slightly wider than the prothorax, elongate-oval in form, fully three-fourths longer than wide, the punctures rather sparse and moderately coarse ; body blackish-piceous in color (♀). Length 1.65–1.9 mm.; width 0.65–0.78 mm. (♂ ♀). Pennsylvania (Westmoreland Co.)—Mr. Schmitt...................**obscura**, sp. nov.

Antennæ and prothorax throughout nearly as in *obscura*, the body stouter in the female, the male slender, rufo-piceous in color, the elytra scarcely two-thirds longer than wide, more rapidly narrowed behind and relatively more narrowly rounded at tip, the punctures coarser and more conspicuous though equally sparse and likewise irregularly disposed (♀). Length 1.8–2.0 mm. ; width 0.75–0.88 mm. (♂ ♀). Kentucky...... ....**subfuscata**, sp. nov.

Obscura is represented by a large series displaying but little variability, and four others of those described above are also present before me in numbers sufficient to demonstrate the constancy of most of the differential characters stated in the table ; the number of apparently valid species is however unexpected, and, as a rule, they are remarkable similar to each other in general habitus, which causes the taxonomic study of them to be unusually difficult and beset with doubt. *Testacea* of Zimmermann, is omitted from the Henshaw list. The *cryptophaga* of Rietter, I have been unable to identify.

ATOMARIINI.

The Atomariini constitute by far the larger part of the subfamily, and comprise several genera in America. The body is much smaller throughout than in the preceding tribe and seldom or never surpasses 2 mm. in length. The genera before me may be briefly defined as follows : —

Elytra not margined at base ; body always pubescent, the antennæ separated at base by a third of the width of the head or less...2
Elytra margined at base ; antennæ separated at base by nearly half the entire width of the head, though purely frontal as usual ; body minute in size and virtually glabrous..3
2—Body elongate and parallel in form, less convex, the prothorax angulate and foveate at the lateral edges far behind the middle ; antennæ very approximate at base, with the basal joint obconical and feebly arcuate ; first ventral segment behind the coxæ not as long as the next two combined ; prosternal process narrow.
Agathengis
Body oval, more convex, the prothorax rounded or angulate at or before the middle, and generally having the minute fovea, in the edge at the point of angulation. less developed than in *Agathengis ;* antennæ less approximate at base, the basal joint shorter and oblong; first ventral segment behind the coxæ as long as the next two; the posterior segments shorter ; prosternal process generally narrow and not prominent but becoming broader and more prominent in certain aberrent European forms, such as *cephennioides*...**Atomaria**
3—Body oblong-oval, strongly convex, the prothorax rounded at the sides from above and not angulate, the edge minutely beaded and not foveate ; first ventral as long as the next three combined, with a short feeble plate behind the inner part of the coxæ, becoming obsolete externally and gradually confounded with the coxal margin, the posterior segments short ; prosternal process very wide, with acute lateral edges not attaining the apical margin, nearly as in *Ephistemus*...**Tisactia**

The last of these genera is evidently a transition toward the Ephistemini in some respects, but the scutellum is broadly oval as in the others, the body more loosely connected and the prosternal process evidently free and broadly, arcuately obtuse at tip. The basal margin of the elytra will isolate it at once from any other member of the subfamily known to me, causing it to bear somewhat the same relationship to the others, in that respect, as *Tomarus*, does in the Cryptophaginæ.

Agathengis *Gozis.*

This aggregate of species, usually treated as a subgenus of *Atomaria*, satisfies the ordinary definition of a genus in having several constant and purely characteristic structural characters, and is therefore valid. It differs from *Atomaria* in the characters stated in the table, and the habital differences are such that it is seldom a matter of doubt as to the proper genus at the first glance. The body is elongate, generally quite slender and subparallel, convex and subuniformly, sparsely clothed with short and subdecumbent hairs, which become gradually still shorter in a sutural region near the elytral apex. The antennæ and eyes are moderately developed, the former generally rather stout, with more pronounced club than in *Atomaria*, and the joints of the shaft also very conspicuously alternating in length; the eyes are never very prominent and are not very coarsely faceted. The species are numerous in North America, relatively more so, apparently, than in Europe, where they are greatly outnumbered by *Atomaria*. Although easily separable by sight as a rule, they are even more homogeneous in adherence to a fixed type form than in *Atomaria*, and consequently form a difficult study for the taxonomist, as the differences are nearly all comparative. They seem to be quite local in distribution, judging from the material at hand excepting *crassula* which is common to the Atlantic and Rocky Mountain regions, and therefore fall very satisfactorily into primary geographic subdivisions as follows: —

Species of the Appalachian regions and Great Lakes.............:...............................2
Species of the Rocky Mountain system...6
Species of the Pacific Coast regions...15

2—Pronotum impressed at basal only in median half, the basal bead bordered by coarse punctures, especially pronounced within the impression, the latter with clearly defined lateral limits; body small, elongate-oval, convex, polished, piceous, the elytra somewhat paler, the legs and antennæ flavo-testaceous; pubescence short and very sparse; prothorax slightly transverse, the sides broadly arcuate and feebly converging from basal fourth, narrowed at base, the punctures notably sparse and rather coarse; elytra about three-fifths longer than wide, somewhat prominently rounded and subinflated at the middle, then rapidly narrowed, the punctures sparse, strong and moderately coarse. Length 1.25–1.35 mm.; width 0.5–0.55 mm. Michigan......................... **subnitens**, sp. nov.
Pronotum more or less distinctly impressed along the basal margin throughout the width 3
3—Elytra variegated in color, red, a small post-scutellar transverse spot on the suture, a large entire fascia behind the middle, fainter toward the suture, and the apex, black, remainder deep black, the legs and antennæ testaceous; body small, elongate-oval, strongly convex, highly polished; antennæ well developed, half as long as the body; prothorax feebly transverse, nearly as in *subnitens*, finely but deeply, very sparsely punctate, the basal impression stronger toward the middle; elytra feebly though subprominently inflated at the middle, then rapidly narrowed to

the apex, which is narrowly rounded, one-half longer than wide, coarsely, very sparsely punctate, the pubescence short and sparse, but coarse and distinct ; prosternum distinctly carinate along the middle of the intercoxal portion. Length 1.4 mm.; width 0.55 mm. Pennsylvania.........................**carinula**, sp. nov.

Elytra virtually uniform in coloration ; prosternum not, or only very feebly, carinulate along the middle ; elytral punctures fine and more or less close-set............4

4—Antennæ more elongate, fully half as long as the body in the male, the basal joint relatively longer and subequal to the next three combined ; body small, narrow, parallel and less convex, the sides very feebly arcuate, piceous in color, the elytra paler and brownish-testaceous, generally still paler near, but not at, the apex, the pubescence very short but abundant ; prothorax moderately transverse, subparallel, the sides distinctly and evenly arcuate from base to apex, the punctures fine and close-set ; elytra about two-thirds longer than wide, but little wider than the prothorax, not inflated at the middle, somewhat narrowly and parabolically rounded behind, finely and closely punctate. Length 1.25–1.5 mm. ; width 0.5–0.6 mm. Massachusetts to Lake Superior and Iowa.......**pumilio**, sp. nov.

Antennæ less developed, much less than half as long as the body, the basal joint subequal in length to the next two combined ; body more convex......................5

5—Body parallel and feebly arcuate at the sides, the prothorax well developed, black in color, the elytra piceous, the legs and antennæ piceo-testaceous ; pubescence short, moderately abundant but inconspicuous ; prothorax a third to two-fifths wider than long, the sides just visibly convergent and broadly, evenly arcuate from the broadly rounded and margined basal angles to the apex, the punctures fine but deep and only moderately close-set ; elytra elongate, three-fourths longer than wide, the sides rather more arcuate near the middle, moderately narrowed behind, at the middle a fourth to nearly a third wider than the prothorax, the punctures fine and moderately sparse ; hypomera scarcely at all punctured. Length 1.78–2.0 mm. ; width 0.72–0.76 mm. Michigan and Pennsylvania ..**patens**, sp. nov.

Body decidedly obese, with relatively much smaller prothorax, similar in coloration, the pubescence still shorter and quite close ; prothorax nearly one-half wider than long, the sides feebly converging from the rounded basal angles to the apex and very slightly sinuate just behind the middle, the punctures fine and close ; elytra shorter, three-fifths longer than wide, nearly one-half wider than the prothorax, widest at or somewhat behind the middle but without trace of inflation, rather narrowly rounded behind, the punctures fine, only moderately close-set but rather less sparse than in *patens;* hypomera thickly punctured except at base. Length 1.65 mm. ; width 0.75 mm. Pennsylvania (Westmoreland Co.) and North Carolina (Asheville)—the single specimen from the latter locality being wholly pale flavo-testaceous, probably from immaturity—and Colorado.

crassula, sp. nov.

6—Body dark in color, or with the head and prothorax darker than the elytra 7

Body rufo-testaceous in color throughout.....................................11

7—Pronotum impressed along the basal margin in about median half only, the hind angles slightly more than right but exceedingly well marked and not blunt......8

Pronotum impressed from side to side along the basal margin, the hind angles blunt, rounded or very obtuse.......................................9

8—Body regularly elongate-oval and strongly convex, black, the legs and elytra rufo-piceous ; antennæ dark testaceous, the club unusually stout and deep black ; pubescence short and not dense ; basal joint of the antennæ unusually large and stout but of the usual form ; prothorax barely two-fifths wider than long, the sides distinctly converging, broadly and almost evenly arcuate from base to apex and almost continuous with those of the elytra, the punctures fine and moderately close ; elytra oval, two-thirds longer than wide, sensibly widest and feebly inflated before the middle, arcuately narrowed thence to the apex, which is rather narrowly rounded, the punctures very fine and quite sparse. Length 1.7 mm. ; width 0.7 mm. Colorado...............................**capitata,** sp. nov.

Body oblong, parallel and subdepressed, much larger, moderately shining, dark rufo-piceous throughout, the legs and antennæ but little paler ; pubescence short but coarse and sparse, even as usual ; antennæ moderate, the first two joints of the club transverse ; prothorax unusually developed, parallel, but little wider than long, the sides feebly, almost evenly arcuate from base to apex ; punctures fine but strong, moderately close, the basal impression very fine and shallow along the middle of the basal margin, elsewhere obsolete ; elytra three-fourths longer than wide, but little wider than the prothorax, the sides broadly, feebly arcuate, gradually arcuato-convergent from about the middle, the apex somewhat broadly rounded ; punctures very fine and relatively sparse, somewhat disposed to linear arrangement. Length 1.9 mm.; width 0.72 mm. Colorado—Mr. Schmitt.
quadricollis, sp. nov.

9—Sides of the prothorax distinctly sinuate behind the middle and prominent at basal third or fourth, the body, legs and antennæ pale piceo-testaceous in color, the head and pronotum darker piceous ; pubescence very short and somewhat abundant ; antennæ rather slender, the club moderate, sometimes dusky, its first two joints but slightly transverse ; prothorax two-fifths or more wider than long, the sides arcuate and strongly converging at apex and base, the former slightly narrower, the punctures fine but strong and rather close-set ; elytra oblong, two-thirds longer than wide, distinctly wider than the prothorax, narrowing gradually behind the middle (♂), or in apical two-fifths (♀), rather obtusely rounded at apex, the punctures fine but rather close. Length 1.7–1.85 mm ; width 0.7–0.8 mm. Colorado.... ....**constricta,** sp. nov.

Sides of the prothorax not or very obsoletely sinuate behind the middle ; the body smaller and black throughout...10

10—Antennal club shorter and broader, its first two joints strongly transverse ; legs and antennæ rufo-piceous, the club of the latter blackish ; pubescence moderately abundant and short but coarse and distinct ; prothorax less than one-half wider than long, narrowing slightly only very near the base, the sides obviously converging, broadly and evenly arcuate thence to the apex ; disk unusually tumid at the middle near the base and just before the impressed margin, the punctures rather fine but deep, close-set and conspicuous ; elytra rather elongate, distinctly wider than the prothorax, parallel anteriorly, gradually narrowed behind the middle, moderately obtuse at tip, the punctures only moderately fine, deep, close-set and distinct. Length 1.65 mm.; width 0.65 mm. Colorado.
tenebrosa, sp. nov.

Antennal club larger, narrow, not darker in color, its first two joints not transverse; body shorter, the legs and antennæ testaceous; prothorax strongly transverse, more than one-half wider than long, strongly narrowed in basal third or fourth, where the sides are very slightly prominent, thence feebly converging and broadly arcuate to the apex; disk not tumid in the middle subbasally, the basal impression very feeble throughout, the punctures fine and moderately close; elytra short, parallel, gradually rounding behind, about three-fifths longer than wide, rather distinctly wider than the prothorax, the punctures very fine and rather close, the pubescence very short but somewhat abundant. Length 1.45–1.55 mm.; width 0.65 mm.   Colorado......................**coloradensis**, sp. nov.

11—Pronotum broadly, or deeply and very obviously, impressed at base..............12

Pronotum not impressed at base, or only with an extremely fine line extending along the basal bead; pubescence extremely short and moderately dense, the punctures fine; sides of the prothorax without trace of sinus behind the middle..14

12—Sides of the prothorax distinctly sinuate for a short distance behind the middle and prominently rounded at basal third or fourth; body elongate-oval, moderately convex, pale rufo-testaceous throughout, the elytra more flavate, polished, the pubescence very short, sparse and inconspicuous; antennæ moderate, about two-fifths as long as the body, with the club moderate, the first two joints moderately transverse (♂), or very short, stouter, with the club joints more transverse(♀); prothorax moderately transverse, strongly narrowed behind the lateral prominences, the apex not distinctly narrower than the base; disk finely, strongly somewhat closely punctate, the basal impression confined to median half of the width; elytra two-thirds longer than wide, gradually parabolically rounded toward apex, but little wider than the prothorax in either sex, the punctures fine but strong and rather sparse, sometimes inclined to serial arrangement. Length 1.6–1.9 mm.; width 0.65–0.75 mm. Idaho (Cœur d'Alène).

stricticollis, sp. nov.

Sides of the prothorax without an obvious post-median sinus, rather strongly converging and broadly, almost evenly and strongly arcuate from base to apex; antennæ moderate in length, rather slender, the club not stout; integuments shining....13

13—Body narrowly oval, dark rufo-testaceous in color throughout; prothorax but slightly transverse, strongly, evenly convex, finely but strongly, sparsely punctate, the basal impression wide and strong, coarsely punctate and confined to the median regions; elytra three-fifths longer than wide, the sides parallel and evenly arcuate, slightly, though obviously, wider than the prothorax, rather narrowly rounded behind, finely and sparsely punctate, the pubescences sparse, moderately long and coarse and distinct. Length 1.3–1.6 mm.; width 0.55–0.65 mm. Colorado......................**lucida**, sp. nov.

Body more broadly oval, equally convex, pale flavo-testaceous throughout; prothorax nearly one-half wider than long, evenly convex, finely and very sparsely punctate, the basal impression broader and more feeble than in *lucida* and stronger in median half or more, subobliterated toward the sides; elytra short, not more than one-half longer than wide, but little wider than the prothorax, narrowed behind from about the middle, the apex moderately obtuse; punctures very fine and sparse, the pubescence short, very sparse and inconspicuous. Length 1.5 mm.; width 0.68 mm.   Arizona?......................**luculenta**, sp. nov.

14—Antennæ thick, moderate in length, the club unusually broad, with its first two
joints distinctly transverse ; body very elongate and moderately convex ; prothorax
moderately transverse, less than one-half wider than long, the sides distinctly
converging and almost evenly, moderately arcuate from base to apex, the latter
distinctly narrower than the base ; disk feebly but almost evenly convex, finely
but strongly, rather closely punctured, with a more or less distinct impunctate
median line ; elytra oblong, parallel, elongate, fully three-fourths longer than wide,
only slightly wider than the prothorax, rather obtusely rounded in apical third,
finely but distinctly, rather closely punctured. Length 1.7–1.8 mm.; width 0.62–
0.65 mm. New Mexico (Coolidge) and Colorado..........**forticornis**, sp. nov.
Antennæ slender, the club narrow, with its first two joints but little wider than long,
moderate in length ; body smaller and less elongate but similar in coloration and
general characters to *forticornis* ; prothorax rather strongly transverse, fully one-
half wider than long, otherwise similar to *forticornis* but still more closely and
more finely punctured ; elytra three-fifths to two-thirds longer than wide, scarcely
wider than the prothorax (♂) or distinctly so (♀), the punctures fine but strong
and very close-set, the pubescence very short but coarse. Length 1.5 mm.; width
0.55 mm. Arizona (Williams)—Mr. Wickham..................**macer**, sp. nov.
15—Pronotum distinctly impressed at base, the impression abruptly limited to about
median half of the width, the surface before the impression never tumid at the
middle..16
Pronotum distinctly impressed but more broadly and indefinitely, the impression extending from side to side..17
Pronotum not impressed at base, other than the fine line rendered apparent by the fine
basal bead..24
16—Body very elongate-oval, narrow, rather convex, polished, pale flavo-testaceous
throughout, the pubescence very short, sparse and inconspicuous; antennæ rather
short and thick, the club broader than usual, short, with its first two joints strongly
transverse ; prothorax not quite one-half wider than long, the sides prominent
near basal third, strongly convergent thence to the base, and, feebly, nearly to
the apex, where they are rounded ; punctures fine, but strong and rather sparse ;
elytra parallel, elongate, moderately obtuse at apex, slightly wider than the prothorax, very finely and quite sparsely punctate, the punctures strongly tending to
serial arrangement. Length 1.75 mm.; width 0.7 mm. California (Sonoma
Co.)..**ochronitens**, sp. nov.
Body much smaller, elongate-oval and narrow, less shining and rather less convex,
dark rufo-testaceous throughout ; antennæ moderately developed, more slender,
the club narrow, with its two basal joints but feebly transverse ; pubescence very
short, moderately sparse ; prothorax moderately transverse, the sides distinctly
sinuate just behind the middle, but not very prominent behind the sinus, thence
converging strongly to the base, converging and rounded at apex, the latter but
little narrower than the base ; disk rather depressed, quite strongly and closely
punctate, the basal impression strong ; elytra rather less elongate, about two-
thirds longer than wide, slightly wider than the prothorax, gradually rather obtusely rounded behind, the punctures strong and moderately close-set, not tending to linear arrangement. Length 1.5 mm.; width 0.65 mm. California
(Siskiyou Co.)..**undulata**, sp. nov.

17—Pronotum evenly convex, not at all tumid at the middle before the basal impression, the surface sloping steeply to the impression, which is very fine; body narrowly oblong-oval, rather strongly convex, shining, black, the elytra feebly picescent, the legs and antennæ dark rufo-testaceous ; pubescence short but coarse and rather abundant; antennæ slender and rather more than half as long as the body, the basal joint well developed and nearly as long as the next three combined, the club rather small and narrow, with its two basal joints but slightly transverse; prothorax slightly transverse, evenly and strongly convex, fully as wide as the elytra, the sides subparallel, broadly, feebly arcuate, more convergent at base and apex, the latter but little the narrower ; punctures only moderately coarse but strong and somewhat close-set ; elytra barely three-fifths longer than wide, obtusely rounded behind, parallel, convex, the punctures strong, close-set and distinctly sublinear in arrangement.   Length 1.4 mm.; width 0.6 mm. California (Lake Co.).................................. ....**melas,** sp. nov.
Pronotum more or less feebly, though perceptibly, tumid at the middle before the basal impression ; body never entirely black...18
18—Elytral punctures conspicuously coarse and well separated..........................19
Elytral punctures not very coarse, though always strong, and very distinct, generally less sparse...21
19—Sides of the prothorax with a broadly rounded feeble sinus just behind the middle, a little more rounded but scarcely prominent near basal fourth, becoming apparently even and broadly arcuate throughout in *puella*.........................20
Sides of the prothorax strongly converging and broadly arcuate from basal fourth to apex with a very short feeble rounded sinus just before the slight prominence at basal fourth, from which point to the base they are strongly convergent ; body oblong-oval, moderately stout and convex, shining, black, the elytra, legs and antennæ piceo-rufous ; pubescence very short and moderately abundant ; antennæ well developed, the club rather stout, with its two basal joints strongly transverse, prothorax fully one-half wider than long, the apex very much narrower than the base, the punctures not very coarse but deep and only moderately close ; elytra scarcely more than one-half longer than wide, fully a fourth wider than the prothorax, parallel, obtusely rounded in apical two-fifths, the punctures quite coarse and moderately close-set.   Length 1.3 mm.; width 0.55 mm.   California (Siskiyou Co.)...**soror,** sp. nov.
20—Body moderately stout, elongate-oval and convex, shining, blackish, the elytra, legs and antennæ testaceous ; pubescence short and sparse ; antennæ rather long and slender, nearly half as long as the body, the club moderately wide, rather long and loose, its two basal joints transversely obtrapezoidal, the eighth joint subquadrate and but slightly narrower than the seventh ; prothorax well developed and transverse, fully two-fifths wider than long, narrowed only slightly in basal fourth, the apex but little narrower than the base ; surface almost evenly convex, finely but strongly, somewhat closely punctate ; elytra parallel and broadly arcuate at the sides, obtusely rounded at apex, three-fifths longer than wide and but little wider than the prothorax, the punctures coarse and moderately sparse. Length 1.65 mm. ; width 0.7 mm.   California (Humboldt Co.).
                                                    cribripennis, sp. nov.

Body smaller and narrower, the elytra parallel and only very slightly arcuate at the sides, similar in coloration and vestiture to *cribripennis*, the antennæ shorter, distinctly less than half as long as the body, the club shorter, more compact and unusually broad, its two basal joints strongly transverse, the eighth joint very small and much narrower than the seventh ; prothorax narrower and less transverse, a third wider than long, otherwise similar, except that the sides are perceptibly convergent and the apex distinctly, although not greatly, narrower than the base ; elytra two-thirds longer than wide, obtusely rounded behind, distinctly wider than the prothorax, the punctures coarse and sparse, becoming gradually fine and rather close posteriorly. Length 1.28-1.55 mm. ; width 0.45-0.6 mm. California (Monterey)............ ....**dispersa,** sp. nov.

Body still smaller and narrower, piceous, the elytra pale testaceous, with a large piceous cloud on the suture behind the middle ; antennæ slender, fully half as long as the body, the club rather small and narrow, its two basal joints moderately transverse, the eighth joint but little narrower than the seventh ; prothorax moderately transverse, nearly as in *cribripennis* but more arcuate at the sides and with the punctures very sparse ; elytra three-fifths longer than wide, parallel and broadly, distinctly arcuate at the sides, obtuse at apex, only very slight wider than the prothorax, the punctures coarse and quite sparse, notably less close-set than in either of the preceding. Length 1.1 mm. ; width 0.48 mm. California.

                                                          puella, sp. nov.

21—Antennæ moderately developed, rather stout, dark testaceous, the club rather broad and shorter, its two basal joints moderately, though very distinctly, transverse ; body narrowly oblong-oval, rather strongly convex, polished, blackish-castaneous, the elytra slightly paler and more rufescent toward base ; pubescence short, sparse and inconspicuous ; prothorax moderately developed, a little less than one-half wider than long, the sides feebly convergent and broadly, almost evenly, arcuate from base to apex, with a short and scarcely visible sinus just behind the middle ; surface strongly, almost evenly, convex, rather strongly, conspicuously and somewhat closely punctate, the basal impression quite finely and feebly impressed ; elytra nearly three-fourths longer than wide, parallel and very feebly arcuate at the sides, broadly and obtusely rounded at apex, distinctly wider than the prothorax and about three times as long, the punctures rather fine but very deep, perforate and distinct, not very close-set and clearly inclined to serial arrangement. Length 1.45 mm. ; width 0.58 mm. California (Sonoma Co.).................... ... .**castanea,** sp. nov.

Antennæ rather slender, the club longer and narrow, with its first two joints not, or but very slightly, transverse...22

22—Antennæ only moderate in length, distinctly less than half as long as the body, the latter rather stout, oblong-oval, moderately convex, the pronotum rather less shining than the elytra, black, the legs piceous, the elytra and antennæ paler and piceo-testaceous ; pubescence very short and sparse ; prothorax one-half wider than long, almost fully as wide as the elytra, very slightly narrowed at apex and abruptly and distinctly so near the base, the median tumidity before the basal impression very obvious ; punctures not coarse but deep, perforate and close-set ; elytra three-fifths longer than wide, parallel and nearly straight at the sides, parabolically obtuse in apical two-fifths, the punctures rather fine but deep, impressed, moderately close and not very distinctly inclined to serial arrangement. Length 1.7 mm. ; width 0.7 mm. Alaska (Kenai)............**vespertina** *Mäkl.*

Antennæ more slender and almost half as long as the body ; pubescence short and inconspicuous though rather abundant...23

23—Body piceous, the elytra paler and piceo-testaceous, the antennæ testaceous, with the club slightly dusky ; prothorax small, but little more than a third wider than long, the sides distinctly, but obtusely, prominent at basal third, thence strongly arcuate and convergent to the base, very feebly and indefinitely sinuate before the prominence, and thence feebly convergent and broadly arcuate to the apex, which is subequal in width to the base; surface broadly convex, finely but strongly, densely punctate ; elytra rather short and broad, parallel, somewhat narrowly parabolic behind from slightly behind the middle, three-fifths longer than wide and nearly a third wider than the prothorax, the punctures rather fine but strong, subimpressed, only moderately close and arranged wholly without trace of order. Length 1.5 mm. ; width 0.63 mm. California (Siskiyou Co.).

        parvicollis, sp. nov.

Body similar in coloration to the preceding but somewhat more narrowly oblong-oval ; prothorax more transverse, two-fifths wider than long, the sides nearly similar but scarcely at all prominent at basal third, the punctures less distinct and much less close-set, the basal impression finer and feebler ; elytra rather shorter, but little more than one-half longer than wide, very obtusely rounded behind, parallel at the sides and less arcuate and only about a fourth wider than the prothorax, the punctures similar but very close-set and arranged in conspicuously even, very close-set rows almost throughout. Length 1.4 mm. ; width 0.6 mm. California (Siskiyou Co.) ....**parvicollis**(♂)?

24—Antennæ rather short, scarcely two-fifths as long as the body, the club somewhat robust, with its two basal joints distinctly, though not strongly, transverse ; body oblong-elongate and parallel, piceous, the elytra and antennæ slightly paler and testaceous ; pubescence very short and rather abundant ; prothorax two-fifths wider than long, parallel, feebly sinuate at the middle point of the sides, obtusely prominent behind the sinus and thence narrowed to the base, the punctures strong and close-set ; elytra oblong, obtusely parabolic behind, three-fourths longer than wide, distinctly, though not greatly, wider than the prothorax, the punctures rather fine but strong and close-set, irregular in arrangement. Length 1.8 mm.; width 0.75 mm. California (Siskiyou Co.).........**subrecta**, sp. nov.

Antennæ slender, the club narrow, with the two basal joints not notably transverse..25

25—Elytra much more than one-half longer than wide and but very slightly wider than the prothorax, the body narrow and elongate-oval..............................26

Elytra short and broad, about one-half longer than wide and fully two-fifths wider than the prothorax, the latter relatively very small..............................27

26—Body black, the elytra and antennæ pale rufo-testaceous, the pubescence short and sparse ; prothorax shorter, two-fifths wider than long, distinctly and rather acutely prominent at the sides near basal third, narrowed gradually thence to the apex, which is slightly narrower than the base, with a small and almost imperceptible sinus just before the prominence, the punctures fine but strong and only moderately close-set ; elytra two-thirds longer than wide, evenly rounded in apical two-fifths, parallel toward base, the punctures moderately fine, strongly

impressed and somewhat close-set.   Length 1.3–1.65 mm.; width 0.55–0.6 mm.
California (Siskiyou Co.)..**nigricollis**, sp. nov.
Body piceous, the elytra and antennæ slightly paler and rufous, the pubescence short
but rather abundant ; prothorax well developed, scarcely more than a third wider
than long, parallel and broadly, feebly sinuate at the middle of the sides, arcuately and equally narrowed at apex and base, the surface rather strongly convex,
only moderately finely, deeply and very closely perforato-puncta e ; elytra well
developed, fully three-fourths longer than wide, parabolically rounding in apical
two-fifths, parallel and almost straight thence to the base, the punctures rather
fine but deep, close-set and arranged without order.   Length 1.6–1.9 mm.;
width 0.6–0.72 mm.   California..............................**longipennis**, sp. nov.
27.—Body evenly dark rufo-testaceous throughout, moderately convex and shining, the
pubescence very short but somewhat abundant ; antennæ moderate ; prothorax
nearly one-half wider than long, the sides conspicuously and rather narrowly
prominent at basal third, thence strongly converging to the base and more gradually so and nearly straight to the apex, which is not distinctly narrower than the
base ; disk more convexo-declivous near the basal margin than in the three preceding species but not properly impressed, the punctures fine and rather cnevenly
close-set, becoming very minute and sparse broadly along the middle ; elytra
short and broad, the punctures very fine but distinct and moderately close-set, altogether irregular in arrangement.   Length 1.4 mm.; width 0.68 mm.   California..**subdentata**, sp. nov.

Ochronitens quite strongly resembles *stricticollis*, but differs in the
more slender form of the body and in the very much more minute
and sparse punctuation.   The species described under the name *parvicollis* is represented by a unique, as is also the form with seriately
punctured elytra which I have surmised to be its male ; more material
is necessary to decide this rather puzzling point, as the difference in
elytral sculpture is certainly very marked.   I have, however, noticed
at times a slight sexual difference in density and arrangement of punctures elsewhere in the family.   *Fuscicollis* of Mannerheim, and *planulata*
of Mäklin, I have not seen, the latter is described as oblong, depressed, fusco-testaceous, finely and densely punctate with the legs
and elytra rufo-testaceous.

Atomaria *Steph.*

The species of this genus are less numerous in America than *Agathengis*, and for the most part present a rather monotonous appearance.
The body is generally oblong-oval and convex, shining and sparsely
clothed with short subdecumbent hairs.   The antennæ are usually
slender, moderate in length, with the basal joint short and oblong
or more developed internally toward base than externally ; the joints

of the funicle are alternately shorter and longer as usual in the tribe, but they are somewhat more widely separated at base than in *Agathen-gis*; the first two joints of the club are generally about as long as wide or longer, and seldom at all transverse. The eyes are larger and more coarsely faceted as a rule than in *Agathengis*. The prothorax is narrowed anteriorly and generally more or less distinctly angulate at the middle—not nearer the base as in *Agathengis*—and the marginal fovea at the point of angulation is not so marked a character as it is in that genus ; the edge is finely beaded and frequently feebly crenulate from the angulation to the base ; the disk is evenly convex, becoming broadly concave along the very finely margined or simple transverse base. The prosternal process is narrow, but in certain species, such as the European *turgida* and *cephennioides*, becomes wider, more prominent and more strongly margined along the sides— a divergence in the direction of the remarkable genus *Sternodea*. There seem to be, in fact, several quite well defined subgenera among the species of the European fauna having for types such forms as *turgida*, with medially lobed thoracic base, stout antennæ and broader and more prominent prosternal process, *cephennioides*, with large and broadly truncate prothorax, broad prosternal process, stout antennæ and very small eyes, and *unifasciata*, which is perfectly con·generic with our species and might be regarded as *Atomaria* proper.

The elytra are finely, irregularly punctured, frequently subinflated before the middle, truncate at base and slightly impressed within the humeral callus. The abdominal sutures are straight, the first segment as long as the next two combined and the fourth shorter than the second or third. The legs and tarsi are slender.

The species before me may be identified as follows :—

Elytra conspicuously ornamented or bicolored, the lines of demarcation more or less
    well defined..2
Elytra unicolorous or nubilously darker toward base................................7
2—Elytra pale, a dark fascia just before the middle extending from side to side......3
Elytra black or blackish, the apical third or fourth abruptly, and the basal regions nu-
    bilously pale...4
Elytra black, abruptly pale in apical two-fifths to half, the base not paler, excepting,
    rarely, the humeral callus...5
3—Suboval, strongly convex, piceous, the antennæ, legs and elytra pale luteo-flavate,
    the latter each with a large, broadly oval oblique blackish spot from basal third
    at the sides, the two mutually tangent on the suture at the middle ; antennæ
    slender, half as long as the body ; prothorax strongly, longitudinally convex in
    profile, moderately transverse, strongly rounded at the sides and narrowed per-

ceptibly more at apex than at base, the basal impression deep and rather narrow, the punctures minute and sparse ; elytra widest before the middle, narrowly
rounded at apex, distinctly wider than the prothorax and but two and one-half
times as long, the punctures very fine but subperforate and sparse. Length
1.4–1.5 mm.; width 0.65–0.7 mm. Vermont and Rhode Island to Iowa and
Colorado..................... ....**ephippiata** *Zimm.*

Var. A—Similar to *ephippiata* in form and size but more narrowly elongate-
oval and with the antennæ shorter, the elytral spots broadly uniting on the
suture, and with the elytral punctures quite coarse, deeply impressed and ap-
parently-denser. Washington State (Spokane Falls)...**hesperica**, v. nov.

Suboblong oval, convex, much smaller in size, pale and rufo-testaceous throughout,
the head blackish, each elytron with a narrower oblique black band from basal
third externally to the suture at the middle, and spreading longitudinally on the
flanks ; antennæ moderate, less than half as long as the body, the prothorax
moderately transverse, but little narrower at apex than at base, broadly rounded
at the sides and more closely punctulate than in *ephippiata*, the basal impression
confined to median two thirds of the width ; elytra nearly similar to the preceding in form but only very slightly wider than the prothorax and much more
closely, though equally finely, punctate. Length 1.1–1.22 mm.; width 0.52–
0.58 mm. California to Washington State (Spokane Falls)..........**lætula** *Lec.*

4—Elongate-oval, convex, shining, rufo-testaceous throughout, the under surface generally piceous, the elytra shaded with blackish from near the base to apical third
or fourth ; antennæ slender, half as long as the body, a little shorter in the
female ; prothorax moderately transverse, rather strongly narrowed from base to
apex, the sides broadly and feebly subangulate at the middle, the base broadly,
feebly arcuate ; basal impressions strong, extending almost to the sides, gradually evanescent laterally ; disk only moderately convex longitudinally, the punctures strong, moderately coarse and well separated ; elytra at least three times as
long as the prothorax and distinctly wider, very much so in the female, more declivous toward apex in profile, the sides parallel and almost evenly arcuate ; apex
rather obtusely rounded ; punctures fine but distinct, moderately sparse. Length
1.4–1.6 mm.; width 0.65–0.72 mm. California (Los Angeles to Monterey)
nubipennis, sp. nov.

5—Species of the Pacific coast regions. Body somewhat broadly oval, strongly convex, polished, black throughout, the antennæ and legs testaceous ; elytra pale in
apical two-fifths to half, the margin of the pale area broadly and posteriorly angulate ; pubescence ashy and distinct ; antennæ slender, nearly half as long as the
body, shorter in the female ; prothorax moderately transverse, both this and the
elytra strongly arcuate in profile as in *ephippiata*, the apex distinctly narrower
than the base, the sides broadly subangulate at the middle, the basal impression
strong, not extending beyond lateral fourth or fifth, the punctures fine but deep
and quite close-set ; elytra much wider than the prothorax and not quite three
times as long, the sides almost evenly arcuate from the humeri to the sutural
angles, but little wider before the middle than at base, the apex ogival ; punctures very fine but strong and perforate, quite close-set. Length 1.3–1.4 mm.;
width 0.68–0.72 mm. California (Monterey to Humboldt Bay) and Nevada
(Reno)...**postpallens**, sp. nov.

Species of the Atlantic slope..6

6—Rather narrowly oval, strongly convex, polished, dark rufo-testaceous throughout, the sterna of the hind body and basal half of the elytra black, the
pale apex more advanced on the suture ; pubescence very sparse ; antennæ slender, not quite half as long as the body ; prothorax not more than three-fifths
wider than long, subangularly inflated at the middle, the apex but little narrower
than the base, the basal impression rather feeble and medial ; disk strongly,
longitudinally convex in profile, the punctures small and very sparse ; elytra two
and one-half times as long as the prothorax and distinctly wider, the sides evenly,
strongly arcuate, the apex rather acute, punctures fine and very-sparse.   Length
1.35 mm.; width 0.63 mm.   District of Columbia...........**distincta**, sp. nov.

More broadly oval, strongly convex, nearly similar in coloration to the preceding but
paler testaceous, the elytra black in basal half, less on the suture ; antennæ
slender and half as long as the body ; prothorax shorter and more transverse,
three-fourths wider than long, the sides strongly arcuate, the apex but little narrower than the base, similarly feebly impressed at base, the punctures fine and
distinctly less sparse ; elytra distinctly wider than the prothorax and about three
times as long, subangularly inflated and widest at basal two-fifths, the apex
acutely rounded ; punctures fine and rather sparse but much less so than in
distincta, the pubescence similarly very short.   Length 1.4 mm.; width 0.7 mm.
Iowa (Independence)........ ....**divisa**, sp. nov.

7—Second antennal joint subequal in length to the third, both elongate ; base of the
prothorax transverse and rectilinear or very feebly arcuate......... 8

Second antennal joint much longer and generally thicker than the third, frequently
as long as the third and fourth combined ; base of the prothorax variable, sometimes distinctly lobed in the middle...16

8—Base of the prothorax, at the middle, finely beaded and frequently abruptly though
feebly elevated...9

Base of the prothorax reflexed but not beaded at the middle of the base.   Body piceous, the elytra paler, sometimes wholly pale, rather stout, oval, strongly convex,
highly polished and very sparsely clothed with short recurved pubescence ; head
subimpunctate, the eyes rather small ; antennæ slender, a little less than half as
long as the body ; prothorax scarcely two-thirds wider than long, widest and
broadly angulate at the sides just before the middle, the apex very much narrower than the base ; surface strongly convex, minutely and very sparsely
punctate, the basal impression broadly concave and gradually evanescent laterally ; elytra at base slightly wider than the base of the prothorax, widest, but
not inflated, before the middle, where they are a third wider than the prothorax,
not quite three times as long as the latter, oval, rather pointed behind, strongly
but very sparsely impresso-punctate.   Length 1.6–1.75 mm.; width 0.72–0.88
mm.   New Jersey...**gilvipennis**, sp. nov·

9—Body oval in form and strongly convex............... 10

Body oblong or oblong-oval in form and subparallel at the sides......................15

10—Juxtahumeral impressions at the base of the elytra very large and conspicuous,
though shallow.   Body short and very stout, dark rufo-piceous in color throughout ; legs and antennæ pale, the latter moderately slender and nearly half as
long as the body ; prothorax well developed, strongly convex, fully three-fourths

wider than long, widest and broadly angulate at the sides at the middle, the
apex much narrower than the base ; punctures fine but deep and perforate,
separated by two or three times their diameters, with a narrow impunctate median
line not attaining the base, the basal impression well developed. the basal bead
very feeble and flat ; elytra subinflated and widest at two-fifths, where they are a
fourth wider than the prothorax, about three times as long as the latter, the
humeri narrowly exposed at base, the apex narrowly roun led ; punctures very
small but perforate and moderately close-set. Length 1.45 mm.; width 0.75
mm.   Canada (Ottawa)..**saginata**, sp. nov.
Juxtahumeral impressions very small and feeble..11
11—Prothorax rather rounded than angulate at the sides at or before the middle.....12
Prothorax conspicuously, though broadly, angulate at the sides at or slightly before
the middle, where it is much wider than at base, the sides rapidly convergent
from the angle to the apex and straight or broadly sinuate; elytral humeri distinctly exposed at base..13
12—Prothorax relatively long and narrow, scarcely three-fourths as wide as the elytra
and but little wider at the submedian dilatation than at base, very strongly narrowed at apex, the latter much narrower than the base ; body elongate-oval,
shining, testaceous throughout, the antennæ distinctly less than half as long as
the body in the female ; prothorax strongly convex, finely but deeply and closely
perforato-punctate, the basal impression deep and narrowly impressed ; elytra distinctly wider at two-fifths than at base, where they are scarcely wider than the base
of the prothorax, two and two-thirds times as long as the latter, acutely rounded
at tip, the punctures fine but distinct, rather impressed and moderately close-set,
at least three times as sparse as those of the prothorax.   Length 1.6 mm.; width
0.78 mm.   District of Columbia..**ochracea** *Zimm*.
   Var. A— Similar in color but with the head and prothorax slightly piceous, the
   latter equally distinctly and very closely punctate, the punctures separated by
   only their own diameters, and, as usual, coarser toward the sides and basal
   angles ; basal impression much feebler and less acutely impressed ; elytra
   strongly and rather sparsely impresso-punctate ; body more narrowly oval.
   Length 1.65 mm.; width 0.75 mm.   Lake Superior.........**lacustris**, v. nov.
   Var. B—Body nearly similar in form but slightly smaller and more rapidly attenuate at the extremities, blackish-piceous to dark testaceous in color, polished, the pronotum finely and rather sparsely punctate, the basal impression
   much feebler and more broadly impressed than in *ochracea ;* elytra minutely
   and rather feebly, moderately sparsely punctate. the punctures but little more
   widely separated than those of the pronotum.   Length 1.2-1.5 mm.; width
   0.58-0.75 mm.   Mountains of Pennsylvania (Westmoreland Co.).
                                              pennsylvanica, v. nov.
Prothorax shorter and decidedly more transverse, two-thirds to three-fourths wider
than long, much narrower at base than at the med an inflation, at least four-fifths
as wide as the elytra, distinctly, though less markedly, narrower at apex than at
base, strongly convex, finely and rather sparsely punctate, the basal impression
rather deep and acutely impressed ; elytra shorter and less obviously narrower at
base than at the feeble inflation two-fifths from the base, the humeri more widely
exposed at base, less than three times as long as the prothorax and narrowly

rounded at tip, the punctures fine but rather strongly impressed, moderately sparse and slightly more widely separated than those of the pronotum; antennæ slender, half as long as the body in the male, a little shorter in the female; body piceo-testaceous in color, shining. Length 1.25-1.45 mm.; width 0.63-0.75 mm. Canada (Ottawa) ....**curtula,** sp. nov.

Var. A—Similar but less stout and much smaller, piceous in color, the legs and antennæ pale luteo-flavate, the latter slender, fully half as long as the body; eyes slightly larger, convex and well developed; pubescence finer and a little closer; prothorax similar in form, deeply and rather acutely impressed at base, very finely and moderately closely punctate; elytra very finely, feebly, rather inconspicuously and moderately closely punctate, otherwise nearly similar. Length 1.2 mm.; width 0.6 mm.   Iowa............**pumilio,** v. nov.

13—Prothorax small, at its widest part not quite as wide as the base of the elytra, the latter less than three times as long as the prothorax. Body rather broadly oval, moderately convex, polished, pale ferruginous in color throughout, the antennæ slender, not quite half as long as the body in the male, shorter and a little stouter in the female; prothorax nearly two-thirds wider than long, angulate at the middle, the sides thence to the base distinctly convergent, feebly arcuate and minutely serrulate, more strongly convergent and feebly sinuate to the apical angles, which are somewhat prominent, the apex distinctly narrower than the base; disk finely but deeply, moderately closely punctate, the impression along the base rather deep; elytra about one-half longer than wide, subinflated at two-fifths and nearly a third wider than the prothorax, acute at tip, finely but strongly and somewhat sparsely impresso-punctate, the punctures at least twice as sparse as those of the pronotum. Length 1.4-1.6 mm.; width 0.7-0.75 mm. New York—Mr. H. H. Smith.. **gonodera,** sp. nov.

Prothorax small, nearly as in the preceding but with the converging sides anteriorly not sinuate and the apical angles not so acute. Body similar but shorter and more convex, black or blackish in color, the legs and antennæ dark testaceous; surface polished; antennæ slender, moderate in length; pronotum similarly punctured and impressed; elytra a little shorter and relatively broader, strongly rounded at the sides, acute at apex, more finely and sparsely punctate. Length 1.5 mm.; width 0.73 mm.   Delaware to Florida............ ....**riparia,** sp. nov.

Prothorax more developed, at its widest part fully as wide as the base of the elytra, the latter fully three times as long as the prothorax.................................14

14—Oval, moderately convex, subalutaceous in lustre, piceous, the elytra slightly paler; legs and antennæ pale; eyes rather well developed, convex; antennæ moderately slender, with the club rather stout, distinctly less than half as long as the body in the female; prothorax nearly four-fifths wider than long, the sides prominently inflated before the middle, thence converging and arcuate to the base and strongly convergent and feebly sinuate to the apex, the latter not more than three-fourths as wide as the base; punctures small but deep and distinct, notably dense, the basal impression strong; elytra fully three times as long as the prothorax, widest, inflated and a fourth wider than the prothorax at two-fifths, rapidly narrowed thence to the acutely rounded apex; punctures fine, moderately close-set, twice as sparse as those of the prothorax. Length 1.6 mm.; width 0.72 mm. (♀) Rhode Island (Boston Neck)..............**subalutacea,** sp. nov.

Oval, rather more convex, shining, black, the antennæ and legs dark testaceous ; antennæ slender, less than half as long as the body in the female ; prothorax more than three-fourths wider than long, somewhat obtusely angulate and inflated at the middle, the sides thence straight and moderately convergent to the base and strongly so and straight to the apex ; punctures fine and somewhat sparse, the impression deep ; elytra nearly as in the preceding but more than a fourth wider than the prothorax, the punctures very fine and quite sparse—though but slightly sparser than those of the pronotum. Length 1.5 mm.; width 0.7 mm. (♀). Colorado...**incerta**, sp. nov.

15—Body oblong, parallel, only moderately convex, polished, piceous, the entire elytra and legs pale flavo-testaceous, antennæ pale with the club infuscate, somewhat slender and slightly less than half as long as the body ; prothorax small, much narrower at any part than the base of the elytra, angularly inflated slightly before the middle, the sides nearly straight and converging to base and apex, the latter quite distinctly narrower than the base ; disk convex, two-thirds wider than long, finely, rather sparsely punctate, the basal impression strong, extending evanescently to the very obtuse basal angles ; elytra oblong, parallel and almost straight at the sides, broadly rounded at apex, more than three times as long as the prothorax and about a third wider, the humeri widely exposed at base, rather coarsely but not densely impresso-punctate and somewhat rugose by oblique illumination. Length 1.55 mm.; width 0.72 mm. Colorado.

brevicollis,sp. nov.

Body oblong, the elytra feebly inflated, polished, black or blackish, the elytra very dark piceo-testaceous throughout ; antennæ and legs dark testaceous, the former slender and nearly half as long as the body ; prothorax large, three-fifths wider than long, fully as wide at base as the closely fitting base of the elytra and nearly as wide as at the very feeble subangular dilatation slightly behind apical third, the apex slightly narrower than the base ; disk convex, finely and quite sparsely punctate, the basal impression moderate ; elytra parallel and broadly arcuate at the sides, widest and distinctly wider than the prothorax at the middle, not more than two and one-half times as long as the latter, rapidly narrowed behind the middle and acutely rounded at tip, the punctures rather coarsely impressed, moderately close-set, not materially sparser than those of the pronotum. Length 1.4 mm. ; width 0.65 mm. Alaska.................................**aleutica**, sp. nov.

16—Prothorax subangularly dilated at, or a little before, the middle, narrowed toward base and still more strongly toward apex, the elytral humeri exposed at base 17

Prothorax, viewed vertically, rounded at the sides from the base, more strongly narrowed toward apex and widest perceptibly behind the middle.....................18

Prothorax, viewed vertically, parallel at the sides from the base to or beyond the middle, then strongly narrowed to the apex......................................19

17—Body, legs and antennæ uniform pale ochreo-testaceous throughout, the latter rather short and stout, but little longer than the head and prothorax, the eyes moderate ; surface shining, the pubescence short, fine, ashy and rather abundant ; prothorax four-fifths wider than long, not quite as wide as the base of the elytra, the sides broadly angulate at apical two-fifths, the apex distinctly narrower than the base, the basal impression rather feeble ; punctures very fine and

quite close-set; elytra oblong, feebly arcuate at the sides, obtusely rounded at
apex, barely three times as long as the prothorax and about a fourth wider, the
punctures very fine and moderately close-set. Length 1.5 mm.; width 0.75 mm.
Colorado..**oblongula**, sp. nov.
Body deep black throughout, the legs and antennæ piceo-testaceous; surface polished,
the pubescence short and inconspicuous; antennæ moderately slender, distinctly
longer than the head and prothorax, the latter convex, two-thirds wider than
long and fully as wide as the base of the elytra, dilated and strongly rounded
laterally just before the middle, the converging sides thence nearly straight to
the base and apex, the latter but little narrower than the base, the basal impression moderate, extending throughout the width but feeble at the sides, the punctures fine but deep and strong and not very close-set; elytra short, oblong, two
and one-half times as long as the prothorax and barely a fifth wider, parallel and
broadly arcuate at the sides and obtusely rounded at apex, rather strongly and
moderately closely impresso-punctate. Length 1.3 mm.; width 0.6 mm. Iowa.
                                                                    crypta, sp. nov.
Body very small, parallel, pale testaceous, the elytra gradually shaded blackish toward base; integuments shining; antennæ slender, nearly half as long as the body;
prothorax short and strongly transverse, about as wide as the elytra, angulato-
dilated at the middle, the apex distinctly narrower than the base; punctures
minute and moderately dense, the basal impression rather acutely impressed;
elytra parallel, feebly arcuate at the sides, obtusely rounded at tip, three times as
long as the prothorax in the female but obviously shorter in the male, the punctures very fine and moderately close-set. Length 0.88–1.05 mm.; width 0.35–
0.48 mm. Europe and Northeastern America....................**pusilla** *Schönh.*
18—Body oblong-suboval, parallel, convex, polished, blackish in color, the elytra gradually pale posteriorly and the humeral callus also slightly paler; legs and antennæ
testaceous; the latter moderately slender, nearly half as long as the body, the
club rather thick; prothorax moderately transverse, rounded on the sides, more
strongly toward base, from above, but broadly subangulate at the middle when
viewed sublaterally, the punctures fine and moderately close-set, the apex but
little narrower than the base, the basal impression distinct; elytra parallel,
broadly arcuate at the sides, but little wider at the middle than at base and only
slightly wider than the prothorax, nearly three times as long as the latter, moderately obtuse at tip, the humeri evidently exposed at base; punctures fine but
strong and distinct, scarcely sparser than those of the prothorax. Length 1.2
mm.; width 0.5 mm. California .... **fallax**, sp. nov.
19—Species of the Atlantic regions. Body oval, rapidly attenuate at the extremities
and very convex, shining, black or piceous-black, the legs and antennæ testaceous, the latter notably stout, nearly half as long as the body in the male; prothorax small, less transverse than usual, three-fifths wider than long, the sides
strongly converging anteriorly, the apex only two-thirds as wide as the base, the
latter with a feeble but distinct arcuate lobe in median third; punctures rather
strong and close-set, the impression somewhat feeble; elytra oval, subinflated,
and, at two-fifths, very much wider than at base and a third wider than the prothorax, the base of the latter scarcely at all narrower than the base of the elytra,
the humeri not exposed at base; apex narrowly rounded, the punctures fine but

distinct, rather sparse, two or three times sparser than those of the pronotum. Length 1.22-1.4 mm.; width 0.6-0.73 mm.   Canada, New York, Pennsylvania, and Iowa..**ovalis**, sp. nov.
Species of the Pacific coast and Alaska.................................... 20
20—Larger species, oblong-oval in form, rather stout, convex, polished, black, the entire elytra bright testaceous ; legs piceous; the antennæ pale, with the club rather stout ; prothorax but little more than one-half wider than long, subangularly rounded at the sides slightly before the middle, then strongly narrowed to the apex, finely, rather sparsely punctate, as wide at base as the base of the elytra, the latter parallel and broadly arcuate at the sides, rather narrowly rounded at apex, finely but strongly, rather sparsely impresso-punctate.   Length 1.6 mm.; width 0.78 mm.   Alaska (Kodiak Island)............................**fulvipennis** *Mann*.
Small species, shining, rather narrow and elongate-oval in form........................21
21—Black, the elytra picescent, the legs paler ; antennæ testaceous, moderately stout, two-fifth as long as the body ; prothorax short, three-fourths wider than long, very slightly narrower than the base of the elytra, the sides parallel almost to apical third, then strongly convergent to the apex, the punctures strong, deep and close-set, dense toward the sides, the basal impression moderate, not attaining the sides ; elytra parallel and broadly arcuate at the sides, ra her obtuse at apex, fully three times as long as the prothorax, finely and rather sparsely punctate, the punctures much less close than those of the pronotum.   Length 1.25 mm.; width 0.6 mm.   California (Mendocino Co.)............................**inepta**, sp. nov.
Black, the elytra suffusedly paler toward tip, frequently pale ferruginous throughout, the antennæ pale, rather stout, two-fifths as long as the body ; prothorax rather small, three-fifths wider than long, slightly narrower than the base of the elytra, the sides parallel for three-fifths the length, then moderately converging to the apex, the punctures very fine and rather close set, not materially denser laterally, the impression rather fine and moderately deep ; elytra parallel, broadly arcuate at the sides, somewhat obtuse at apex, widest at the middle, not quite three times as long as the prothorax and fully a fourth wider, the punctures fine but strong and moderately close-set.   Length 1.2 mm.; width 0.55 mm.   California (Hoopa Valley, Humboldt Co.)..**nanula**, sp. nov.

The species in the neighborhood of *ochracea* form a very difficult study, and my treatment of them above must be regarded as provisional.   *Fallax* bears some resemblance to *nanula*, but the antennæ are more approximate in insertion upon the front, being separated by a third of the total width in the latter.   *Kamtschatica* Mots., is quoted by Mannerheim as occurring in Alaska, but I hav not seen it ; it is ovate, black, with the elytral humeri and apex testaceous and the prothorax arcuately dilated at the middle.   The species *lepidula* of Mäklin, from Sitka, is also unknown to me ; it is described as oval, slightly convex, shining, testaceous, with the prothorax slightly rounded at the sides and deeply, the elytra finely, punctate, and the antennæ not approximate at base ; it must be an unusually large species, as its length is given " 1 line," and it is said to be extremely rare.

Tisactia, gen. nov.

Although bearing a certain general resemblance to *Atomaria*, this genus is really profoundly different in several structural characters, and it may be readily recognized by the marginal bead at the base of the elytra ; it also differs in having the pronotum perfectly even and unimpressed at base, in its widely separated frontal antennæ and in its broad prosternal process, margined at each side by an acute cariniform edge. The head is rather deeply inserted, the eyes well developed and rather coarsely faceted but not very convex, and the clypeus, which is slightly prolonged and expanded before the antennæ, is separated from the front by an impressed straight suture extending between the antennal foveæ. The antennæ are nearly as in *Atomaria*, the first joint relatively still smaller but subsimilar in form, and the club parallel, loosely 3-jointed and well developed. The legs and tarsi are slender, the latter filiform, moderately short and pentamerous, the mesosternum moderately wide and unimpressed between the coxæ, and the deep-set anterior coxæ are oblique and much more transverse than in *Atomaria* approaching *Ephistemus* in this respect, the cavities sharply angulate externally. The scutellum is moderate in size and transversely oval. The single species is the following :—

Body oblong-oval, very convex, black or blackish, the legs and antennæ paler, testaceous, the club of the latter blackish, apparently glabrous, each puncture, however, with an excessively minute hair ; punctures throughout very fine and sparse, not denser on the prothorax, which is moderately transverse, very convex and deep on the flanks, the base distinctly wider than the apex, transverse, finely beaded and feebly lobed at the scutellum ; sides broadly arcuate from above, the lateral edges finely but acutely reflexo-beaded and nearly straight from a sublateral viewpoint ; elytra slightly longer than wide, ogival at tip, widest and distinctly wider than the prothorax slightly before the middle, the base equal to the base of the latter, the humeri not at all exposed at base, the sides arcuate, the suture not margined, minutely dehiscent at apex as usual. Length 0.9 mm.; width 0.55 mm. Indiana ....**subglabra**, sp. nov.

Two specimens are before me, one much damaged.

EPHISTEMINI.

This is one of the more highly specialized tribes of the family, composed of very minute, broadly oval and convex glabrous species, feebly represented in the nearctic, but moderately abundant in the palæarctic provinces. It is distinguished from the other tribes of the Atomariinæ, excepting the Sternodcini, by the structure of the pro- and mesosterna, and in the close juncture of the prothorax with the hind body, and, in the extremely specialized *Ephistemus*, also by a form of anterior coxa, antennal club, antennal clefts of the proster- num and form of scutellum which are wholly foreign to the rest of the family. The post-coxal plates of the first ventral segment, though feebly developed, should also be alluded to as an important distinguishing character. In the general structure of the body, legs, palpi and tarsi it is however a perfectly normal member of the sub- family Atomariinæ. The elytra are never margined at base, the pro- notum is always unimpressed, and the deeply seated anterior coxæ are transverse and subcylindrical and attached near the sides of the body, the cavities acutely angulate externally. The species before me may be assigned to the two following widely differentiated genera :—

Antennæ free, the club more slender and bilaterally symmetric, the grooves before the eyes and prosternal clefts wholly obsolete ; scutellum transversely oval as in Atomariini ; prothorax more transversely truncate at base, feebly arcuate at the middle ....* **Curelius**
Antennæ partially received in repose within narrow deep grooves before the eyes and in a broad shallow cleft and excavation between the prosternum and hypomera, the club rather more developed, parallel, loose and asymmetric, the joints being more developed on the inner side ; scutellum still more minute, cordate, pointed behind and as long as wide or longer ; prothorax broadly angulate at base.
Ephistemus

These genera are related to the Atomariini through the singularly synthetic genus *Tisactia* described above, which has the unimpressed pronotum, broad, flat and laterally margined prosternal process, an- tennal insertion and sensible, though somewhat differently formed, post-coxal plates of the Ephistemini, the loosely connected body and prothorax and free prosternal process of the Atomariini, and a strongly margined elytral base, which very exceptional character is foreign to both but existent to a well-developed degree in *Tomarus* of the Cryp- tophaginæ.

<h3 style="text-align:center">Curelius, gen. nov.</h3>

This genus is founded upon the *Ephistemus dilutus* of Reitter, and *exiguus* of Erichson, and, as far as known to me, is exclusively European. Although abundantly distinct from *Ephistemus*, it does not seem to have been recognized thus far by Reitter and other European authors.

Ephistemus *Steph.*

In this genus, as in the preceding, the body is evenly oval and rather pointed behind, the sides of the elytra and prothorax being perfectly continuous and without a reëntrant angle at the contiguous bases. The surface is virtually glabrous, having only a few extremely minute hairs visible under high amplification, and is feebly and sparsely sculptured. Our single representative is the following :—

Oval, convex, polished, black or piceous-black the elytra gradually rufo-testaceous posteriorly almost in apical half, the legs and antennæ paler ; surface impunctate ; prothorax moderately transverse, the sides convergent and rather strongly, almost evenly arcuate ; elytra rather less than three times as long as the prothorax and about a third to nearly half wider, widest at two-fifths, the sides strongly, almost evenly arcuate, converging behind, the tip narrowly rounded. Length 1.1 mm. ; width 0.72 mm. New York, New Jersey, Pennsylvania and Indiana.

apicalis *Lec.*

Almost perfectly resembles the European *dimidiatus*, but rather stouter and much larger ; the latter species seems to be distinct from *globulus*, with which it is united as a variety in the European catalogue of Heyden, Reitter and Weise.*

TRITOMIDÆ.

MYCETOPHAGIDÆ *Auct.*

It matters but little what name is used to designate a genus, and consequently perhaps, a family, provided it be the oldest properly published name, and that there be unanimity of opinion in regard to the points at issue. The Geoffroyian name *Tritoma* has been adopted in the most complete European catalogue, presumably after proper investigation, for the familiar *Mycetophagus*, and, as arbitrary dissent from this decision would only tend to perpetuate ambiguity in the fundaments of nomenclature, I am ready to take any course which

* The following is a new species from the European fauna, recently received from Mr. Reitter :

Narrowly oval, polished, blackish throughout, the elytra obscurely rufescent,blackish toward base, the legs and antennæ pale, extremely minutely, feebly and sparsely punctate; prothorax rather short and strongly transverse, the sides converging and arcuate as usual; elytra relatively long, rather more than three times as long as the prothorax but only about a fourth to a third wider, subinflated between a third and two-fifths from the base and narrowly rounded at tip. Length 0.95 mm. ; width 0.6 mm. Russia (Caucasus)..................   ....**reitteri**, sp. nov.

Distinguishable at once from *globulus* or *dimidiatus* by its narrower and less ovate form, the elytra in the species referred to being from two-fifths to a half wider than the prothorax.

may tend to bring about permanent agreement, assuming that it is never too late to correct a mistake, however repugnant it may be to our spirit of conservatism. The name *Triplax* is therefore to be re-established in the Erotylidæ.

The present family is taken up for investigation at this time, primarily to draw attention to the inharmonious and composite scope which has been given to it hitherto by our systematists. Of the genera which have been included within its limits by LeConte and Horn, *Diplocœlus* and *Biphyllus* are assigned by Heyden, Reitter and Wise to the Cryptophagidæ, which disposition of them is eminently appropriate. *Hypocoprus* forms a subfamily of Cucujidæ near the Monotominæ, and is also to be removed.

Again, as an important fact because affecting both the European and American scope of the family, it should be stated that *Berginus* is in no wise allied to the Tritomidæ, but belongs near *Lyctus*, in fact only distinguishable from that genus by the obliquely truncate maxillary palpi.*

Finally, but by no means least, it is to be remarked that the European *Triphyllus* does not occur in America, the species assigned by LeConte and Horn to that genus forming in reality two purely heteromerous genera in the vicinity of the malandryid *Tetratoma*. The Tetratomini are distinguished from other Malandryidæ by the 3- or 4-jointed antennal club, and will be alluded to in more detail near the close of the present paper.

The present family is evidently closely related to the Trixagidæ

---

* The following is an interesting new species of *Berginus* :—

Very slender, convex, blackish, the under surface, legs and antennæ paler; head and pronotum coarsely and closely punctured, the elytra with approximate series of similar coarse and close-set but well-defined punctures, each puncture throughout bearing a very small recurved squamiform hair; prothorax as long as wide, slightly narrower than the elytra and a little wider than the head, the sides arcuate and parallel ; eyes small and prominent ; antennæ slender, the two basal joints larger and the club 2-jointed; under surface coarsely, sparsely punctured, except the last four segments of the abdomen which are finely and longitudinally strigato-punctate, the first segment as long as the next three combined; legs short, the femora stout, the tibiæ and tarsi slender. Length 0.9–1.2 mm.; width 0.32–0.42 mm. Bahama Island (Eleuthera)—Mr. Wickham.

bahamicus, sp. nov.

Differs from *pumilus* in its smaller size, more slender form, evenly seriato-punctate elytra, even pronotum and general habitus. I have taken *pumilus* in abundance at San Diego, California ; it has an almost entire longitudinal impression at each side of the pronotum, which exists in the European *tamarisci* only as a minute basal impression, and in *bahamicus* is wholly wanting ; it was described from Pennsylvania, but perhaps this may be an error.

(Byturidæ) and Dermestidæ, and is quite out of position in the catalogue of Heyden, Reitter and Weise. Its general characters have been sufficiently presented by LeConte and Horn and need not be repeated at the present time. The tarsi are filiform and 4-jointed and the anterior in at least the first subfamily, are 3-jointed, more or less dilated and pubescent beneath in the males ;* the basal joint is generally elongated. The anterior coxæ are large, obliquely ovoidal and prominently convex in the first subfamily but smaller in the second, narrowly separated, with the cavities widely open or closed. The ornamentation of the elytra in many species is remindful of the Attagenini, but the eyes are coarsely faceted—in marked contrast to the Dermestidæ. The Trixagidæ are intermediate between the two families in this respect.

The Tritomidæ of America consist of two subfamilies which differ greatly from each other in general habitus, and are sufficiently defined by the following characters :—

Anterior coxæ large and convexo-prominent, the cavities widely open behind ; bases of the prothorax and elytra equal in width, the scutellum well developed ; sides of the prothorax defined by a thin acute edge ; hind coxæ narrowly separated.
TRITOMINÆ.
Anterior coxæ small and more deep-set, oblong-oval, the cavities broadly closed behind ; base of the prothorax much narrower than that of the elytra, its lateral edges obtuse and not acutely defined ; scutellum small ; hind coxæ rather widely separated..MYRMECHIXENINÆ.

The latter of these is represented by a single isolated genus common to Europe and America.

TRITOMINÆ.

The body is oblong-oval, convex or moderately depressed and always clothed with coarse and sparse pubescence. The four American genera before me may be separated by the following primary characters :—

Basal angles of the prothorax well defined...2
Basal angles broadly rounded ; body very minute...5
2—Epipleuræ horizontal and flat...3
Epipleuræ concave and rapidly descending externally...4
3—Eyes transverse, sinuate anteriorly................................. ....**Tritoma**
Eyes more rounded, not sinuate...**Typhæa**

---

* The anterior tarsi are said to be 4-jointed in both sexes in the Myrmechixeninæ, but my four examples seem to be females and I cannot, therefore, confirm this.

4—Eyes nearly as in *Tritoma ;* body much smaller and more oval...........**Litargus**
5—Epipleuræ flat and horizontal, not extending much behind the middle.
Thrimolus
All of these genera are common to the Atlantic and Pacific districts, except the last, which has been taken thus far only in Texas.

Tritoma *Groff.*
Mycetophagus Hellw.

The species are oblong-oval in form, moderately convex and clothed rather sparsely with short stiff reclined pubescence, the elytra generally ornamented with a pale design upon a darker ground ; they are moderately numerous and the American forms may be defined as follows :—

Antennæ gradually incrassate toward tip, the outer joints sometimes feebly subserriform, the prothorax widest at base, with the sides more or less strongly convergent and broadly arcuate thence to the apex, the two subbasal foveæ deep and distinct ; body broadly oblong-oval. [**Tritoma**, in sp.]............. 2
Antennæ with a very feebly differentiated subparallel 5-jointed club ; prothorax but little wider at base than at apex, more or less serrulate at the sides, much wider near the middle, the sides strongly arcuate, the subbasal pits deep and distinct ; body narrowly elongate-oval, the elytral intervals each with a series of semi-erect hairs. [**Ilendus**, sg. nov.] 8
Antennæ with a feeble parallel 4-jointed club ; body shorter and moderately broadly oblong-oval, the prothorax with the sides but feebly converging from the base and broadly arcuate, the subbasal pits distinct. [**Parilendus**, sg. nov.]............11
Antennæ with a 3-jointed club ; body rather broadly oblong-elongate, the prothorax widest before the base, with the subbasal pits feeble or obsolete. [**Gratusus** sg. nov.]..12
2—Last joint of the antennæ elongate, distinctly longer than the two preceding combined ; punctures rather coarse, not dense ; elytra blackish, with a large reddish-yellow design involving the suture from fifth to three-fourths, extending obliquely to the humeri, and, transversely at its posterior limit, nearly to the side margin, the apices also yellow. Length 4.5–5.7 mm. ; width 2.2–2.6 mm. New York, Indiana and North Carolina ....**punctata** *Say*
Last joint of the antennæ shorter, never longer than the two preceding combined ; body smaller in size...3
3—Elytral striæ impressed, strongly punctured and distinct almost throughout. Atlantic regions...4
Elytral striæ scarcely at all impressed, very finely punctured and almost completely obliterated behind the middle. Pacific coast...7
4—Pale design of the elytra somewhat as in *punctata*, involving the suture from basal fifth or sixth to slightly behind the middle, extending obliquely to the humeri, near which there is a projection from each side of the ramus, extending obliquely outward also at its posterior limit to the middle of the width and with

a subdisconnected transverse lateral spot more posterior, the apex also maculate. Length 2.8-4.0 mm.; width 1.4-1.8 mm. New York, North Carolina, Indiana, Lake Superior and Montana; [*bimaculata* Melsh.] ....**flexuosa** *Say*

Pale design of the elytra never involving any part of the suture.........................5

5—Side edges of the prothorax finely serrulate, the punctures not very dense, unequal as usual, moderately coarse toward the sides, which are narrowly explanate; elytra blackish-piceous, each with seven pale spots, one, quadrate, at the humeri, one smaller, rounded, at inner third and basal fifth, one small rounded, at inner fourth just behind the middle, one elongate, near the median line at four-sevenths, one very small, subattached to the last at outer and basal third, one transverse, near the margin at three-fifths and one rather large, involving the apex. Length 3.6 mm.; width 1.7 mm. Virginia........................**serrulata**, sp. nov.

Side edges of the prothorax even, not at all serrulate, the sides more or less narrowly explanate; abdomen finely and closely punctate...6

6—Body more elongate-oval, larger and more convex; elytra each with a large sub-quadrate humeral pale spot not involving the callus, another, large and slightly elongate-oval, very near the suture at basal sixth and narrowly connected with the humeral, a narrow irregular spot near the center, extending along and scarcely broader than the sixth interval from three-sevenths to slightly behind the middle and then obliquely extending internally nearly to the suture at four-sevenths, two submarginal spots, the anterior minute at two-fifths, the posterior larger and transverse at three-fifths and a moderate subapical spot. Length 4.2-4.3 mm.; width 1.9 mm. New York..**picta**, sp. nov.

Body oblong-oval, rather depressed, black, the prothorax scarcely paler and more transverse than in *picta*; elytra of the male each with two large coalescent subbasal pale spots in oblique line and one, smaller, sublateral at one-fourth from the base and frequently obsolete, also an oblique irregular fascia at or near apical third, sometimes obsolete or existing as two minute pale spots, and, finally, a large subapical spot; in the female the inner of the two subbasal spots is wholly obsolete, only the humeral and subapical remaining, or, sometimes, with the two minute pale spots in oblique line near apical third in addition. Length 3.7-4.2 mm.; width 1.75-1.9 mm. Indiana and North Carolina.

subdepressa, sp. nov.

7—Body oblong-elongate, rather depressed and shining, the punctures finer than usual; elytra blackish, each with a large oblique subbasal spot, from the humeri nearly to the suture at basal fourth, and a smaller transverse spot at apical fourth, not attaining the suture or margin; subapical pale spot wholly obsolete. Length 3.4-4.2 mm.; width 1.65-1.8 mm. Washington State to California.

californica *Horn*

8—Elytra more than twice as long as wide; abdomen sparsely punctured; pronotum coarsely, sparsely and equally punctate...9

Elytra not more than twice as long as wide, the abdomen more closely punctured; pronotum less coarsely, more densely and somewhat unequally punctate.........10

9—Body black or piceous-black throughout above, the under surface, legs and antennæ testaceous, the latter becoming blackish in outer half; elytra maculate with pale spots, of which two on each, elongate-oval, disposed in oblique line near the base and one transverse, discal and anteriorly angulate at apical third or fourth,

1. punctain

2

9

3. mm.

3.

10

var.

3 mm

4 pucta

(5w) ♂

8 = 5a

(5b) ♀

2.? mm.

3

9

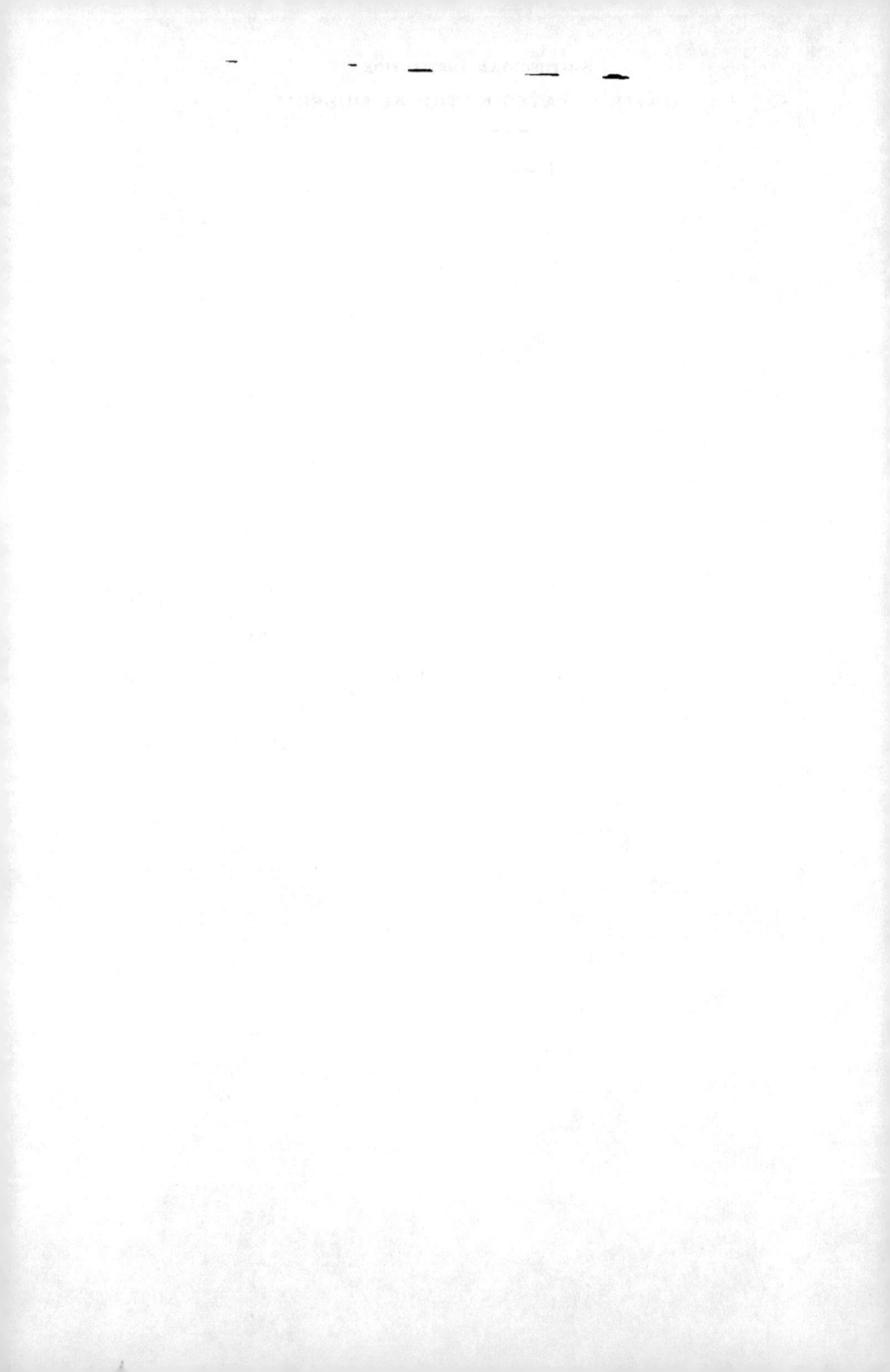

are most conspicuous, a small elongate spot, just before the middle and near the side margin, is also generally evident ; striæ strongly punctured, feebly impressed and distinct very nearly to the tip.   Length 4.3 mm.; width 1.45 mm.   Texas.
    melsheimeri *Lec.*
Body nearly similar in form and sculpture but rather less slender, the under surface and legs more dusky-testaceous, the antennæ blackish throughout, except near the base, distinctly longer than in *melsheimeri* and almost two-fifths as long as the body in the male, the elytra black throughout, without trace of paler spots ; intromittent organ of the male very slender and nearly straight.   Length 4.6 mm.; width 1.6 mm.   Indiana ; [Georgia—LeConte]..................**obscura** *Lec.*
10—Body very narrow, piceous or blackish, the elytra with numerous small flavo-testaceous spots, the striæ rather distinctly impressed but somewhat finely punctured and obliterated well before the tip ; under surface, legs and antennæ pale, the latter dusky distally as usual.   Length 3.4 mm.; width 1.3 mm.   Indiana.
    pluripunctata *Lec.*
Body distinctly broader but almost similarly sculptured, the prothorax less transverse, relatively narrower at apex and somewhat more coarsely punctured ; elytra evenly piceous-black throughout, without trace of paler maculation ; under surface, legs and antennæ pale testaceous.   Length 3.75 mm.; width 1.6 mm.   North Carolina (Asheville)...**pini** *Zieg.*
11—Body oblong-oval, rather strongly convex, the sides very feebly arcuate ; antennæ testaceous throughout, short and rather thick, not as long as the head and prothorax, the latter dark piceous-brown, three-fourths wider than long, the sides very feebly convergent from base to apex and very slightly arcuate ; disk convex, coarsely, densely and unequally punctured throughout, the edges minutely serrulate ; elytra dark, finely, densely punctulate, the striæ feebly impressed, finely punctate and obliterated toward tip, each with a suffused humeral pale spot and another, transverse and discal, near three-fifths ; each interval with a single series of suberect hairs ; abdomen finely and densely punctate.   Length 2.9–3.5 mm.; width 1.3–1.5 mm.   Massachusetts, New York, Indiana, Iowa and Nebraska.
    bipustulata *Melsh.*
Body oval, piceous, sparsely pubescent ; elytra densely punctulate, with a feebly striate arrangement at the middle near the base, maculate with large yellow spots somewhat as in *flexuosa*.   Length [4.5 mm.].   Colorado ....**confusa** *Horn*
12—Subbasal impressions of the pronotum distinct but in the form of short narrow canaliculations, the punctures moderately coarse, deep, not very dense, equal and evenly distributed throughout, the sides broadly, evenly arcuate, very feebly convergent, the apical angles broadly rounded ; disk widest behind the middle ; elytra piceous, with pale humeral, post-humeral and post-median maculation, the striæ scarcely at all impressed, rather finely and not conspicuously punctured and obliterated toward tip, the pubescence short and even ; abdomen finely, densely punctate.   Length 4.3–4.8 mm.; width 1.7–1.9 mm.   California (Truckee and Lake Tahoe)...**pluriguttata** *Lec.*
Subbasal impressions small and feeble but rounded and foveiform.....................13
Subbasal impressions wholly obsolete...14
13—Elytra immaculate, except some very minute widely scattered pale spots which are clothed with paler pubescence, of which there is on each one at base at each

side of the scutellum, one at outer fourth and one-sixth from the base, two at inner fourth in line with the basal spot at two-sevenths and one-half from the base, and one, transverse, near the margin at three-fifths; body oblong-oval, piceous-black throughout, the prothorax a little less than twice as wide as long, with the sides feebly convergent and feebly arcuate, only slightly wider behind the middle than at base, the punctures rather fine, close-set and unequal, with coarse punctures intermingled toward the sides; elytral striæ feebly impressed but strongly punctured, distinct nearly to the extreme apex; abdomen finely but not densely punctate; legs blackish, the tarsi paler. Length 5.0 mm.; width 2.1 mm. British Columbia.................................... **notatula**, sp. nov.

Elytra with very narrow sinuous bands of grayish pubescence at basal third and behind the middle, and also an apical spot, the posterior band bifurcating near the middle of each elytron, sending one branch forward the other backward to the side margin; body otherwise nearly similar to *notatula*. Length [5.0 mm.]. New Hampshire (White Mts.)............................... **tenuifasciata** *Horn*

14—Body moderately stout, oval, strongly convex, piceous-black, the legs and antennæ paler; prothorax distinctly less than twice as wide as long, the sides strongly convergent throughout, but very feebly arcuate, obsoletely serrulate, obliquely convergent near the base to the basal angles, which are very obtuse; disk convex, very coarsely, almost equally punctate, not densely toward the middle but closely laterally; elytra sparsely pubescent, with impressed entire striæ of coarse deep punctures, each with an irregular pale oblique subbasal fascia from the humeri nearly to the suture, another, oblique in contrary sense, at three-fifths and not attaining the suture and an oval, margino-median and a subapical spot. Length 4.5 mm.; width 2.0 mm. Virginia...................... ...**obsoleta** *Melsh.*

Confusa and *tennifasciata* I have not seen, and the characters are drawn from the descriptions. *Pluriguttata* is a very aberrant species, with the 3-jointed club very much feebler than in the others of that section and with a very complex male intromittent organ, consisting of a gradually narrowed thin basal piece, arcuate in plane, with an apical appendage curved sharply in contrary sense, and having two posteriorly diverging, rapidly and finely acuminate basal alæ and a terminal asymmetric button.

Typhæa *Curtis.*

Closely related to *Tritoma* and distinguished by the much smaller size of the body and the form of the eyes. The single species seems to be cosmopolitan :—

Narrowly oblong-oval, moderately convex, pale flavo-testaceous throughout, the elytra rarely piceous; antennæ with a 3-jointed club, distinctly shorter than the head and prothorax, the latter about twice as wide as long, with the apex but little narrower than the base and the sides arcuate, the punctures fine, subequal and rather close-set; elytra finely punctate, with unimpressed series of fine punctures

becoming obliterated toward tip, the pubescence short, moderately dense ; each
strial interval with a single series of suberect hairs. Length 2.25–2.7 mm.; width
0.85–1.15 mm. Vermont to Washington State, Florida and Texas

fumata *Linn.*

The single specimen with dark elytra is from Palm Beach, Florida,
and seems to have the prothorax slightly less transverse and the antennal club a little thicker ; additional material may ultimately prove
it to represent a variety or closely related species.

Litargus *Erichs.*

This genus differs profoundly from the two preceding in the form
of the epipleuræ, but the eyes are nearly as in *Tritoma* and the antennæ have a loose 3-jointed club as in *Typhæa*. The ornamentation of
the elytra is similar to that of *Tritoma*, and the body is very small in
size. The species are rather less numerous than in *Tritoma* and may,
as far as discovered, be separated by the following characters :—

Elytra with the pubescence short and sparse but stiff, pale in color and arranged
    throughout in even approximate series, piceous to blackish in color, each with a
    large transversely oval discal spot near basal and apical fourth, the posterior ap-
    proaching more closely to the suture ; punctures sparse throughout, the body
    rather broadly oval, convex and shining, the pronotum not impressed at base but
    with the basal sinuation at each side of the middle distinct ; last antennal joint
    short, rounded, the labrum small ; epistomal suture wholly obsolete. Length
    1.7–2.0 mm.; width 0.85–1.2 mm. Rhode Island to Texas and Lake Superior.
    [**Tilargus,** sg nov.]..4-**spilotus** *Lec.*
Elytra with the pubescence in general confusedly arranged...............................2
2—Elytra with fine dark pubescence, closely punctulate and with widely separated
    single series of longer semi-erect and paler hairs, each with a small subbasal spot
    at three-fifths from the suture, a larger triangular subsutural spot at two-sevenths
    and an obliquely oval subsutural spot at five-sevenths, the pale spots clothed also
    with pale hairs ; pronotum finely, not very densely punctate, the punctures simple
    and not asperate, feebly biimpressed at base ; body elongate-oval and depressed ;
    last antennal joint short, narrowly rounded at tip ; labrum large and very trans-
    verse. [**Litargus** in sp.; type *connexus*]. Length 2.2 mm.; width 1.1 mm.
    Illinois and Kansas...6 **punctatus** *Say*
        Var A—Similar but with the rows of erect paler hairs only evident toward the
        sides of the elytra, the basal spot more oblique, the anterior subsutural
        smaller, rounded and more distant from the suture, the elytra relatively less
        elongate, the spots subobsolete occasionally. Length 1.8–2.2 mm.; width
        0.9–1.2 mm. New Jersey and Indiana..................**obsolescens,** v. nov.
Elytra without widely separated series of pale hairs 3
3—Epipleuræ strongly concave and deeply descending, the epistoma trapezoidal
    as usual...4

Epipleuræ much narrower, almost flat and but slightly descending externally; epistoma rounded, the suture fine but rather more distinct; punctures granulato-asperate; last antennal joint short and transverse, somewhat obliquely but broadly rounded at tip; pronotum not at all impressed at base and with the sinuations very feeble. [**Paralitargus**, sg. nov.]..7

4 – Last antennal joint elongate, the tip obliquely and rectilinearly truncate; pronotal punctures simple. [**Alitargus**, sg. nov.]..5

Last antennal joint short and very broadly, subobliquely arcuato-truncate at apex; pronotal punctures minute, slightly elevated and subannulate. [**Litargellus**, sg. nov.]..6

5—Rather narrowly oval, moderately convex, shining, piceous or darker, finely, rather closely punctate, the pronotum with two feeble subbasal impressions, the basal sinuations small but evident; elytra with humeral, post-scutellar, subsutural and transverse post-median paler maculation and also with a very feeble paler spot at the side margin at two-sevenths, the paler spots clothed with paler pubescence. Length 1.75-1.9 mm; width 0.9-1.0 mm. Texas to California (Los Angeles). **balteatus** *Lec.*

Var A—Similar but larger, more elongate-oval and more depressed, the body generally darker, the subbasal impressions of the pronotum feebler and less linear, the elytra similarly maculate. Length 2.2-2.4 mm., width 1.0-1.15 mm. California (San Francisco)...............................**transversus** *Lec.*

6—Body oval and strongly convex, shining, the punctures sparse, the pubescence fine and rather sparse, closely decumbent and even; prothorax about two and one-half times as wide as long, the sides strongly converging from base to apex and moderately arcuate, flavo-testaceous, sometimes transversely clouded with piceous in the central part; elytra flavo-testaceous, each with three incomplete narrow piceous fasciæ, the two posterior anteriorly arcuate, the subbasal less obvious. Length 1.5-1.75 mm.; width 0.73-0.83 mm. New York and Pennsylvania to New Mexico (Las Cruces)...............................**nebulosus** *Lec.*

7—Elytra with the post-median pale fascia transverse, or, to a slight degree, posteriorly oblique toward the suture. Body narrowly oblong-oval, rather convex, not coarsely but strongly, somewhat closely and asperately punctate, blackish throughout above, the elytra each with an oblique pale spot from the humeri nearly to the suture at two-sevenths and a more or less narrow fascia at four-sevenths, which is virtually entire. Length 1.6-1.9 mm.; width 0.7-1.0 mm. Rhode Island and Illinois to Florida (Palm Beach)......**didesmus** *Say*

Elytra with the posterior pale area median and anteriorly oblique toward the suture. Body narrowly oval and convex, not densely but strongly, evenly and asperately punctate, the pubescence shorter and sparser but coarse and rather pale in color; integuments piceous above, the elytra each with an oblique subbasal pale spot nearly as in *didesmus* and also having an equally broad and conspicuous pale spot extending from the sides, just behind the middle, almost to the suture well before the middle and near the apex of the subbasal spot. Length 1.6 mm.; width 0.78 mm. Dakota—Mr. Wickham...............................**asperulus**, sp. nov.

In *6-punctatus* and its varieties the epistoma of the male is clothed densely with an extremely fine short pale pubescence, which is want-

ing in the female, and the labrum is larger than in any other species, extending to the extreme limits of the epistomal truncature. *Infulatus*
of LeConte, I have not seen ; it is said by Horn to be a synonym
of *balteatus*.

Thrimolus, gen. nov.

This genus is composed of a single exceedingly minute species,
differing radically from those which precede in the broadly rounded
basal angles of the prothorax. The body is oblong-oval, moderately
convex, clothed rather sparsely with coarse and moderately long reclined hairs, with other longer erect setæ serially bristling from the
elytral flanks. The head is large, transverse and well developed, the
eyes moderately large, basal, not very prominent, somewhat transversely oval, entire and much less coarsely faceted than usual ; the
clypeus is rather short and broad, with the suture transversely rectilinear, not impressed and very feeble.   Antennæ moderate in length,
11-jointed, with a compactly cylindric stout and 3-jointed club, the
joints six to eight gradually increasing in width and decreasing in
length, the latter as wide as the base of the club.   Prothorax broadly
arcuate and very finely beaded at base.   Scutellum well developed,
broadly subtriangular or parabolic.   Anterior coxæ large, obliquely
suboval, very convex and narrowly separated.   Basal segment of the
abdomen as long as the next two combined ; two to four relatively
shorter than usual and gradually diminishing somewhat in length, the
hind coxæ very narrowly separated.   Legs slender, coarsely, sparsely
herissate with moderately long hairs, the tarsi extremely slender, filiform, much shorter than the tibiæ, with the basal joint but little longer
than the second, normally 4-jointed throughout, the claws small and
very slender ; tibial spurs small and much less developed than usual.

The antennæ are bilaterally symmetric, shorter and more compact
than in *Typhæa* and the elytral punctures are altogether irregular in
distribution.   The type may be briefly defined as follows :—

Body dark luteo-testaceous in color throughout, the legs and antennæ still paler, the
club of the latter very feebly infuscate, shining ; head and pronotum subimpunctate, the latter short and strongly transverse, more than twice as wide as long, the
sides converging, broadly arcuate and subcontinuous with those of the elytra, the
disk wholly devoid of basal foveæ or impression ; elytra oblong-oval, rather convex, slightly longer than wide, broadly and obtusely rounded conjointly at tip,
much wider than the prothorax and more than three times as long, the sides
broadly arcuate ; punctures very fine, sparse and subasperate ; under surface
shining, sparsely clothed with coarse inclined hairs.   Length 0.78 mm.; width
0.45 mm.   Texas...**minutus,** sp. nov.

The single example before me is so frail that I am unable to dismount it to better observe the structure of the mouth, the trophi however appear to be in perfect homology with the rest of the family.

MYRMECHIXENINÆ.

This subfamily is evidently assigned properly to the Tritomidæ by LeConte and Horn, although the facies departs conspicuously by reason of the small prothorax and wide elytra, the latter rather sparsely clothed with an even decumbent vestiture, finer and less conspicuous than in Tritominæ. The single genus is as follows:—

Myrmechixenus *Chev.*

Our single species occurs throughout the more southern parts of the United States, from the Atlantic to the Pacific, and may possibly be identical with some European form; it may be briefly defined as follows:—

Body narrowly oblong, convex, rather shining though finely, deeply and very closely punctured throughout, reddish-brown in color, the legs and antennæ paler; head subtriangular, the eyes well-developed, moderately convex, coarsely faceted as usual; antennæ moderate, the club loosely 5-jointed, joints six to eight increasing gradually in width; prothorax distinctly wider than the head, slightly transverse, widest near apical third, the sides parallel, rounded, the base and apex equal and feebly arcuate; elytra between two and three times as long as the prothorax and about two-fifths wider, the humeri exposed at base; sides parallel and broadly arcuate, the apex obtusely rounded; abdominal segments convex, gradually and but slightly decreasing in length, as usual in the Tritomidæ, the last partly exposed dorsally. Length 1.7–1.8 mm.; width 0.75–0.78 mm..**latridioides** *Crotch*

This species does not appear to be very common; the specimens in my cabinet are from South Carolina, El Paso, Texas, and Riverside, California, the latter sent to me by Mr. H. C. Fall. The basal joint of the hind tarsi is much elongated, as in normal members of the family, which is an additional reason for believing that it is correctly placed in the Tritomidæ.

DERMESTIDÆ.

The Dermestidæ are a small family of clavicorn beetles, which, in their notably varied structural characters, seem to constitute one of the old synthetic types of Coleoptera, having some philogenetic relationship with both the Geodephaga and Serricornia. They have the anterior coxal cavities open behind, the tarsi simple and 5-jointed, the claws unmodified, the sternal side pieces very wide and the hind coxæ lamellate and transversely excavated. The antennæ are extremely varied in structure and may or may not be received within protecting pits or excavations, and the legs may be free or strongly retractile. In considering the depression for the protection of the antennæ, a distinction should be drawn between a large and vaguely limited concavity of the hypomera—or inflexed side of the prothorax— as in *Dermestes*, and a closely circumscribed and sharply defined pit ; the former characterizes most of the genera in some form, and becomes a true protective fossa in a few genera, but the latter only occurs in *Anthrenus*.

The genus *Trixagus* (*Byturus* Lat.), is evidently allied to the Dermestidæ, but differs in so many radical characters, such as the closed anterior acetabula, lobed tarsi, dentate claws, narrow sternal side pieces and structure of the mesosternum, that the position assigned it by Reitter as a distinct family is probably as satisfactory as any, and I have therefore not considered it in the following revision. As thus restricted, the American Dermestidæ may be assigned to five distinct tribes characterized as follows :—

Head without ocellus ; anterior coxæ large, contiguous, the prosternum not visible between them, the mesosternum between the coxæ moderately wide, ogival and not sulcate ; antennæ 11-jointed, with a 3-jointed club, similar in the sexes and not received within sharply circumscribed pits ; hypomera concave anteriorly ; epipleuræ strongly defined, wide and inflexed toward base ; body clothed with short hairs ....DERMESTINI

Head with a single ocellus 2

2—Prosternum visible between the coxæ ; metacoxal lamina not extending to the sides of the body. 3

Prosternum not visible between the coxæ ; metacoxal lamina extending to the sides of the body..5

3—Metacoxal plate extending laterally half way across the parapleuræ ; prosternal process impinging upon the exposed surface of the mesosternum between the coxæ ; epipleuræ well developed toward base ; legs in great part free ; body clothed with short hairs...ATTAGENINI

Metacoxal plate only extending laterally to and abutting against—squarely in Trinodini, obliquely in Anthrenini—the inner boundary of the parapleuræ...............4

4—Epipleuræ subobsolete ; lateral margin of the prothorax entire as usual ; antennal club received within deep fossæ at the apical thoracic angles ; body compact, clothed with decumbent scales, the legs all very closely retractile ; coxæ large ; scutellum very minute..ANTHRENINI

Epipleuræ narrow but strongly delimited and inflexed toward base ; lateral thoracic margins obliterated at apex ; legs and antennæ perfectly free, excepting, as usual,

the hind femora ; coxæ small ; body clothed with long sparse erect coarse and
bristling hairs ; scutellum large .... TRINODINI

5—Anterior coxæ contiguous at apex over the prosternum, which has the form of a
transverse pointed plate ; antennæ 11-jointed, the club 3-jointed, not received
within abruptly excavated pits, the hypomera biconcave ; legs very closely retractile ; body glabrous, the epipleuræ distinct toward base ; scutellum well developed.
ORPHILINI

Except the small and isolated tribe Trinodini, which is confined to
the Atlantic and Sonoran regions, all of these groups are very general
in distribution.

DERMESTINI.

The genus *Dermestes* differs so greatly from the other types of the
family in the absence of the very characteristic vertexal ocellus and
contiguous anterior coxæ, that it is necessary to regard it as a distinct
tribe. The metacoxal lamina is narrow, extending only to the parapleuræ, and is notably elongate internally, the tibiæ seriate with short
stout spinules and the tarsi rather stout, with the basal joint shorter
than the second, generally very markedly so, but sometimes only
slightly as in *lardarius*.

Dermestes *Linn.*

The species of *Dermestes* are rather numerous and are the largest
of the family. They can be readily classified by the form of the
inner marginal suture of the abdomen toward base, and by the form
and vestiture of the prothorax, as follows : —

Inner lateral suture of the first abdominal segment inflexed at base to the outer limit
of the hind coxæ, becoming deeply excavated at the basal margin 2
Inner lateral suture straight, not inflexed basally and distant at base from the outer
limit of the coxæ ; pronotum not deeply declivous laterally, the margin visible
throughout from above 10
2—Pronotum clothed densely throughout with variegated black and fulvous pubescence, except in *medialis*, the flanks deeply declivous ; male with the third and
fourth segments foveolate at the middle...3
Pronotum clothed with dense cinerous pubescence laterally, leaving a large triangular
or parabolic discal area sparsely clothed with almost uniform pubescence, the
flanks rather deeply declivous..7
Pronotum somewhat sparsely or inconspicuously and quite uniformly pubescent
throughout, the flanks less declivous, the lateral margin visible from above
throughout the length ; third and fourth ventrals foveolate in the male.............9
3—Pronotum having, as a marked feature of the vestiture, three widely separated
points of pale pubescence arranged transversely at about the middle of the
length..4
Pronotum without the three points of paler pubescence........................6

4—Large species, 10 mm. or more in length ; vestiture cinereous to ochreous, the pale points of the pronotum cinereous, sometimes ochreous and less distinct ; elytra with a large oblong area of dense pubescence at each side, extending two-fifths, and elsewhere marmorate with black and cinereous or ochreous hairs ; ventral segments each with the usual lateral dark spot, that of the basal segment very large ; median fovea of the male very small. Length 10.0–12.5 mm.; width 3.9–5.3 mm. Texas to California.................................**marmoratus** *Say*
Smaller species, always distinctly less than 10 mm. in length............................5
5—Pronotal punctures fine and more distinctly separated ; body moderately large, the elytra marmorate with cinereous and black, usually subtransversely, and with certain parts of the surface uniformly clothed with the pale hairs, the abdomen densely clothed with whitish pubescence, with black lateral spots ; ventral foveolæ of the male much larger than in *marmoratus*. Area of uniform pale pubescence subquadrate, extending from near the base to basal third and from the side margin nearly to the middle. Length 7.5–8.4 mm.; width 3.5–3.75 mm. Atlantic Coast from Canada to Florida (Palm Beach); [*nubilus* Say, *dissector* Kby., and *murinus* Lec. nec Linn.]...**caninus** *Germ.*
    Var. A—Area of pale pubescence extending from near the base scarcely to basal third, but prolonged transversely to or near the suture. Length 7.0–8.3 mm.; width 3.0–3.75 mm. Pacific Coast.................**mannerheimi** *Lec.*
    Var. B—Area of pale pubescence extending entirely across the elytra and prolonged to about apical third. Length 6.5–8.0 mm.; width 3.2–3.7 mm. Iowa (Keokuk) to Florida ; [*nubilus* Lec. nec Say]...**nubipennis**, v. nov.
    Var. C—Area of pale pubescence extending at the lateral margin from the base for two-fifths and dilated internally subbasally nearly to the middle, the entire sutural region also clothed with a very large preponderance of pale hairs. Length 8.3 mm.; width 3.65 mm. Texas (Galveston).
                                         compactus, v. nov.
Pronotal punctures quite coarse and more close-set ; body very small in size, the elytra clothed to the tip with dense cinerous pubescence, with a few small black spots, especially at base ; ventral pubescence much less dense, especially toward tip. Length 5.0 mm.; width 2.2 mm. Oregon...................................**rattus** *Lec.*
Pronotal punctures coarse, deep and narrowly separated ; body small in size, elongate-oval and strongly convex, deep black throughout, the pronotum clothed rather sparsely and almost uniformly with dusky pubescence, with three small widely separated spots of pale pubescence arranged transversely ; scutellum transverse, densely clothed with coarse pale yellowish hairs ; elytra rather coarsely and quite closely punctured and clothed uniformly throughout with short blackish inconspicuous hairs ; abdomen densely clothed with white pubescence only in the middle third of the two basal segments, elsewhere more sparsely clothed with a mixture of white and fuscous hairs ; femora annulated. Length 5.6 mm.; width 2.4 mm. California.... ....**medialis,** sp. nov.
6—Elytra transversely marmorate with black and cinereous pubescence, the pale hairs generally forming a condensed transverse fascia behind the base, the portion thence to the basal margin having some fulvous hairs intermingled ; body larger and more broadly oval. Length 6.8–7.5 mm.; width 3.0–3.5 mm. Wyoming to New Mexico (Fort Wingate) ... .**fasciatus** *Lec.*

Elytra finely and more uniformly variegated with black, cinereous and fulvous hairs throughout; body narrowly and evenly ellipsoidal, rather small in size. Length 5.0–6.7 mm.; width 2.35–3.0 mm. Idaho (Cœur d'Alène), Nevada (Reno) and California (San Francisco and Monterey)..............................**talpinus** *Mann.*

7—Elytra black, rufo-piceous toward the humeri, where there is a small post-humeral area of fulvous pubescence, elsewhere marmorate subtransversely with black and cinereous pubescence; body rather small and stout, the abdomen very densely clothed with white hairs, the black marginal spots very small; male with the third and fourth segments foveolate. Length 6.5–6.9 mm.; width 2.9–3.25 mm. Arizona; [*mucoreus* Lec.]..**carnivorus** *Fabr.*

Elytra uniform in color and uniformly clothed with a mixture of black and paler hairs; fourth ventral alone foveolate in the male, at least in *vulpinus*...........8

8—Elytra piceous, uniformly and rather sparsely clothed with a mixture of black and fulvo-cinereous hairs in almost equal proportions, dense whitish pubescence toward the sides of the pronotum not maculate at base; body elongate. Length 5.8–8.9 mm.; width 2.4–3 6 mm. Indiana, Florida, California and Guadalupe Island............ ....**vulpinus** *Fabr.*

Elytra black, sparsely clothed with black hairs, among which longer yellowish-cinereous hairs are uniformly but sparsely intermingled; densely pubescent lateral area of the pronotum with a small rounded dark spot at base; body stouter and more oval. Length 7.5–8.5 mm.; width 3.5–3.75 mm. New Jersey, Virginia (Fort Monroe) and Iowa (Keokuk)..**frischi** *Aug.*

9—Pubescence throughout above and on the abdomen uniform and yellowish-cinereous, somewhat sparse, not concealing the sculpture, the abdomen without trace of quasi-denuded dark spots at any part; body very elongate, subparallel, the pronotum with two pronounced basal impressions and the elytra with feebly impressed longitudinal lines extending almost to the base. Length 8.9 mm.; width width 3.6 mm. Indiana..**elongatus** *Lec.*

Pubescence, thoracic impressions and elytral lines as in *elongatus*, the vestiture of the abdomen even less conspicuous and dark fulvous in color, with two marginal and two discal series of rounded subdenuded spots, the two male foveolæ small; legs not annulated; body shorter and less parallelo-subcylindric than in *elongatus*. Length 7.5 mm.; width 3.3 mm. Florida (Key West)......**cadaverinus** *Fabr.*

10—Elytra densely cinereo-pubescent in basal two-fifths or more, each with three small nigro-pubescent points in transverse posteriorly arcuate series at or near basal fourth; male with two ventral foveolæ..11

Elytra black, pale and fulvo-pubescent at base for a short distance, not maculate; male with two ventral foveolæ...12

Elytra piceous, uniform in color and vestiture throughout; male with a single ventral foveola on the fourth segment as in *vulpinus*..13

11—Pronotum closely punctured throughout and uniformly clothed with blackish hairs, with small clusters of yellowish-cinereous hairs interspersed; basal pubescent area of the elytra not extending to the middle and sharply delimited, the hairs of the remainder being entirely black. Length 6.8–7.7 mm.; width 2.75–3.4 mm. United States and Europe..**lardarius** *Linn.*

Pronotum finely and sparsely punctured toward the middle, clothed uniformly throughout with longer fulvo-cinereous pubescence, the elytra rufo-piceous throughout, the densely pubescent basal area extending well beyond the middle and not sharply defined, the pubescence of the remaining parts being in large part similar in color but sparser.   Length 5.7–7.4 mm.; width 2.5–3.2 mm.   Vancouver Island and New Mexico (Fort Wingate)..................................**signatus** *Lec.*

12—Body oblong-oval, more depressed than usual, pale rufo-ferruginous throughout above and beneath and clothed with rather sparse fulvous pubescence, the elytra black, except at the basal margin and along the sides to basal fourth or more, the black parts clothed uniformly with inconspicuous blackish pubescence ; abdomen without quasi-denuded spots.   Length 6.3 mm.; width 3.0 mm.   Illinois.

pulcher *Lec.*

13—Narrow and convex ; body and legs throughout uniform dark piceous-brown in color, the pronotum rather finely, not very densely punctate, deeply and narrowly bisinuate at base, broadly biimpressed at the basal margin, with rounded hind angles, the vestiture uniform throughout and consisting largely of fulvo-cinereous hairs ; elytra clothed rather sparsely with dark pubescence, with fulvo-cinereous hairs sparsely and uniformly interspersed throughout ; pubescence of the under surface denser and uniformly flavo-cinereous, the abdomen without quasi-denuded spots.   Length 6.7 mm.; width 2.7 mm.   Texas (El Paso).

angustus, sp. nov.

Sobrinus of LeConte, I have been unable to identify amidst the material accessible to me.   *Rattus* and *signatus* are by no means varietal forms, but perfectly valid and very interesting species ; *mannerheimi* seems, however, to be a variety of the very widely distributed *caninus ;* it is wholly different from *marmoratus*, as I have previously pointed out (Bull. Bk. Ent. Soc.).   The identity of *mucoreus* and *carnivorus* rests upon the authority of the Hanshaw List.   Say described his *nubilus* from Florida and Pennsylvania, and the characters given coincide entirely with those of *caninus* and not with the more pubescent form named *nubipennis* above.

ATTAGENINI.

This is the largest tribe of the family, and contains a considerable number of genera having the legs more or less free throughout.   The laminate portion of the hind coxæ extends about half way across the end of the parapleuræ ; the epipleuræ are distinct and generally strongly defined toward base, and the prosternal process is visible, though generally narrow, between the coxæ, its free tip resting in an apical pit of the mesosternum which is frequently prolonged to the apex of the latter as a well-defined sulcus or fossa.   The antennæ are of varied structure, and the antennal fossa may be traced in successive stages of development through the genera in an instructive and interesting manner.   In the first four or possibly five genera of the tribe, the hypomera are merely flat or concave, without trace of an enclosed antennal fossa, but in *Trogoderma* the fossa appears in one of its primitive stages, and may be conceived to be the result of retractility of the anterior femora. The crural fossæ are deep and defined anteriorly by a strongly elevated acute cariniform line, extending obliquely to the hind angles of the prothorax, and forming the posterior boundary of the hypomeral concavity. To suggest that this latter concavity has not been evolved primarily as a shelter for the antennæ as in *Anthrenus*, for example, it may be observed that it is equally large and well formed in both sexes, although the antennæ differ sexually to a great degree, and it is only in the male that it is in any way completely utilized or compactly filled by that organ; in the female, where the antennæ are comparatively very feebly developed, these organs lie in repose along the bottom of the concavity, which is much too large to form a secure shelter. In *Trogoderma* the fossa occupies the entire length of the prothorax, but in *Cryptorhopalum* while having a general form which undoubtedly betrays a development from that of *Trogoderma*, it has become smaller and forms a secure shelter for the antennæ, these having become similar in the sexes and assuming a form so radically different from those of *Trogoderma* that it is difficult to trace any philogenetic relationship, and in *Thaumatoglossa*, the modification is carried still further, the two closely connected club-joints of *Cryptorhopalum* becoming a single very large joint. *Acolpus* appears to be a very satisfactory intermediate between the non-fossate genera and *Trogoderma*, and it is possible that more careful observation may there show the antennal fossa in a still more incipient stage of formation. The American genera may be defined as follows :—

Basal joint of the hind tarsi very short, much shorter than the second ; antennal fossa not defined ; legs free, the hind femora retractile as usual.............................2

Basal joint elongate, generally but little shorter than the next two combined ; antennæ 11-jointed in both sexes... 3

2—Antennæ 11-jointed in both sexes, the two basal joints of the male club short and transverse, the last greatly elongated ; mesosternum between the coxæ longer than wide, not sulcate, the anterior coxæ narrowly separated ; metacoxal lamina greatly elongated internally............ ....**Attagenus**

Antennæ 10-jointed in the male, 11-jointed in the female, the two basal joints of the male club much elongated and the last joint relatively much less so ; mesosternum between the coxæ very narrow and elongate, not sulcate, the prosternal process extremely narrow ; metacoxal lamina as in *Attagenus*, the epipleuræ less inflexed and less strongly defined ; body with denser and more variegated pubescence..**Novelsis**

Antennæ 9-jointed in both sexes, the club oval, compact and dilated in the male, with its two basal joints very short and transverse ; mesosternum between the coxæ rather narrow, divided longitudinally throughout by a narrow shallow sulcus ; anterior coxæ narrowly separated ; hypomera feebly concave anteriorly ; metacoxal lamina short, gradually and very slightly longer internally ; epipleuræ narrow but distinct..**Dearthrus**

3—Hypomera indefinitely concave as usual, without antennal fossa......................4
Hypomera with a deep concavity which is well defined internally by acute edges...5

4—Antennal club 3-jointed in both sexes, formed nearly as in *Attagenus* but with the last joint less elongate in the males ; mesosternum between the coxæ moderately narrow, divided throughout by a very shallow longitudinal impression and deeply emarginated behind by the tip of the metasternal process ; anterior coxæ rather narrowly separated ; epipleuræ strongly defined ; metacoxal lamina scarcely at all longer internally...**Perimegatoma**
Antennal club of the male 6-jointed and serriform, nearly as in *Trogoderma ;* hypomera concave ; metacoxal plates only attaining the parapleuræ ; mesosternum as in *Trogoderma* [Jayne]..**Acolpus**

5—Antennæ stout, claviform and usually serrate in the male, with the subbasal joint small, generally very small and with a narrow 4-jointed club in the female ; mesosternum very short and wide between the coxæ and completely divided longitudinally by a deep broad sulcus ; anterior coxæ rather narrowly separated ; metacoxal lamina short, gradually, feebly and rectilinearly longer internally, as in *Dearthrus ;* epipleuræ rather feebly inflexed and not coarsely delimited ; anterior femora retractile, the crural cavities separated from the antennal fossæ by a thin cariniform interval..**Trogoderma**
Antennæ with a large oval and compactly 2-jointed club, securely and closely fitting in repose within deep fossæ, which are separated by a flat interval from the crural cavities in both sexes ; mesosternum as in *Trogoderma*, the anterior coxæ more widely separated ; epipleuræ feebly inflexed, rather well defined ; metacoxal lamina short, with its hind margin transverse......................**Cryptorhopalum**
Antennæ with a male club consisting of a single very large subsecuriform joint, closely fitting in repose within hypomeral fossæ ; remaining characters nearly as in *Cryptorhopalum* ; [*Axinocerus* Jayne]..................................**Thaumatoglossa**

If the metacoxal plates only attain the parapleuræ in *Acolpus*, as stated by Jayne, this genus forms a remarkable exception to the entire tribe, and I strongly suspect that the author is mistaken. Neither this genus nor *Thaumatoglossa* is represented before me at present, and I am therefore unable to consider them below. The species are all pubescent, generally with nubilous variation in density, usually elongate or oblong-oval in form and of less compact build than in the Anthrenini or Orphilini, but similar in this respect to the Dermestini and Trinodini.

Attagenus *Latr.*

The prosternal process is wider between the coxæ than in *Novelsis*, though still very narrow, and the species are larger, stouter, more ob-long and almost uniformly clothed with rather sparse dark and incon-spicuous pubescence. The species are somewhat numerous but closely allied among themselves, those forms which are apparently worthy of distinctive names may be defined as follows :—

Elytra deep black throughout, the head and pronotum concolorous.. 2
Elytra rufous to piceous-black in color, the anterior parts frequently darker than the
  elytra... 4
2—Elytra each with a small spot of white pubescence at the middle of the length and
  at inner fourth of the width ; pronotum with three small and widely separated
  areas of pale pubescence at base ; third joint of the male antennal club black,
  as long as the entire remainder of the antenna and rather more than four times
  as long as the two basal joints of the club combined. Length 4.0 mm.; width
  2.0 mm. Rhode Island...**pellio** *Linn.*
Elytra without paler pubescence at any part; pronotum without pale hairs at the
  base..3
3—Pronotum coarsely and closely punctate, without subbasal impressions; antennæ
  of the male nearly as in *pellio* ; pubescence fulvo-piceous in color. Length 3.3
  –4.0 mm.; width 1.6–1.9 mm. Indiana and California ; [*megatoma* Fabr.].
                                                                       piceus *Oliv.*
Pronotum very finely and less closely punctured, with three widely separated subbasal
  impressions ; body more broadly oblong-oval, shining ; legs piceous, the tarsi
  ferruginous, pubescence blackish ; last joint of the female club more than one-
  half longer than the two preceding combined ; male not observed. Length 3.8
  mm.; width 2.0 mm. Idaho (Cœur d'Alêne).................**schæfferi** *Herbst*
4—Pronotum not impressed, or clothed with paler pubescence along the basal sinuations... 5
Pronotum feebly impressed along the broadly rounded basal sinuations, the impressed
  margin clothed with finer and pale pubescence....................................9
5—Entire upper surface dark piceous-brown to piceous-black in color ; pronotum
  with a feeble subbasal impression before the scutellum...............................6
Elytra bright red, sometimes narrowly infuscate along the suture, the head and pronotum black and much more closely punctured, the ante-scutellar impression
  not visible................ 8
6—Last joint of the male antennal club black, about as long as the entire preceding
  part of the antenna, which is testaceous, and slightly more than three times as
  long as the two preceding joints combined ; body moderately stout, oblong-oval,
  legs ferruginous throughout. Length ♂ 3.2, ♀ 3.7 mm.; width ♂ 1.7, ♀ 1.85
  mm. Pennsylvania..**extricatus**, sp. nov.
Last joint of the male antenna much shorter than the entire preceding part...........7
7—Male club stout, the last joint two and one-half times as long as the two preceding
  combined ; prothorax of the male fully twice as wide as long. Length ♂ 3.25,
  ♀ 3.4–3.7 mm.; width ♂ 1.6, ♀ 1.65–1.85 mm. New York, District of Columbia and Virginia (Norfolk)...............................**deficiens**, sp. nov.

Male club relatively still shorter, the last joint but slightly more than twice as long
as the two preceding combined ; prothorax less transverse and less strongly and
densely punctured, not quite twice as wide as long, the male narrower and the
female larger than in *deficiens*. Length ♂ 2.8–3.1, ♀ 3.8–4.5 mm.; width ♂
1.4–1.65, ♀ 2.0–2.4 mm. Iowa (Keokuk) and Nebraska ; [*spurcus* Lec., *flor-
icola* and *obscurus* Mels., i. litt.]...**cylindricornis** Say

8—Body narrowly oval or oblong-oval, the head and pronotum strongly and moderately closely punctured, the elytra unusually sparsely and much less coarsely so ;
hypomera but feebly concave ; last joint of the male antennæ black, longer than
the entire preceding part and four times as long as the two preceding joints combined ; female club black ; under surface piceous, the legs ferruginous throughout. Length ♂ 2.9, ♀ 3.6–4.1 mm.; width ♂ 1.35, ♀ 1.7–2.1 mm. Cali-
fornia to Utah...**rufipennis** Lec.

9—Head and pronotum generally blackish-piceous, the elytra rufous, the entire body
sometimes testaceous ; hypomera deeply concave ; punctures moderately dense,
those of the pronotum finer than the elytral ; male club very elongate and
slender, the last joint contorted distally, longer than the entire preceding parts.
Length ♂ 3.2, ♀ 4.9 mm.; width ♂. 1.65, ♀ 2.4 mm. Iowa (Keokuk);
[*dichrous* Lec.]..**bicolor** G. & H.

Head and pronotum blackish, the elytra somewhat, but not very noticeably, paler
piceo-rufous; in body and antennæ nearly similar to *bicolor*, the former obviously
narrower and relatively more elongate-oval. Length ♂ 2.9–3.4, ♀ 4.4 mm.;
width ♂ 1.4–1.7, ♀ 2.1 mm. Nebraska to Utah.......**elongatulus,** sp. nov.

As may be inferred from the detailed measurements given in the
table, the female is generally very much larger than the male, but in
extricatus and *deficiens* there is greater equality in this respect, judging
from the material accessible to me. The discriminative work hitherto
bestowed upon this comparatively monotonous, and consequently less
interesting, genus, has been very superficial, and detailed study of the
male antennæ reveals a variety of
structure too great apparently to be
the result of fortuitous variation ;
some of the names proposed by
LeConte must therefore be restored
to specific weight ; *rufipennis* is, in
fact, quite isolated as a species—more
so than *pellio* when compared with
piceus for example. The diagrams
given in the accompanying cut will

FIG. 3.

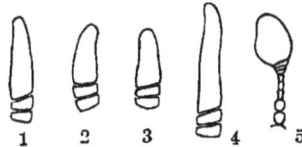

1. Antennal club of *Attagenus extricatus* ♂ ; 2, same of *A. cylindricornis;* 3
same of *A. deficiens;* 4, same of *A. elongatulus;* 5, antenna of *Dearthrus longulus.*

serve to show some of the variations in the club of the male antennæ,
and, although some variability in an organ so over-developed is to
be expected, it will be probably granted that such extreme variations,

especially when accompanied by differences in the form, color and sculpture of the body, must, until further evidence, be held to have specific weight.

Novelsis, gen. nov.

This genus is comparatively local, occurring only in the Sonoran provinces, and is distinguishable at once from *Attagenus* by the structure of the antennal club and hypomera and the 10-jointed male antenna as well as by the complex vestiture. The few species before me may be identified as follows :—

Hypomera nearly horizontal, not concave and with the outer edge rather obtuse and not at all descending ; mesosternum very narrow between the coxæ. [**Novelsis**, in sp.] 2

Hypomera concave and strongly descending, the outer edge very acute ; mesosternum wider between the coxæ. [**Paranovelsis**, sg. nov.]...6

2—Elytra without distinct paler pubescent maculation behind the middle..............3

Elytra with transverse paler pubescent spots or bands in apical half................ 5

3—Elytra with the suture, external margin in basal two-fifths, and an oblique line connecting the latter with the pale sutural line at basal third or less, pale testaceous and clothed with coarser fulvo-cinereous hairs, the remainder blackish and clothed with shorter blackish pubescence ; head and pronotum blackish, the basal margin of the latter testaceous ; last joint of the male antennal club much longer than the preceding. Length 3.2 mm. ; width 1.5 mm. Arizona....**horni** *Jayne*

Elytra piceous to testaceous in color and almost uniform throughout, the pubescence dense and less variegated, a condensed oblique spot near basal third generally more or less distinct...4

4—Subbasal spot of condensed cinereous pubescence posteriorly angulate at inner third or fourth of the width ; body stouter ; sides of the prothorax strongly convergent and distinctly arcuate. Length 2.7–3.4 mm. ; width 1.55–1.75 mm. Arizona.

byturoides, sp. n. (Cr. MS)

Subbasal spot straight and oblique, frequently suffused and indistinct; body narrower and much smaller in size, the prothorax less narrowed at apex, the sides very broadly and feebly arcuate from base to apex ; last joint of the male antenna three times as long as wide and distinctly shorter than the two preceding combined. Length 2.4–2.65 mm. ; width 1.15–1.35 mm. Utah (southwestern)—Mr. Weidt.

uteana, sp. nov.

5—Body narrow and elongate-oval, convex, piceous-black above and beneath, the legs testaceous ; pubescence very dense, rather short, subdecumbent, the longer semi-erect hairs not conspicuous, uniform, brownish-cinereous on the pronotum and pale areas of the elytra, of which there is, on each, a large transverse basal spot, an oblique fascia between basal third and fourth, separated from the spot by a short transverse darker interval, a narrow and irregularly sinuous transverse band near apical third, and a straight transverse fascia very near the apex prolonged to the apical angles along the suture ; male antennal club extremely long, the last joint nearly as long as the two preceding combined and as long as the width of the head. Length 3.25 mm. ; width 1.65 mm. Arizona (Riverside)— Mr. Wickham ....**picta,** sp. nov.

6—Body much broader, oblong-oval, more sparsely pubescent, the sub-erect hairs longer, abundant and conspicuous ; elytra piceous, variegated with paler and with three transverse fasciæ of pale hairs, the second usually divided into two spots on each elytron, and the third broadly interrupted at the suture, also with a spot of paler pubescence at each side of the scutellum. Length 3.4 mm. ; width 1.8 mm. Arizona ....**varicolor** *Jayne*

Perplexa of Jayne, I have not seen, but it is evidently allied to *byturoides,* differing in the relatively shorter last joint of the antennal club. *Byturoides* was considered by Dr. Jayne as the female of *horni,* but this is not the case, as I have both male and female of that species as well as the allied *uteana.*

Novelsis differs from *Lanorus* in antennal and hypomeral structure, and from *Telopes* in the structure and armature of the legs in addition.

Dearthrus *Lec.*

This genus is allied to *Attagenus* but differs in having the meso-sternum completely divided by a narrow shallow sulcus, in the 9-jointed antennæ and in the shorter, less inwardly postero-extended metacoxal lamina. The single species may be defined as follows from the male :—

Narrowly oblong-oval, moderately convex, piceous-black in color ; prothorax twice as wide as long, strongly narrowed from base to apex, with the sides evenly and feebly arcuate, the base broadly and feebly lobed, feebly oblique and sinuate laterally, the surface rather strongly but not densely punctate, with a fine exca-vated median line not attaining base or apex ; elytra three-fourths longer than wide, rather strongly but not very closely punctured ; under surface black, the legs rufo-piceous ; pubescence throughout dark in color, uniform, short and not conspicuous ; antennæ as figured under *Attagenus.* Length 2.4 mm.; width 1.15 mm. Indiana...**longulus** *Lec.*

Apparently rare ; I have before me only a single specimen in rather poor state of preservation.

Perimegatoma *Horn.*

In this exclusively western genus, which belongs to an important section of the Attagenini differing from those above considered in the elongate basal joint of the tarsi, the antennal club is 3-jointed, with its two basal joints transverse and the last elongate, though to a less degree than in *Attagenus.* The prosternum is strongly deflexed at tip to form a protection to the mouth in repose, as in most of the other genera of the family, the process between the coxæ moderately narrow, the mesosternum narrow and divided throughout by a relatively wide parallel sulcus. The hypomera are moderately and indefinitely concave, and the metacoxal lamina short. *Belfragei*, which is assigned to the genus by Jayne, undoubtedly forms the type of a distinct genus because of the 5-jointed antennal club ; it is therefore not considered in the following table, which comprises all the species known to me : —

Last joint of the male antennal club short, scarcely one-half longer than the two preceding combined.. 2

Last joint much longer, nearly twice as long as the two preceding ; body narrower...9

2—Last joint conical, pointed at apex..3

Last joint ovo-conoidal, rounded at apex..8

3—Body in great part black or piceous-black in color.......................... 4

Body wholly rufo-ferruginous, stout.......... 7

4—Pubescence rather persistent ; zig-zag testaceous bands at basal third and apical fourth very narrow and frequently indistinct.....................................5

Pubescence readily denuded, the rufous bands very wide, the anterior broadly interrupted at the suture............... 6

5—Vestiture rather fine, largely black, the suberect bristle-like hairs rather inconspicuous ; elytral punctures close-set ; body moderately stout. Length 3.5–4.6 mm.; width 1.65–2.1 mm. California (San Francisco to Calaveras).
                                                                        jaynei, sp. nov.

Vestiture much coarser, the sub-erect bristles conspicuous, the hairs sparser, largely fulvous and whitish, the darker much less numerous ; body less stout and more elongate, the elytral punctures sparse. Length 4.0 mm.; width 1.9 mm. Guadalupe Island......... ....**guadalupensis**, sp. nov.

6—Black areas of the elytra clothed with nearly uniform short blackish pubescence, the rufous bands with sparse uniform fulvous hairs; body broad, feebly convex, oblong, the elytral punctures rather fine and sparse. Length ♂ 3.9, ♀ 6.0 mm.; width ♂ 1.7, ♀ 2.8 mm. California....................................**ampla** sp. nov.

7—Oblong-oval, convex, the vestiture short but abundant, much variegated, in great part fulvous and white, the suberect black bristles distinct, the white hairs generally forming a distinct cluster at basal and inner third and three detached spots at apical fourth in the zig-zag paler band. Length 4.7 mm.; width 2.2 mm. California ....**variegata** *Horn*

8—Body rather narrowly oblong-oval, moderately convex, black, the elytra rufous along the lateral edges and rufo-piceous in two narrow obscurely evident bands at the usual positions, the vestiture persistent, nearly as in *jaynei*, but with the whitish hairs more abundantly interspersed. Length 3.4 mm.; width 1.5 mm. Nevada (Reno)....................................**nevadica**, sp. nov.

9—Body black, the elytra with the usual two rufous bands clothed with paler, denser and more persistent pubescence, the latter elsewhere readily denuded, the anterior pale areas more impressed than usual ; punctures of the elytra fine but deep, perforate as usual and somewhat sparse ; prothorax about twice as wide as long in the male. Length 3.6 mm.; width 1.5 mm. Utah (southwestern)—Mr. Weidt...**impressa**, sp. nov.

Body black, more depressed, the elytra more strongly and closely punctured, without distinct rufous areas, almost evenly clothed with subdecumbent fulvous pubescence, with very narrow and scarcely noticeable zig-zag bands of more cinereous hairs in the usual positions; prothorax of the male more transverse, more than twice as wide as long; under surface black, the legs and antennæ piceous-black. Length 3.65 mm.; width 1.5 mm. Wyoming (Laramie)...**monticola**, sp. nov.

Cylindrica of Kirby (Saskatchewan), and *angularis* Mann., (Alaska), are not known to me at present, the former is said to be distinguished by its uniform elytral vestiture and was assigned by Kirby to *Attagenus;* it was considered to be the same as *piceus* by Gemminger and Harold, but is probably different, as it is said by the author to rese mble a *Cryptophagus*. The *Attagenus angularis* of Mannerheim, seems by the description to be uniformly pubescent, except toward the hind angles of the prothorax, where the hairs become whitish and condensed ; it cannot be the same as *jaynei*, of the above list, which latter was considered to be *cylindrica*, var. C, by Horn. The *falsa* of Horn, is evidently a rare and local species, entirely unknown to me, having the male antennal club slightly longer than the funicle, with its first joint " extremely short "—language which will not apply to any other species known to me—and the last joint more than twice as long as the two preceding together and pointed at tip ; it occurs at and near Sta. Barbara, California.

The pronotum throughout the genus is coarsely and very closely punctured, and there are generally two small and very shallow subbasal fovea at outer fourth, in which the punctures become still more crowded and coalescent. The species are difficult to identify, as there is a strong mutual resemblance throughout. *Ampla*, however, is a very striking species, differing enormously in the relative size of the sexes ; the females are the largest by far of the entire genus. Generally the divergence of the sexes in this respect is not quite so noticeable as in *Attagenus*, although the paucity of material before me will not allow of definite statement in this regard.

Trogoderma *Latr.*

In this genus the body is oblong-oval, less elongate than in *Perimegatoma* but almost similarly clothed with variegated pubescence. The species described by Dr. Jayne under the name *Trogoderma simplex*, seems to have a somewhat unusual construction of the side pieces of the prosternum, and it should therefore form the type of a distinct genus ; it is unknown to me.

The antennæ are of a different type of structure from that prevailing elsewhere in the tribe, the club being 6- to 8-jointed and
generally loose and serriform in the males, and 4-jointed and regular
in the females. The prosternum is not so strongly deflexed at apex
as in *Perimegatoma*, and the process between the coxæ is wider, the
mesosternum between the coxæ very much wider, transverse and
divided throughout by a broad deep sulcus.

Dr. Jayne was mistaken in his diagnosis of the species of the
sternalis group in two important particulars. The mesosternum is as
completely and widely divided by the median sulcus as in the others,
but the metasternal process is rather more arcuate, and the broad flat
marginal bead usually extends along the apex throughout the width;
this misled the author in determining the true anterior limit of the
metasternum. The author also failed to observe the true structure of
the male antennæ, the very minute third joint giving rise to the appearance of a 10-jointed condition, which is alluded to as a general
fusion of the tenth and eleventh joints in the male (Proc. Am. Phil.
Soc., XX, p. 363).

The species are quite numerous and those before me may be thus
briefly characterized :—

Eyes entire, the inner frontal margin not sinuate; antennæ serrate in the male 2
Eyes sinuato-emarginate at about the middle of their inner frontal edge; male antennæ
    compact, not serrate, the third and fourth joints subequal and transverse; prono-
    tum minutely, sparsely punctate, becoming strongly and more densely so toward
    the sides................................ 14
2—Male antennæ with the third and fourth joints equal in size............................3
Male antennæ having the third joint minute and very much smaller than the fourth..10
3—Body more elongate in form, the elytra nearly one-half longer than wide..........4
Body stout and broadly oblong-oval, the elytra one-fourth longer than wide or even
    less.............. 9
4—Submedian testaceous band of the elytra crossing the suture at the middle of the
    length; species small and inhabiting the Eastern and Gulf States...................5
Submedian testaceous band crossing well behind the middle of the length; species
    much larger and inhabiting the Pacific States..................................... 8
5—Pronotum strongly and rather closely punctate, especially toward the sides; pubes-
    cence persistent.. 6
Pronotum very minutely and sparsely punctate throughout, the pubescence readily de-
    nuded... 7
6—Elytra black, with the usual pattern of fine irregular rufescent bands clothed with
    paler hairs; vestiture of the pronotum much variegated. Length 3.0 mm.; width
    1.65 mm. Iowa (Keokuk); [*pusilla* Lec.]...!.......................**orŋata** *Say*
Elytra and pronotum almost similarly colored, and with the variegated pubescence

nearly similar but finer, the subapical irregular band emitting a fine spur anteriorly at inner two-fifths ; body narrower ; club of the male antennæ beginning with the fourth joint.   Length 2.9 mm.; width 1.45 mm.   Texas..**serriger**, sp. nov.

7—Body nearly similar in ornamentation and color to the preceding, the basal lobe of the pronotum not so distinctly marked with white pubescence ; serrate antennal club of the male beginning with the sixth joint.   Length 1.6-2.7 mm.; width 0.8-1.3 mm.   Massachusetts, New York (Long Island) and Virginia (Norfolk).

tarsalis *Mels.*

8—Body large, elongate-oval, black, the elytra with irregular anastomosing bands of testaceous nearly as in the preceding, the pale vesture of the rufous areas rather long and fulvous, that of the black areas short, dark and inconspicuous ; pronotal punctures fine and sparse, those of the elytra coarser but sparse ; legs ferruginous throughout.   Length 4.0 mm.; width 2.0 mm.   California.

pollens, sp. nov.

9—Epipleuræ of the elytra flat ; body stout, black, the elytra variegated with paler areas which are clothed with paler pubescence, nearly as in the preceding species ; pronotum in the female not quite twice as wide as long, rather strongly but sparsely and evenly punctured throughout, the sides moderately convergent.   Length 2.7 mm.; width 1.6 mm.   Texas. ♀ ....**complex**, sp. nov.

Epipleuræ deeply concave ; body in coloration and sculpture nearly as in *complex*, the pale areas of the elytra larger and more suffused and the variegated vesture shorter ; pronotum in the female much more transverse, more than twice as wide as long, the sides very strongly converging from base to apex : legs pale, the femora black.   Length 3.2 mm.; width 1.78 mm.   California (Shasta Co.)

variipes, sp. nov.

10—Pronotum minutely punctate, the punctures simple and perforate.   Pacific Coast.............. 11

Pronotum strongly and closely punctate, the punctures simple and perforate.   Atlantic and Sonoran... 12

Pronotum strongly but sparsely punctate, the punctures rugose.   Sonoran............13

11—Body black, the elytra with broken transverse pa'er bands clothed with the usual paler pubescence, the sutural portions of the submedian band far in advance of, and detached from, the lateral portion ; prothorax of the male with the median lobe of the base rather broadly rounded, the sides evenly convergent and broadly, almost evenly arcuate from base to apex.   Length 2.4 mm.; width 1.2 mm.   California...**sternalis** *Jayne*

Body as in the preceding, the submedian band of the elytra finer, almost continuous, the sutural crossing but little more advanced than the lateral part : basal lobe of the prothorax smaller, more narrowly rounded and more abruptly formed, the sides strongly rounded basally, becoming thence much more strongly convergent and almost straight to the apex in the male, the base somewhat wider than the base of the elytra ; size very small.   Length 1.8 mm.; width 1.0 mm.   California (Los Angeles.)....................................... ....**simulans**, sp. nov.

12—Body black, with variegated white and fulvous bands nearly as in *sternalis*, the sutural part of the submedian band far in advance of the lateral angulation and detached from it ; prothorax at base equal in width to the elytra, very strongly transverse, in the male distinctly more than twice as wide as long, the sides very strongly convergent toward apex, more rounded toward base.   Length 2.65 mm.; width 1.4 mm.   Virginia [Fort Monroe]..................**virginica**, sp. nov. Body black and with variegated pubescence nearly as in the preceding species,the submedian band of the elytra almost continuous, transverse, the sutural part not much in advance of the lateral, forming a broad even arc in more than inner half of each elytron ; prothorax of the male much less transverse, scarcely twice as wide as long, the sides less convergent and more even in curvature ; size much smaller.   Length 2.2 mm.; width 1.18 mm.   Texas [El Paso].

oblongula, sp. nov.

13—Body black, with variegated white and fulvous elytral bands nearly as in *virginica*, except a distinct sutural rhombus included within the subapical band, which is wanting in that species, the submedian band much broken ; prothorax at base not quite as wide as the elytra in the female, the punctures deep, well separated and strongly annulo-rugose, much less than twice as wide as long ; elytral punctures rather strong but twice as sparse as in *virginica*, the pubescence very much sparser than in that species or *oblongula*.   Length 2.5 mm.; width 1.3 mm.   Arizona.

aspericollis, sp. nov.

14—Form very short and broad, oblong, the elytra in both sexes scarcely a fourth longer than wide..**15** Form narrower and more elongate in both sexes, the pattern of elytral ornamentation obsolete or partially so..**16**

15—Elytra black, with narrow anastomosing paler bands nearly as in *ornata*, which are clothed sparsely with whitish hairs, the subapical transverse band enclosing a transverse rhombus on the suture ; elytral punctures sparse and rather fine ; prothorax of the female twice as wide as long, the sides evenly and moderately arcuate.   Length 2.9 mm.; width 1.7 mm.   Pennsylvania ; [*pallipes* Zieg.].

inclusa Lec.

Elytra as in the preceding, with the pale anastomosing markings broader and clothed in great part with fulvous pubescence, the punctures somewhat stronger and slightly less sparse, the subapical band not forming a distinct sutural rhombus ; hairs of the pronotum sparse, suberect and black, becoming paler laterally toward base ; prothorax of the male more than twice as wide as long.   Length 2.3 mm.; width 1.35 mm.   California (San Francisco)..............**brevis**, sp. nov.

16—Elytra parallel and feebly arcuate at the sides, rounded and narrowed only at the apex, black, with a narrow testaceous bisinuate band clothed with paler pubescence near the base, and a few small spots of pale pubescence posteriorly, notably one on each at the suture at the middle, and at the side slightly behind the middle of the length, and one at the middle of the width at apical fourth. Length 2.4 mm.; width 1.28 mm.   Indiana?—Cab. Levette.

obsolescens, sp. nov.

Elytra narrowed slightly from the rather pronounced humeral swelling to the rounded apex ; body pale testaceous throughout, the head and pronotum slightly piceous ; pubescence sparse and not at all varied, pale in color ; surface of the elytra rugose, sparsely punctate.   Length 1.9 mm.; width 0.9 mm.   Arizona.

advena, sp. nov.

Unlike nearly all the other genera of Dermestidæ, the present seems to be very rare in individuals, and it is seldom that more than a single one it taken at any one time ; most of the species, which appear however to be abundantly distinct among themselves, are therefore represented at present by unique types. *Perimegatoma* resembles it in this respect to some extent. The pale coloration of *advena* may be due to immaturity, at least partially. In the adjoining diagram the antenna of *advena*, which is representative of that entire section of the genus, is drawn in a contracted state, but the insect has the power to separate the joints slightly, when they are seen to be deeply concave at their apices ;

FIG. 4.

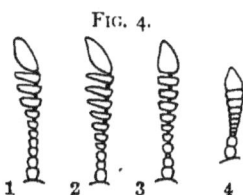

1 Antenna of *Trogoderma tarsalis* ♂ , 2 same *T. serriger*. 3 same of *T. oblongula*. 4 same of *T. advena*.

they are mutually attached by short stipes or pedicels as in the others, but differ in being virtually symmetrical and not eccentric. These antennal differences, although marked, are not indicative of subgeneric groups, as the general structure of the under surface, and particularly of the hypomera, is indentical throughout.

Cryptorhopalum *Guér.*

The body in this genus, which is the most extensive of the American Dermestidæ, becomes more oval and compact than in any other of the present tribe, but in anatomical structure it is evidently homologous with *Trogoderma*. The species are small to quite minute in size, of sober color and generally uniformly clothed with short dark pubescence, which, in some forms, becomes slightly variegated as in most of the other genera. The species before me are the following :—

Pubescence of the elytra variegated............................ 2
Pubescence uniform throughout... 9
2—Elytra with irregular or interrupted transverse bands of dense paler pubescence...3
Elytra without transversely fasciate pubescence, but with a spot of dense pale and
    coarser hairs near the apex of each ; last ventral of the female unmodified.......8
3—Last ventral segment of the female with two small, widely separated and rounded
    discal erosions ; elytra not paler posteriorly..4
Last ventral of female with two small, rounded, flat and entirely unexcavated scar-
    like spots ; elytra paler in apical third..7
4—Pubescence of the pronotum dusky, sparse and inconspicuous but becoming pale
    and conspicuous toward the sides and on the basal lobe..............................5
Pubescence of the pronotum uniform or nearly so, coarse, denser, pale and conspicu-
    ous throughout..6

5—Pale pubescent bands of the elytra at basal and apical third entire, the anterior irregular, the posterior narrowly interrupted at the middle ; body narrowly oval, more or less pale piceo-testaceous in color, finely punctate, the pronotum minutely and sparsely ; tarsi very slender, the posterior as long as the tibiæ in the female and distinctly longer in the male ; antennal club of the latter stout, not twice as long as wide, the second joint slightly shorter than the first.   Length 1.9–2.25 mm.; width 1.1–1.3 mm.   California (Sta. Barbara)...........**filitarse**, sp. nov.

Pale pubescent bands subentire but composed of short, sparse hairs and mutually separated by a distance equal to that of the anterior band from the base ; body castaneous, sparsely punctured ; legs testaceous, the femora picescent ; posterior tarsi slightly shorter than the tibiæ ; antennal club of the female rather small, stout, one-half longer than wide, with the second joint distinctly longer than the first—a reversal of the general rule.   Length 2.15 mm.; width 1.3 mm. New Mexico (Fort Wingate—Dr. Shufeldt).................**reversum**, sp. nov.

Pale pubescent bands broken up into small sparse spots, a spot also behind the humeri and another near the apical angle of each elytron ; body much larger, elongate-oval, darker in color, castaneous, the punctures a little coarser and rather more close-set than in *filitarse ;* hind tarsi nearly similar.   Length 2.5–2.65 mm.; width 1.4–1.55 mm.   Arizona (Cañon of the Colorado River)—Dr. T. Mitchell Prudden...**pruddeni**, sp. nov.

6—Elytra feebly narrowed posteriorly from the humeral callus, the pale pubescent bands cinereous and almost entire, separated mutually by a distance which is equal to that of the anterior band from the basal margin : apical spot of pale pubescence concolorous or nearly so, the spots and bands rather poorly defined, and with the pubescence largely cinereous toward base throughout the width, joining the first band at the suture ; hind tarsi quite distinctly shorter than the tibiæ in the female.   Length 2.65 mm.; width 1.6 mm.   Arizona.......**balteatum** *Lec.*

Elytra rapidly narrowed behind from the humeral callus, the apex more narrowly rounded, body smaller, convex, and relatively stouter, cas'aneous in color, the bands of coarser pale yellowish-cinereous pubescence narrower, subentire and better defined, the two mutually much more distant than the first from the base, the apical spot fulvous in color ; basal regions with a large proportion of pa'e hairs ; hind tarsi very slightly in the male, distinctly in the female, shorter than the tibiæ ; male antennal club stout, not twice as long as wide, the second joint a little shorter than the first, the cavities extending to basal third.   Length 1.9–2.25 mm.; width 1.2–1.4 mm.   Texas (Brownsville)—Mr. Wickham.

festivum, sp. nov.

7—Body oval, blackish, the elytra rufous in apical third or more, with subhumeral an-
• nulus and two transverse bands of short, rather sparse pale hairs ; pronotum with pale hairs toward the sides and basal lobe ; joints of male antennal club sub-equal, the second but slightly shorter than the first, nearly similar, but a little smaller, in the female ; tarsi slender and elongate.   Length 1.8–2.5 mm.; width 1.2–1.5 mm.   Missouri, Kansas and Texas..................**hemorrhoidale** *Lec.*

8—Rather broadly suboblong-oval, black, the elytra gradually and suffusedly rufescent toward tip, the pubescence short, dark, sparse and inconspicuous, becoming pale and distinct, though sparse, toward the sides and basal lobe of the pronotum and toward the sides, and more densely, near the apices, of the elytra ; pronotal lobe

rather broadly, rectilinearly truncate ; legs testaceous, the femora blackish, except
toward tip, the hind tarsi shorter than the tibiæ; male antennal club extending
three-fifths of the thoracic length, with the second joint three-fifths as long as the
first, in the female smaller, with the second joint slightly shorter than the first
Length 2.0–2.8 mm.; width 1.23–1.8 mm.   Oregon, California (Humboldt to
San Diego) and Nevada (Reno)..**apicale** *Mann.*

9 – Body broadly oval, the thoracic lobe broadly truncate ; joints of the antennal club
   very unequal...10

Body more or less narrowly oval, the thoracic lobe much narrower.......................11

10—Body deep black throughout, the elytral punctures sparse and coarse, the pubescence sparse, fine, blackish in color, uniformly distributed and very inconspicuous ; antennal club of the male slender, two and one-half times as long as wide,
   extending to basal third, its second joint relatively very short, much less than
   half as long as the first, the latter twice as long as wide, of the female much
   smaller, extending to the middle, the second joint much shorter than the first.
   Length 2.1–2.8 mm.; width 1.4–1.8 mm.   Arizona...**dorcatomoides**, sp. nov.

Body piceous-brown in color, the elytra coarsely and less sparsely punctured, the
   pubescence uniform, more abundant, short, coarse, fulvo-cinereous in color and
   distinct ; male club not extending quite to basal third, the second joint more than
   half as long as the first, the latter not twice as long as wide.   Length 2.5–2.7
   mm.; width 1.6–1.7 mm.   Texas (Austin)....................**obesulum**, sp. nov.

11—Thoracic punctures sparse, at least toward the middle......... 12

Thoracic punctures rather close-set throughout,... 16

12—Elytra coarsely, though rather sparsely, punctate.   Sonoran and Pacific re-
   gions,.. 13

Elytra very finely and rather less sparsely punctate.   Atlantic regions.. 15

13—Pubescence of the elytra longer, coarse, yellowish-cinereous and distinct ; body
   very small, somewhat narrowly oblong-oval, black or piceous-black ; male an
   tennal club extending beyond basal third, elongate-oval in form, relatively large,
   more than twice as long as wide, the second joint three-fifths as long as the first,
   the latter much longer than all the preceding portion together, the funicle very
   short, not as long as the two globular basal joints combined.   Length 1.65 mm.;
   width 1.0 mm.   Arizona...**granum**, sp. nov.

Pubescence short, fine, dark in color and less conspicuous.................. 14

14—Body deep black in color, the elytral pubescence blackish and not at all fulvous ;
   joints of the antennal club in the female less unequal, the second four-fifths as
   long as the first.   Length 2.3 mm.; width 1.3 mm.   Arizona...**anthrax**, sp. nov.

Body piceous-black, polished, sparsely punctured and unusually sparsely pubescent,
   the hairs fulvo-piceous in color and more distinct ; joints of the antennal club in
   the female very unequal, the second about two-thirds as long as the first, the latter
   longer than the entire funicle ; legs ferruginous.   Length 2.0–2.6 mm.; width
   1.25–1.6 mm.   California (Lake and Sonoma Cos.). ....**affine**, sp. nov.

15—Narrowly oblong-oval, black or piceous-black, shining, the pubescence very short,
   dark in color and inconspicuous ; antennal club pale as usual, large and evenly
   oval, in the male not twice as long as wide, the second joint very much shorter
   and narrower than the first.   Length 1.73 mm.; width 1.0 mm.   Georgia.
                                                                ruficorne *Lec·*

16—Pubescence coarse, pale, ashy-cinereous and distinct, rather sparse but denser toward the sides of the prothorax ; elytra coarsely, rather sparsely punctured ; male antennal club more than twice as long as wide, the joints very unequal, the second scarcely more than one-half as long as the first but only a little narrower, the first as long as the entire preceding parts, the funicle fully as long as the two basal joints combined ; club of the female much smaller but with the joints unequal.   Length 1.65-2.2 mm.; width 0.9-1.28 mm.   Arizona (Benson).

fusculum *Lec.*

Pubescence dark fulvo- or piceo-cinereous and less distinct...............................17
17—Pronotal punctures moderately close-set but very fine, not dense and very inconspicuous.   Sonoran regions.. 18
Pronotal punctures small but strong, dense and very distinct.   California coast regions... 19
18—Pubescence fulvo-cinereous, coarser and distinct, moderately dense ; body oval, black or piceous-black, less elongate ; sides of the male pronotum strongly convergent and almost evenly arcuate throughout ; antennal club rather dark brownish-ferruginous in color, narrowly oval and two and one-half times as long as wide in the male, with the second joint three-fourths as long as the first ; hind tarsi distinctly shorter than the tibiæ.   Length 1.9-2.25 mm.; width 1.1-1.3 mm.   Texas (Brownsville)....................... ....**modestum**, sp. nov.
Pubescence finer, piceous and much less distinct ; body black, narrower, more parallel ; sides of the pronotum in the male strongly convergent anteriorly, subangularly rounded behind the middle and thence parallel and straight to the base, the edges more widely subexplanate ; antennal club black or blackish, nearly similar in form in the male but with the joints less unequal, the second four-fifths as long as the first.   Length 1.7-1.8 mm.; width 1.05-1.15 mm.   Texas and Utah (southwestern)..**fusciclave,** sp. nov.
Pubescence blackish, nearly as in the preceding but much denser ; body black, stouter, the elytra coarsely and unusually closely punctured ; legs ferruginous, the femora piceous ; antennal club pale rufo-testaceous the joints only slightly unequal in the female ; hind tarsi about as long as the tibiæ in the latter sex.   Length 2.0 mm.; width 1.25 mm.   Arizona......................................**pumilum,** sp. nov.
19—Body narrowly oblong-oval, black, the elytra more closely punctured, the pubescence blackish, fine, rather dense but short and very inconspicuous ; antennal club black or blackish, extending to basal fourth or fifth in the male, elongate-oval, with the second joint nearly four-fifths as long as the first ; legs piceous, the hind tarsi slightly shorter than the tibiæ, notably so in the female.   Length 2.4 mm.; width 1.3 mm.   California (Humboldt to Sta. Barbara); [*nigricorne* Lec.].. ....**triste** *Lec.*

There are a few other apparent species indicated by inadequate or poorly preserved material, and the genus is evidently a large one.   In striking contrast to *Trogoderma*, individuals are abundant when discovered, and most of the species are represented by good series.   The species *fusculum* of LeConte, which is entirely valid, is said by Dr. Jayne to inhabit the Atlantic regions ; it is however Sonoran, and was not correctly identified, and *triste* is not an Atlantic, but a Pacific, species. One female of *apicale* in my cabinet has the two joints of the antennal club equal in length : as it is not in very good condition, I cannot state whether it differs specifically. The remarks made by Dr. Jayne in regard to the female of *balteatum* are erroneous, as the antennal cavity is normal in form. The same author gives "California" as the locality of *ruficorne*, whereas it is confined in reality to the southern Atlantic States. *Picicorne*, described by LeConte from the southern Atlantic regions, is unknown to me, but is probably a valid species.

ANTHRENINI.

The distinguishing characters of this tribe are the compact body, very retractile legs and the deep and acutely defined fossæ for the antennal club. The tarsi are short and rather slender, the basal joint of the posterior distinctly shorter than the second, the next three subequal or progressively decreasing slightly in length. The mouth parts are completely protected in repose by the deflexed prosternum. The antennæ vary in the number of joints, but these divergencies do not indicate more than subgenera, as the structure otherwise is quite homogeneous. There is but one genus :—

Anthrenus *Geoff.*

The eyes may be sinuato-emarginate within or entire as in *Trogoderma*, and are finely faceted as usual. The prosternal process is rather narrow, impinging upon the transverse, deeply sulcate mesosternum, also as in that genus. The species are moderately numerous, and number among them some of the most destructive enemies of dried insects preserved in cabinets ; those before me may be easily identified as follows :—

Eyes emarginate ; antennæ 11-jointed, the club broadly oboviform and composed of three closely connate joints of which the two basal are strongly transverse and much shorter than the last; body clothed with broad scales. [**Anthrenus** in sp.]..2
Eyes entire...6
2—Pronotum having a large well-defined lateral spot of pale scales not inclosing a darker spot...3
Pronotum with a large pale lateral spot, as above, but inclosing a small darker spot at nearly its central point. Pacific Coast...4
Pronotum clothed throughout with a mixture of white and brown scales. Sonoran regions..5

3—Elytra having the suture clothed throughout with whitish or rufescent scales, the vitta dilated laterally near base and apex and at the middle, also with a transverse area of pale scales just behind the middle and seldom attaining the sutural area, a subbasal and subapical marginal pale area and a basal ring at each side of the scutellum. Length 2.7–3.6 mm.; width 1.75–2.25 mm. New Jersey and Europe ....scrophulariæ *Linn.*

Elytra similar but with a large uniform area of white scales extending from fourth to three-fifths from the base, and from the margin to inner third or fourth. Length 2.3–3.4 mm.; width 1.5–2.25 mm. Texas................ ....thoracicus *Melsh.*

4—Elytra clothed with black scales, with clearly limited areas clothed with whitish scales nearly as in *scrophulariæ*, but with the sutural vitta generally interrupted at apical third and the transverse marginal spot behind the middle rarely extending beyond the median line, the oblique marginal fascia at basal third or fourth sometimes enlarged internally and forming with the basal sutural white regions a large irregular white spot covering a third of the entire area ; pale scales of the elytra always white. Length 2.3–3.0 mm.; width 1.5–2.0 mm. California (Sta. Cruz and Lake Cos.)........................... ....occidens, sp. nov.

Var. A—Similar to *occidens* but more narrowly oblong-oval, the scales of the subbasal sutural area yellow and not white ; enclosed black spot within the lateral pale area of the pronotum very near the inner edge of the latter. Length 2.8 mm.; width 1.85 mm. Nevado (Reno).....nevadicus, v. nov.

Var. B—Similar to *occidens*, except that the large subbasal area on the suture is clothed with dark fulvo-ferruginous scales, and the enclosed thoracic spots are composed of fulvous, and not black, scales, the formation nearly as in *lepidus* and its varieties. Length 2 5 mm.; width 1.7 mm. California ....pictus, v. nov.

Elytra variegated nearly as in the preceding but with a sprinkling of brown scales ; enclosed dark spot within the lateral white areas of the pronotum never black as in *occidens* but clothed with fulvous-brown scales ; body smaller and less dilated. Length 2.25–2.5 mm.; width 1.6–1.7 mm. California (San Diego)

lepidus *Lec.*

Var. A—Body similar in form to the preceding, the pronotum less transverse, densely clothed throughout above with ochreo-fulvous scales, replacing the black scales of *occidens;* black scales wholly wanting at any part. Length 2.7 mm.; width 1.75 mm. California...............obtectus, v. nov.

Var. B—Similar to *lepidus* but with the scales of the paler areas more suffused and dispersed, the body more broadly oval, the prothorax larger, with the sides less convergent ; antennæ longer, the club broader. Length 2.4–2.7 width 1.6–1.8 mm. California (Lake Co.)...................suffusus, v. nov.

Var. C--Similar to *lepidus* but smaller and still narrower, the scales of the elytra black and fulvous, confusedly intermingled, with some feeble whitish sutural and external areas remindful of *lepidus*. Length 2.15 mm.; width 1.4 mm. California ;(San Diego)—Mr. Dunn........conspersus, v. nov.

5—Broadly and evenly elliptical, convex, blackish-piceous, the legs paler ; antennæ moderate, ferruginous throughout; upper surface clothed with relatively very large white and brown scales, confusedly mottled on the pronotum and elytra, but with the white scales forming two tolerably distinct suboblique fasciæ on the

latter behind the middle; on the under surface white throughout.   Length 1.8–1.95 mm. ; width 1.15–1.25 mm.   Texas (El Paso).............**parvus**, sp. nov.
6—antennæ 11-jointed, the club subparallel, consisting of three connate joints, the two basal slightly transverse; scales elongate.   [**Nathrenus**, sg. nov.]..........7
Antennæ 8-jointed, the club consisting of two closely connected joints.   [**Florilinus**]..8
Antennæ 5-jointed, the club consisting of a single very elongate claviform joint.   [**Helocerus**] 10
7—Oblong-oval, moderately convex, black, clothed with yellow, black and white scales, largely black on the median parts of the pronotum, the basal lobe always with whitish scales, the elytra with a transverse zig-zag pattern of pale scales, largely white bordered with yellow in two fasciæ.   Length 1.8–2.8 mm. ; width 1.25–1.75 mm.   Europe and Eastern United States ; [*varius* Fabr.]
                                                                            verbasci *Linn.*
    Var A—Similar but more narrowly oblong-oval, the yellow scales still narrower, more elongate and more dispersed over the entire surface, the pattern of *verbasci* scarcely traceable and the scales more isolated among themselves.   Length 2.2 mm. ; width 1.3 mm.   Virginia (Norfolk)...............**pistor**, v. nov.
    Var B—Nearly similar to *verbasci* but larger and more broadly oblong, the yellow scales entirely covering the pronotum, the elytral pattern nearly similar but with the yellow scales more dispersed, the white patches similar in position but larger.   Length 2.8–3.0 mm. ; width 1.8–2.15 mm.   Indiana ....**vorax**, v. nov.
    Var C—Almost similar to *vorax* but very much smaller, the yellow scales densely clothing the entire surface, except where replaced by the equally dense white scales in patches similar in position to those of *verbasci ;* form more broadly rounded than in the European *nebulosus ;* scales broader than in *verbasci*.   Length 1.7–2.2 mm. ; width 1.15–1.6 mm.   Iowa (Keokuk).
                                                                            destructor *Melsh.*
    Var D—Similar to *verbasci* but larger and more broadly oblong-oval, the zigzag pattern of the elytra equally well marked but with the post-median fascia more sharply anteriorly angulate near the suture, the surface when denuded showing feebly impressed longitudinal lines.   Length 3.0 mm. ; width 2.2 mm.   Central America..**substriatus**, v. nov.
8—Basal joint of the antennal club subquadrate, the second nearly twice as long as wide; body piceous, rather sparsely clothed with scales which are less decumbent, elongate and with oval cross-section and concave apices, the pale scales less numerous than the darker ones and irregularly disposed, the dark scales apparently flatter and more decumbent ; elytral punctures fine but rather deep, moderately sparse.   Length 2.9 mm. ; width 1.75 mm.   Europe.
                                                                            museorum *Linn.*
Basal joint much smaller, transversely obtrapezoidal, the second relatively shorter and but little longer than wide apparently in both sexes ; scales less elongate, flatter, more decumbent and triangular, the punctures very shallow.   America...........9
9—Antennal funicle moderately stout, the third joint about one-half as thick as the second ; pale scales of the elytra strewn without order toward base but forming two somewhat evident transverse fasciæ behind the middle, the scales all broadly triangular and coarsely strigose ; body castaneous in color, evenly and not very

broadly oval in form and strongly convex, much smaller than *museorum*. Length 2.2 mm.; width 1.3 mm. Pennsylvania and Indiana..**castaneæ** *Melsh.*

Var A—Similar in color, form and size to *castaneæ* but with the scales more narrowly triangular, less coarsely strigose, the paler sparsely dispersed but forming a tolerably evident ring on each at base, very sparse behind the middle, though arranged in two more evident transverse areas nearly as in *castaneæ* but more widely separated. Length 2.25 mm.; width 1.35 mm. North Carolina (Asheville) ....**carolinæ**, v. nov.

Var B—Similar in color to the preceding, the elytra a little paler ; form much more narrowly oblong-oval, the surface more rugulose and alutaceous ; pale scales as in *carolina* but much more abundant than in either of the preceding, scattered without order in basal half but with a large lateral condensation at basal third or fourth, the two transverse fasciæ behind the middle more evident. Length 1.9 mm.; width 1.2 mm. Texas......**angustulus**, v. nov.

Antennæ funicle very slender, the third and fourth joints scarcely a third as wide as the second ; pale scales of the elytra very few in number and sparsely interspersed among the darker ones, more noticeably abundant at basal third or fourth, just behind the middle and near the apex ; body shorter and more broadly rounded than in the preceding, castaneous in color, the tibiæ and tarsi testaceous ; surface shining between the very minute sparse punctules. Length 2.0 mm.; width 1.4 mm. New York.. ......**rotundulus**, sp. nov.

10—Body oval and convex, nearly as in *castaneæ* in color, form and vestiture, the pale s ales of the elytra forming on each a transverse fascia at basal fourth, curving forward internally to the scu ellum, and forming two less evident transverse fasciæ behind the middle. Length 1.9–2.5 mm.; width 1.2–1.4 mm. Europe ; [*claviger* Er.]..................................... ....**fuscus** *Latr.*

A form which I have not seen was described by LeConte, from New York, under the name *flavipes ;* this was supposed by Jayne to be the same as the European *albidus* of Brullé, and may have been founded upon an introduced individual of that species, which in my opinion is distinct from *scrophulariæ*, although inscribed as a variety in the catalogues ; *signatus* and *proteus* appear to be identical and to form a variety of *albidus*, but *senex* may be another distinct species. The two European species *museorum* and *fuscus* are introduced above into the table, although I have never seen any examples taken in this country. Those mentioned by Jayne may have been adventitious importations. From the illustrations given of the antennæ, however, it is probable that Dr. Jayne did not have the true *museorum* before him at all, but mistook the much smaller *castaneæ* for it ; *museorum* might therefore be stricken from the American lists. *Verbasci* and its varieties constitute the chief destructive element of entomological collections in temperate climates, but I have never known of any such habits in *scrophulariæ* or allied species.

TRINODINI.

This tribe includes at present but two very anomalous minute species, differing radically in sternal structure but perfectly homologous otherwise, and inhabiting the palæarctic and nearctic regions respectively. They represent two distinct genera as follows :—

Anterior coxæ narrowly separated, the process feebly carinate, free and received at tip within a deep anterior excavation in the broad mesosternum ; tarsi shorter, the first joint of the posterior but little longer than the second.   Europe
                                                                                    *Trinodes
Anterior coxæ more widely separated, the intercoxal process flat, non-carinate and extending beneath the anterior margin of the still broader mesosternum, which is free, arcuate and feebly deflexed ; prosternum more deflexed at apex, the tarsi longer with the basal joint more elongate.   Eastern America............Apsectus

In both these genera the hypomera are flat, becoming broadly, feebly impressed posteriorly, the antennæ long, with very slender shaft, received in repose within a narrow groove beneath the eyes, extending posteriorly for a short distance along the suture separating the prosternum from its hypomera, the club 3-jointed, with the two basal joints small, the third large and oblong-oval. The legs are slender and free, the posterior retractile, the hind coxal plate very short, but little longer internally and extending only to the wide parapleuræ, which are in a single piece.

Apsectus Lec.

The single species seems to be rare, though rather widely distributed ; its general characters are as follows :—

Oval, convex, piceous-black, polished, sparsely clothed with long erect and bristling fulvo-piceous hairs, each of which completely fills at base a very minute punctule, which, consequently, only become distinctly visible on the removal of the bristles ; prothorax transverse, closely fitted to the elytra and lobed as usual at base, the lateral edges entirely devoid of acute margin anteriorly but with an acute and narrowly reflexed margin in basal two-thirds, the surface with a fine deep groove in outer third, closely paralleling the basal margin ; scutellum large, flat, equilatero-triangular ; elytra wider than the prothorax and more than three times as long, evenly and conjointly rounded behind ; under surface sparsely and very minutely punctulate, the pubescence shorter ; legs and antennæ testaceous throughout, the large terminal joint of the latter blackish.   Length 1.5 mm.; width 0.9 mm.   Texas (Austin)............................hispidus Melsh.

The ocellus is unusually small and feeble in *Apsectus* but is much more distinct in *Trinodes*. I have seen specimens, either of *hispidus* or a species closely allied, collected by Mr. Schwarz in Arizona, but probably in the higher regions.

ORPHILINI.

This tribe is quite as anomalous as the Trinodini, and differs from any other in having the metacoxal plate well developed, almost equal in length throughout the width and extending to the sides of the body. The head rests in repose upon the vertical pointed plate form- ing the prosternum between the coxæ, and the body is glabrous. The legs and head are strongly retractile, the mesosternum transverse and even between the coxæ and the epipleuræ well defined. We have a single genus which is also palæarctic in range :—

Orphilus *Er.*

The body is compact, oblong-oval in form, moderately convex, the elytra impressed along the suture except at base and with rather prominent humeral callus, the prothorax at base as wide as the elytra, to which it is closely fitted, the base broadly lobed in the middle. The scutellum is well developed and ogival in form. The tarsi are slender, glabrous, much shorter than the tibiæ and the two basal joints of the posterior are subequal and each rather shorter than the third or fourth, which also are subequal, the fifth about as long as the first three together. The antennæ are 11-jointed, with a broadly oval compact club composed of three transverse free joints, and the eyes are emarginated by the short post-antennal sides of the front. The species are rather closely allied among themselves, and those repre- sented in my cabinet may be distinguished as follows :—

Integuments deep black, without metallic lustre....................... 2
Integuments black, with bright steel-blue reflection..............................4
2—Elytra finely and sparsely punctured throughout, the punctures toward base separ-
    ated by at least twice their own diameters; pronotum finely and sparsely punc-
    tured throughout; integument highly polished.   Length 3.2 mm.; width 1.8
    mm. Arizona (Cañon of the Colorado River)—Dr. Prudden..**æqualis,** sp. nov.
Elytra coarsely punctate toward base, where the punctures are separated by their own
    diameters or less...3
3—Punctures of the elytra toward base smaller, always clearly separated, those of the
    pronotum fine but rather close-set.   Length 2.8–3.5 mm.; width 1.6–2.1 mm.
    California to Colorado..................................... ....**subnitidus** *Lec.*
Punctures of the basal regions coarser and usually densely crowded so as to become
    more or less distorted in form ; pronotal punctures larger and stronger but rela-
    tively scarcely so close-set; body distinctly smaller in size.   Length 2.3–2.8
    mm.; width 1.28–1.7 mm. Lake Superior to Georgia.................**ater** *Erichs.*
4—Nearly similar in form to *ater*, the elytral punctures not so coarse or deep toward
    base and widely isolated among themselves, the pronotal punctures very fine and
    not close-set.   Length 2.8–2.9 mm.; width 1.7 mm. Idaho (Cœur d'Alène).
                                                        chalybeus, sp. nov.

Individuals of the various species appear to be abundant, and the genus, both in number of species and relative abundance, is much better represented in America than in Europe. *Niger* of Rossi, (= *glabratus* Fabr.), is the only European species, and its occurrence in this country has not been confirmed.

CIOIDÆ.

Maphoca, gen. nov.

The genus based upon the following characters may be placed for the present near *Plesiocis*. The body is narrow, parallel and moderately convex. Head well developed, wider than long, only moderately inclined, the eyes slightly behind the middle, remote from the base, moderate or rather small, entire, convex, relatively rather coarsely faceted, the facets individually strongly convex ; front broadly and evenly arcuate from eye to eye, with a small transversely oval inclosed clypeus defined by a very feeble suture, the labrum small, rounded. Antennæ inserted under the sides of the front immediately before the eyes, short, 9-jointed, with a moderately developed loose parallel 2-jointed club, the two basal joints enlarged ; three to six forming a slender shaft ; third as long as the next two combined ; four to six small, moniliform, the seventh transverse and wider. Maxillary palpi well developed, the last joint large, oval, slightly longer than wide, narrowly truncate at tip, the labial very minute ; buccal opening small, the mentum very minute, longer than wide. Antennal grooves before the eyes rather distinct, the buccal processes almost obsolete ; mandibles short and stout, bifid at tip. Prothorax widest toward apex, the disk even throughout and slightly convex ; prosternum long before the coxæ, broadly truncate, the intercoxal process narrow. Elytra completely enclosing the abdomen, striato-punctate. Scutellum small, transversely oval or broadly angulate behind. Abdomen with five perfectly mobile segments, the sutures straight throughout, the first segment unmodified, as long as the next two combined ; two to four decreasing scarcely visibly in length, the fifth scarcely longer than the fourth and rounded. Anterior coxæ small, very deep-set, transverse, the cavities narrowly open behind and angulate externally ; intermediate and posterior narrowly separated, the latter extending nearly to the sides of the body, the met-episterna extremely narrow. Mesosternum even, transversely convex, the metasternum large. Legs rather short, slender, the femora but slightly dilated, the tarsi much shorter than the tibiæ, 4-jointed, the three basal joints small, the first with a brush of long hairs beneath, the fourth long and notably stout, the claws well developed, divaricate, slender, simple and arcuate. Epipleuræ extending almost to the sutural angles but narrow throughout, scarcely at all dilated but horizontal toward base, inflexed behind the middle.

The extremely minute species having the assemblage of characters given above is one of those aberrant forms continually occurring among the serricorn Clavicornia.   It may be described as follows :—

Body narrow and parallel, testaceous, the elytra blackish and the under surface piceous, the legs and antennæ pale ; surface rather shining ; head nearly three-fourths as wide as the prothorax, the antennæ as long as the width of the head ; prothorax about a fourth wider than long, the sides rather prominently rounded at apical fourth, thence feebly convergent and straight or broadly, feebly sinuate nearly to the basal angles, which are somewhat obtuse ; apex broadly arcuate, equal in width to the base, which is even and subtruncate ; disk feebly convex, declivous at the sides, very minutely and feebly margined at base, minutely and rather sparsely punctate, each puncture with an extremely minute hair ; elytra nearly twice as long as wide, scarcely visibly wider than the prothorax, rather obtusely rounded at tip, the sides parallel and almost straight, the humeral angles right and well defined ; disk with even feebly impressed series of small punctures, the intervals each with a series of extremely minute punctures, each of which bears a very short, stiff erect hair. Length 1.05 mm.; width 0.35 mm.   California (Mokelumne Hill, Calveras Co.)—Dr. F. E. Blaisdell.......**blaisdelli**, sp. nov.

No notes concerning the habits of this species have come to me, but probably they do not differ from those of other members of the family.

MELANDRYIDÆ.

TETRATOMINI.

The definition of this tribe must be enlarged to include all those Melandryids, with simple claws, which have the outer three or four antennal joints abruptly dilated to form a strongly developed loose and parallel club.   The genera may be defined as follows :—

Last four joints of the antennæ dilated ; eyes well developed, emarginate anteriorly ; pronotal foveæ distinct.. 2

Last three joints abruptly and strongly dilated, forming a loose club ; eyes emarginate or sinuate anteriorly, generally less developed ; abdominal sutures moderately fine ; edges of the prothorax subeven..4

2—Pronotal margins not reflexed at the sides ; basal segment of the abdomen about as long as the next two combined ; joints of the antennal club pedunculate, the seventh not much dilated..**Tetratoma**

Pronotal margins rather broadly concave and reflexed, the edge unevenly undulato-crenulate 3
3—Basal segment of the abdomen as long as the next two combined, the sutures very coarse. ....**Abstrulia**
Basal segment but little longer than the second, one to five decreasing gradually in length, the sutures rather fine ....**Incolia**
4—Eyes rather well-developed ; pronotum broadly reflexo-explanate at the sides, transversely truncate and scarcely perceptibly bisinuate at base, the foveæ obsolete ; abdominal segments decreasing uniformly and slowly in length, the first scarcely visibly longer than the second.....................................**Eupisenus**
Eyes smaller, very short and strongly transverse ; prothorax not at all explanate at the sides, the base broadly, arcuately lobed at the middle, the foveæ distinct though not very well developed ; first abdominal segment as long as the next two combined ; body much shorter and more convex.................................**Pisenus**

The last two of these genera were mutually confounded by LeConte and Horn, and both considered identical with the European tritomid genus *Triphyllus*. *Pisenus* resembles the latter considerably in form, and the noting of the 4-jointed hind tarsi, antennæ and pronotal foveæ no doubt led the distinguished authors astray; an inspection of the anterior and intermediate tarsi, which are 5-jointed, would have enabled them to avoid the error.

Tetratoma *Fabr.*

This holarctic genus contains several species in the European fauna, and the two following American species seem to be perfectly congeneric, as far as can be judged by the descriptions : —

Elongate-oval, strongly convex, rufo-testaceous, the head and antennæ black, the elytra steel-blue ; body above polished, glabrous, except that each puncture encloses an infinitesimal hair ; antennæ well developed, nearly two-fifths as long as the body, the club as long as the entire preceding portion, the joints quadrate or oblong, the last a little longer and pointed and all pedunculate at base ; prothorax transverse, as wide at base as the base of the elytra, narrowed moderately from base to apex, the latter scarcely at all sinuate, with the angles broadly rounded, the base very broadly and feebly lobed at the middle and finely margined throughout like the sides, the latter broadly and very feebly irregular or subundulate ; basal angles obtuse but not in the least blunt or rounded ; punctures rather coarse and sparse, the basal foveæ distinct, deep and punctiform ; scutellum moderately transverse, cordiform, finely punctured ; elytra three-fifths longer than wide, a little more than three times as long as the prothorax, rather wider behind the middle than at base, thence rapidly, arcuately narrowed to the subogival apex ; humeral callus obtusely prominent ; punctures coarsely impressed and sparse ; under surface more finely but rather sparsely punctate and sensibly pubescent ; basal joint of the hind tarsi about as long as the last. Length 4.7–5.8 mm.; width 2.1–2.65 mm. Northern Atlantic regions.................**truncorum** *Lec.*

Elongate-oval, convex, finely and sparsely pubescent, shining, piceous with a reddish tinge; legs and base of the antennæ yellowish-brown; head more finely punctured; prothorax and elytra equally punctured, the former transverse, narrower in front, rounded at the sides which are narrowly margined, the base margined like the sides and with a large puncture half way between the middle and the basal angle, the latter obtuse and rounded. Length 4.0 mm. Colorado (Veta Pass)..**concolor** *Lec.*

The latter of these species I have not seen, but, from the originally published characters reproduced above, it would seem to be provisionally attachable to the true *Tetratoma*; the principal differences appear to reside in the pubescence and in the rounded basal angles of the prothorax.

Abstrulia, gen. nov.

The species of this genus differ greatly from *truncorum* in general habitus and in the structure of the sides and base of the prothorax. The irregularly crenulate sides of the latter are prominent just before the middle and at basal third or fourth, and the disk is concave along the basal margin, with the foveæ larger, deep and more impressed or less punctiform, the scutellum smaller and more nearly subquadrate, and the elytra are dark in color with a complex maculation of pale spots, the punctures coarse, impressed and sparse. The surface is sparsely but distinctly pubescent, and the basal joint of the hind tarsi is, as a rule, obviously shorter than the last. The species are mutually closely allied, and the three before me may be thus defined from the male :—

Basal joint of the hind tarsi very much shorter than the last.........................2
Basal joint scarcely visibly shorter than the last......................................3
2—Oblong-oval, convex, polished, pale piceous-brown in color, the legs and antennæ concolorous, the elytra blackish with pale flavous and sharply defined intricate markings, the pale areas together somewhat exceeding the dark, and having as a prominent feature a subsutural obverted C-shaped mark on each extending near the suture to apical two-fifths; antennæ scarcely as long as the head and prothorax, the club distinctly shorter than the stem, cylindric, rather compact, the joints wider than long, the last a little longer than wide and conically pointed, the seventh joint transverse and forming a gradual passage to the club, third as long as the next two combined; prothorax short, nearly twice as wide as long, the base and apex equal in width, the latter transverse, narrowly and feebly sinuate at each side, the middle broadly arcuate and as advanced as the very broadly rounded angles, the base broadly, feebly lobed at the middle, the angles very obtuse but not rounded; punctures coarse and rather sparse, but not as coarse or sparse as those of the elytra, the pubescence distinct; elytra three and a half times as long as the prothorax and equal in width, two-thirds longer than wide, parabolic behind, parallel at the sides; basal angles obtuse, the cal-

lus feeble ; under surface finely and more closely punctured.   Length 3.5 mm.;
width 1.7 mm.   New York...**tessellata** *Mlsh.*
Oblong-oval, more convex, polished, black, the antennæ piceous toward base, the legs
    dark testaceous, the elytral pale markings rufo-testaceous and together not occu-
    pying as much area as the black ground, the subsutural pale spot not extend-
    ing behind the middle and not forming an obverted C-shaped macula ; antennæ
    nearly as in *tessellata* but black and stouter, the third joint very much shorter
    than the next two combined ; prothorax much more convex, distinctly less than
    twice as wide as long, the sides broadly arcuate, converging anteriorly, the prom-
    inence before the middle almost obsolete ; apex much narrower than the base and
    broadly sinuate ; punctures moderately coarse and sparse ; elytra nearly similar
    in form and sculpture, as wide as the prothorax but only three times as long ;
    under surface finely, sparsely punctate.   Length 3.0 mm.;   width 1.45 mm.
    Pennsylvania...**variegata**, sp. nov.
3—Body throughout nearly as in *tessellata* but black, the antennæ concolorous, the
    periphery of the pronotum rather paler, the legs piceo-testaceous, the elytral pale
    maculation nearly similar but less extended, much less in area than the black
    ground, the subsutural C-shaped marks before the middle much shorter and not
    extending distinctly behind the middle ; antennæ nearly similar in structure but
    stouter, and with the third joint very much shorter than the next two combined ;
    prothorax and elytra nearly similar in form, the former a little narrower at apex,
    with the apical angles somewhat more advanced and much less broadly rounded,
    the pubescence longer and more conspicuous ; lateral prominence before the
    middle equally conspicuous and much more so than in *variegata*.   Length 2.9
    mm.; width 1.4 mm.   Indiana.—Cab. Levette...............**maculata**, sp. nov.

In *tessellata* the male has a large and very abruptly limited deep
oval excavation, slightly wider than long, occupying almost median
third of the fifth ventral, and extending from the apex almost to the
base, the bottom of the excavation polished, impunctate and glabrous,
with a very few piliferous punctures posteriorly ; in *maculata* it is
equally deep and abrupt but smaller, occupying about median fourth
and is more distinctly pubescent posteriorly; in *variegata* it is as large
as in *tessellata* or larger, but very much more shallow.

Incolia, gen. nov.

In this genus the body is much more elongate and less convex than
in either of the preceding, and differs greatly in abdominal structure
and somewhat in its finer sculpture ; in the form and structure of the
antennæ and prothorax it is nearly similar to *Abstrulia*.   The single
species may be described as follows from the unique type, which appears to be a female : —

Body elongate, parallel, feebly convex, polished, blackish, the antennæ toward base,
    legs, limb of the pronotum and an indefinite oblique elytral streak, extending for

a short distance from the elytral humeri, dark testaceous; pubescence short, inclined, very sparse and rather inconspicuous; head rather small, not half as wide as the prothorax, the antennæ rather stout, nearly as long as the head and prothorax, the third joint as long as the next two combined, the seventh wider than the sixth, transverse, forming a broader support for the club, which is fully as long as the stem, cylindrical, the joints transverse and rather closely connected, the last oval and pointed; joints of the club much more than twice as thick as three to six; prothorax short, about twice as wide as long, the sides broadly arcuate and coarsely, feebly and irregularly crenulate throughout, more convergent anteriorly, the apex slightly narrower than the base and transversely, rectilinearly truncate; base feebly and arcuately lobed in rather more than median half, the fovea very large and impressed; side margins broadly reflexo-explanate, less widely so anteriorly, the basal angles very obtuse but not rounded, the apical obtusely rounded and not at all advanced; disk not concave along the base, but finely impressed within the basal bead, finely, sparsely punctate, the punctures gradually becoming closer and coarser toward the sides; scutellum slightly transverse, broadly angulate behind, minutely punctate; elytra much elongated, about twice as long as wide and four times as long as the prothorax, just visibly wider at three-fifths than at base and thence rapidly narrowed to the strongly rounded apex; humeral callus decidedly pronounced and elongate, gradually disappearing at some distance from the base; punctures impressed, rather sparse, moderately coarse, gradually becoming very fine posteriorly; sterna strongly, rather closely but not very coarsely, punctured, the abdomen minutely and rather densely so, especially toward the sides; tarsi slender, the first joint of the posterior as long as the last two combined. Length 3.8 mm.; width 1.6 mm. Indiana? ....**longipennis,** sp. nov.

The locality is reasonably certain, but the type bore no label in the cabinet of the late Dr. Levette.   I considered this to be the *concolor* of LeConte, for some time, but the description will not serve, especially regarding the " narrowly margined " sides of the prothorax of *concolor*.

Eupisenus, gen. nov.

This is the only genus of the tribe Tetratomini which has been discovered thus far on the Pacific coast, the others all being inhabitants of the Atlantic districts.   The body is elongate, parallel and moderately convex, with the prothorax relatively narrower than in the preceding genera, and the elytral humeri somewhat exposed at base. The following description of the only known species will bring out other characters which may prove to be generic :—

Parallel, polished, sparsely clothed with short fine and subdecumbent pubescence, black, the legs, antennæ, trophi and elytra pale luteous, the latter indefinitely shaded with piceous at the middle of the flanks and on the suture toward tip; antennæ rather stout, as long as the head and prothorax, the third joint about as long as the next two together, eighth globular and perfectly similar to the seventh, the club very strong, parallel, the joints rather closely connected and strongly transverse, the last pointed and but little longer than wide ; prothorax three-fourths wider than long, not more than two-thirds wider than the head, widest near basal third, the sides broadly arcuate, gradually converging anteriorly and almost even, the apex sensibly narrower than the base and broadly arcuate ; basal angles slightly more than right and not at all rounded ; surface rather coarsely and closely punctate ; elytra parallel, obtusely and broadly rounded behind, four-fifths longer than wide, three and a half times as long as the prothorax and nearly a fourth wider, the punctures moderately fine but deeply impressed, somewhat close-set and nearly similar in size to those of the pronotum ; humeri obtusely rectangular, the callus distinct ; scutellum moderate, transverse ; under surface polished, finely, rather sparsely punctured ; legs slender, the four basal joints of the anterior and middle tarsi short, subequal and together but little longer than the last ; basal joint of the posterior much shorter than the last.   Length 4.5 mm.; width 1.65 mm.   Alaska and southward......... ....**elongatus** *Lec.*

The head has a deep frontal impression at the middle of the line between the antennæ apparently in both sexes.

Pisenus, gen. nov.

The species of this genus may be readily distinguished from the preceding by the shorter, more oval form, greater convexity and much smaller size, as well as by the characters of the table ; the prothorax, also, is as wide at base as the base of the elytra, so that the humeri are not exposed at base, and the sides of both form a virtually continuous arc.   The antennæ are nearly similar in structure.   The two species are the following : —

Body more elongate-oval, shining, clothed sparsely with rather short line subdecumbent pubescence, black, the legs and antennæ dark testaceous, the basal regions of the elytra, especially at the humeri, suffusedly rufous ; head about as wide as the rectilinearly truncate apex of the prothorax, the antennæ stout, fully as long as the head and prothorax, the eighth joint similar to the seventh and the club similar to that of *Eupisenus elongatus* but narrower ; prothorax three-fourths or more wider than long, the sides almost perfectly even and broadly arcuate from the distinct basal angles to the apex, the latter much narrower than the base ; surface rather finely but strongly, moderately closely punctate ; scutellum transverse, broadly angulate behind ; elytra suboval, rather ogivally pointed behind, scarcely at all wider than the prothorax and three times as long, two-thirds longer than wide, the punctures only moderately coarse, impressed, larger than those of the pronotum and somewhat sparse ; under surface finely, rather sparsely punctate ; legs moderately slender, rather short, the tarsi short, with the four basal joints of the anterior and intermediate equal among themselves and together about as long as the fifth, the last joint of the posterior very nearly as long as the first three combined.   Length 2.8–3.1 mm.; width 1.3–1.4 mm.   Pennsylvania, Indiana and northern Illinois ; [*Triphyllus ruficornis* Lec.].....**humeralis** *Kirby*

Body nearly similar to the preceding but shorter and more broadly oval, strongly convex, shining, rufo-testaceous throughout, the pubescence long, coarse, rather
abundant and conspicuous, ashy in color; head smaller, notably narrower than
the apex of the prothorax, the antennæ similar to those of *humeralis* but still
stouter, and with the third joint very much shorter than the next two combined;
prothorax similar but only a little more than one-half wider than long and with
the punctures coarse, deep and densely crowded; elytra sensibly wider than the
prothorax and two and a half times as long, the sides slightly arcuate toward
base, the apex gradually, rather narrowly rounded, one-half longer than wide,
the punctures rather smaller than those of the pronotum and somewhat sparse,
moderately coarse toward base, especially externally, gradually fine posteriorly;
under surface finely, very densely punctate; legs rather stouter, the tarsi short
but slender, the last joint of the posterior as long as the first three combined.
Length 2.75 mm; width 1.4 mm. Virginia...............**pubescens,** sp. nov.

In no individual of the Tetratomini that I have seen, is there the
faintest trace of serial arrangement of the always conspicuous elytral
punctures at any part of the surface; the placing of *Tetratoma* near
Triplax, by Redtenbacher, is an unaccountable error for this, as well as
a multitude of other reasons, besides the radically different formation
of the tarsi and palpi.

www.ingramcontent.com/pod-product-compliance
Lightning Source LLC
Chambersburg PA
CBHW021518210326
41599CB00012B/1298